Vehicular Networks

From Theory to Practice

CHAPMAN & HALL/CRC
COMPUTER and INFORMATION SCIENCE SERIES

Series Editor: Sartaj Sahni

PUBLISHED TITLES

ADVERSARIAL REASONING: COMPUTATIONAL APPROACHES TO READING THE OPPONENT'S MIND
Alexander Kott and William M. McEneaney

DISTRIBUTED SENSOR NETWORKS
S. Sitharama Iyengar and Richard R. Brooks

DISTRIBUTED SYSTEMS: AN ALGORITHMIC APPROACH
Sukumar Ghosh

FUNDEMENTALS OF NATURAL COMPUTING: BASIC CONCEPTS, ALGORITHMS, AND APPLICATIONS
Leandro Nunes de Castro

HANDBOOK OF ALGORITHMS FOR WIRELESS NETWORKING AND MOBILE COMPUTING
Azzedine Boukerche

HANDBOOK OF APPROXIMATION ALGORITHMS AND METAHEURISTICS
Teofilo F. Gonzalez

HANDBOOK OF BIOINSPIRED ALGORITHMS AND APPLICATIONS
Stephan Olariu and Albert Y. Zomaya

HANDBOOK OF COMPUTATIONAL MOLECULAR BIOLOGY
Srinivas Aluru

HANDBOOK OF DATA STRUCTURES AND APPLICATIONS
Dinesh P. Mehta and Sartaj Sahni

HANDBOOK OF DYNAMIC SYSTEM MODELING
Paul A. Fishwick

HANDBOOK OF PARALLEL COMPUTING: MODELS, ALGORITHMS AND APPLICATIONS
Sanguthevar Rajasekaran and John Reif

HANDBOOK OF REAL-TIME AND EMBEDDED SYSTEMS
Insup Lee, Joseph Y-T. Leung, and Sang H. Son

HANDBOOK OF SCHEDULING: ALGORITHMS, MODELS, AND PERFORMANCE ANALYSIS
Joseph Y.-T. Leung

HIGH PERFORMANCE COMPUTING IN REMOTE SENSING
Antonio J. Plaza and Chein-I Chang

INTRODUCTION TO NETWORK SECURITY
Douglas Jacobson

PERFORMANCE ANALYSIS OF QUEUING AND COMPUTER NETWORKS
G. R. Dattatreya

THE PRACTICAL HANDBOOK OF INTERNET COMPUTING
Munindar P. Singh

SCALABLE AND SECURE INTERNET SERVICES AND ARCHITECTURE
Cheng-Zhong Xu

SPECULATIVE EXECUTION IN HIGH PERFORMANCE COMPUTER ARCHITECTURES
David Kaeli and Pen-Chung Yew

VEHICULAR NETWORKS: FROM THEORY TO PRACTICE
Stephan Olariu and Michele C. Weigle

Vehicular Networks
From Theory to Practice

EDITED BY

STEPHAN OLARIU
OLD DOMINION UNIVERSITY
NORFOLK, VIRGINIA, U.S.A.

MICHELE C. WEIGLE
OLD DOMINION UNIVERSITY
NORFOLK, VIRGINIA, U.S.A.

CRC Press
Taylor & Francis Group
Boca Raton   London   New York

CRC Press is an imprint of the
Taylor & Francis Group, an **informa** business

A CHAPMAN & HALL BOOK

Cover design by Syed R. Rizvi

Chapman & Hall/CRC
Taylor & Francis Group
6000 Broken Sound Parkway NW, Suite 300
Boca Raton, FL 33487-2742

First issued in paperback 2017

© 2009 by Taylor & Francis Group, LLC
Chapman & Hall/CRC is an imprint of Taylor & Francis Group, an Informa business

No claim to original U.S. Government works

ISBN 13: 978-1-138-11659-7 (pbk)
ISBN 13: 978-1-4200-8588-4 (hbk)

Library of Congress Cataloging-in-Publication Data

Vehicular networks : from theory to practice / editors, Stephan Olariu, Michele
   C. Weigle.
      p. cm. -- (Chapman & Hall/CRC computer & information science series ;
      20)
   Includes bibliographical references and index.
   ISBN 978-1-4200-8588-4 (hardback : alk. paper)
      1. Vehicular ad hoc networks (Computer networks) I. Olariu, Stephan. II.
Weigle, Michele Aylene Clark. III. Title. IV. Series.

   TE228.37.V44 2009
   384.3'3--dc22   2008037102

Visit the Taylor & Francis Web site at
http://www.taylorandfrancis.com

and the CRC Press Web site at
http://www.crcpress.com

Contents

Preface

The past decade has witnessed the emergence of Vehicular Ad-hoc Networks (VANETs), specializing from the well-known Mobile Ad Hoc Networks (MANETs) to Vehicle-to-Vehicle (V2V) and Vehicle-to-Infrastructure (V2I) wireless communications. The importance and potential impact of VANETs have been confirmed by the rapid proliferation of consortia involving car manufacturers, various government agencies, and academia. Examples include, among many others, the Car-2-Car Communication Consortium, the Vehicle Safety Communications Consortium, and the Advanced Safety Vehicle Program. While the original motivation for VANETs was to promote traffic safety, recently it has become increasingly obvious that VANETs open new vistas for Internet access, distributed gaming, and the fast-growing mobile entertainment industry.

The main motivation for offering this handbook stems from the realization that, in spite of its importance and potential societal impact, at present, there is no comprehensive source of information about vehicular networks. We believe there is value in bringing together, in an integrated and cohesive fashion, the state of the art in this important new field that promises to see a phenomenal growth in the next few years.

The handbook consists of six broad sections, each containing a number of chapters contributed by leading experts in the field:

I. Traffic Engineering
II. U.S. and European Initiatives
II. Applications
IV. Networking Issues
V. Simulation
VI. Human Factors

Many of the chapters overlap. This is deliberate, the idea being that such chapters benefit the reader by providing different perspectives on the same or related topics. The breadth of the sections is meant to cover many different areas that impact the design of applications for vehicular networks, not only those related to computer science or computer engineering. We include chapters on traffic engineering to provide a context for traffic safety and traffic monitoring applications that may be developed. We include a chapter on driver distraction to allow designers to consider how their in-vehicle systems might actually affect drivers.

Key features of this handbook include

1. Introducing the first compiled state-of-the-art research, technologies, tools and innovations that provide sufficient reference material for researchers and practitioners in industry and academia
2. Providing a comprehensive technical guide covering introductory and advanced concepts that can serve as a reference material to graduate students in the field of computing, engineering and computer engineering
3. Providing detailed discussions on key research challenges and open research issues that are of great importance to graduate students and practitioners who intend to do work in VANETs
4. Serving as a guideline and giving insights to professional and industrial standards organizations for developing future standards in vehicular networking
5. Providing valuable information on existing experimental studies including case studies, simulation tools and implementation test-beds in industry and academia that can be used for validation of theories and analytical results by researchers and practitioners at various levels

While a number of key aspects of vehicular networks are covered, other equally important ones were deliberately omitted. Indeed a comprehensive discussion of security issues in VANETs, while very important, was left out chiefly because the consensus is to provide security through the use of Public Key Infrastructure (PKI), which is well-described in the academic literature and in *Public Key Infrastructure* by John Vacca. Nevertheless, a brief overview of security and related privacy issues is presented in Chapter 7.

We expect this handbook to be of significant importance to researchers and practitioners in academia, government agencies, automotive industry and transportation science in general. We anticipate the handbook to be a "must have" for all university libraries and an invaluable source of information for all those working in this new and exciting area.

We want to take this opportunity to thank all the authors for their enthusiastic support of this project and for their willingness to share their knowledge by contributing a chapter; we are indebted to the (anonymous) referees for their diligent and painstaking work in ensuring the high quality of the offerings. Last, but certainly not least, we wish to thank Bob Stern from Taylor & Francis for his help, patience and advice throughout the time this handbook was in the making.

Stephan Olariu
Michele C. Weigle
Department of Computer Science
Old Dominion University
Norfolk, Virginia

Contributors

Mahmoud Abuelela
Department of Computer Science
Old Dominion University
Norfolk, Virginia

Mohammad S. Almalag
Department of Computer Science
Old Dominion University
Norfolk, Virginia

Roberto Baldessari
Network Research Division
NEC Laboratories Europe
Heidelberg, Germany

Carryl L. Baldwin
Department of Psychology
George Mason University
Fairfax, Virginia

Azzedine Boukerche
School of Information Technology
    and Engineering
University of Ottawa
Ottawa, Ontario, Canada

Ryan Cheung
Department of Computer Science
University of California
Los Angeles, California

Susan Dickey
Partners for Advanced Transit and
    Highways (PATH)
University of California
Berkeley, California

Andreas Festag
Network Research Division
NEC Laboratories Europe
Heidelberg, Germany

Marco Fiore
Department of Electronics
Polytechnic Institute of Torino
Torino, Italy

Michael D. Fontaine
Virginia Transportation Research
    Council

and

Department of Civil and
    Environmental Engineering
University of Virginia
Charlottesville, Virginia

Mario Gerla
Department of Computer Science
University of California
Los Angeles, California

Khaled Ibrahim
Department of Computer Science
Old Dominion University
Norfolk, Virginia

Arne Kesting
Institute for Transport and Economics
Technical University of Dresden
Dresden, Germany

Long Le
Network Research Division
NEC Laboratories Europe
Heidelberg, Germany

Uichin Lee
Department of Computer Science
    and Engineering
University of California
Los Angeles, California

Cong Liu
Department of Computer Science
Florida Atlantic University
Boca Raton, Florida

Antonio A.F. Loureiro
Department of Computer Science
Federal University of Minas Gerais
Minas Gerais, Brazil

James A. Misener
Partners for Advanced Transit
    and Highways
University of California
Berkeley, California

Eduardo F. Nakamura
Department of Computer Science
Federal University of Minas Gerais
Minas Gerais, Brazil

Stephan Olariu
Department of Computer Science
Old Dominion University
Norfolk, Virginia

Horacio A.B.F. Oliveira
Department of Computer Science
Federal University of Minas Gerais
Minas Gerais, Brazil

Francisco J. Ros
Department of Information and
    Communications Engineering
University of Murcia
Murcia, Spain

Pedro M. Ruiz
Department of Information and
    Communications Engineering
University of Murcia
Murcia, Spain

Juan A. Sánchez
Department of Information and
    Communications Engineering
University of Murcia
Murcia, Spain

Raja Sengupta
Department of Civil and Environmental
    Engineering
University of California
Berkeley, California

Yifeng Shao
Department of Computer Science
    and Engineering
Florida Atlantic University
Boca Raton, Florida

Ivan Stojmenovic
School of Information Technology
    and Engineering
University of Ottawa
Ottawa, Ontario, Canada

Martin Treiber
Institute for Transport and Economics
Technical University of Dresden
Dresden, Germany

Joel VanderWerf
Partners for Advanced Transit
   and Highways
University of California
Berkeley, California

Michele C. Weigle
Department of Computer Science
Old Dominion University
Norfolk, Virginia

Jie Wu
Department of Computer Science
   and Engineering
Florida Atlantic University
Boca Raton, Florida

Gongjun Yan
Department of Computer Science
Old Dominion University
Norfolk, Virginia

Wenhui Zhang
Network Research Division
NEC Laboratories Europe
Heidelberg, Germany

I

Traffic Engineering

1

Traffic Monitoring

Michael D. Fontaine
Virginia Transportation
Research Council
and
Department of Civil and
Environmental Engineering
University of Virginia

Data on traffic conditions are used to make a number of critical transportation decisions. Because of the importance of this data, departments of transportation (DOTs) have developed a series of programs to collect information that allows them to characterize traffic flow conditions on the road system. The goals of these monitoring programs are directly tied to specific functional objectives of the DOT, so the type of data and its level of spatial or temporal aggregation vary depending on the ultimate use of the data. For example, some data are collected to support real-time traveler information and traffic control, whereas other data are collected and used off-line to help characterize typical travel patterns and project future traffic conditions. Examples of some of the uses of traffic data include the following:

1. Predicting where roads should be built or expanded in the future
2. Designing bridges and pavements to withstand predicted traffic loads
3. Analyzing air quality in urban areas
4. Alerting drivers to congestion and accidents
5. Controlling traffic signals

Obviously, traffic data is critical to planning and operating an effective transportation system. One of the challenges when discussing traffic data is that information is collected by a variety of different governmental entities to support very different purposes. There is no master data source of traffic data. Data resides in a variety of different databases that are separately maintained by states, cities, counties, and the federal government. In some cases, data is duplicated and data sources are not always well coordinated. In addition, a great deal of data collected to support real-time traffic control and operations is never archived for later use. This can create a challenging environment for engineers who are seeking data.

Before discussing vehicular networks, it is useful to develop an understanding of the capabilities and limitations of existing methods used to monitor traffic. This chapter is intended to provide some context on how data is currently collected and used by transportation agencies. First, the causes of congestion are discussed. Common types of data collected by DOTs are then reviewed, along with sensors that are typically used to gather this information. Several key uses of sensor and traffic data are also examined. Finally, emerging methods for monitoring traffic are reviewed, and possible future trends in traffic monitoring are examined.

1.1 Causes of Congestion

The highway system touches the life of every person. People and businesses rely on the highway system to move passengers and freight safely, efficiently, and reliably. Data shows that the nation's highways are becoming increasingly congested, which negatively impacts both the quality of life of travelers and the economic competitiveness of U.S. businesses. The Texas Transportation Institute's 2007 *Urban Mobility Report* quantifies the extent of congestion in major urban areas in the United States.[1] In 2005, that report estimated that Americans wasted 4.2 billion hours and 2.9 billion gallons of gasoline while stuck in congestion. This represented an economic loss of over $78 billion dollars, which was an increase of over 70% in the cost of congestion since 1995.

There are several underlying reasons why congestion continues to worsen in the United States. First, roadway and public transportation capacity have failed to increase at the same rate that travel has increased. In the last 20 years, the number of vehicle miles of travel has more than doubled, while roadway and public transportation capacities have only increased by 30 to 45%.[1] These increases in travel are in part generated by increasing car ownership. From 1969 to 2001, the average number of automobiles per household increased from 1.2 to 1.9.[2] Car ownership increases were driven by the relative affordability of automobiles, as well as the growth in suburban development since World War II. Low-density suburban development makes it more difficult to operate public transportation efficiently, creating situations where the automobile is the only feasible way to travel to the store or work. About half of all workers are now estimated to live in a suburb, and the proportion of workers living in the central city and nonurban areas continues to decrease.[3] In essence, people are traveling more due to higher automobile ownership and longer commute distances, and the transportation infrastructure has not expanded fast enough to keep up with these changes. As a result, the term "rush hour" has become inaccurate in many urban areas, where the peak congested periods in the morning and afternoon often last three or more hours.

Two broad categories of congestion can be defined: recurring and nonrecurring congestion. Recurring congestion occurs regularly at the same locations on the roadway system at about the same time every day. This typically includes congestion that is routinely experienced during the morning and afternoon peak periods. Recurring congestion is usually created by a combination of increases in traffic on roadways that do not have sufficient capacity to process the traffic. Generally speaking, recurring congestion cannot be mitigated without expensive projects to expand a road's capacity or significant changes in how or when people choose to travel (e.g., many people shifting to public transportation or telecommuting). Travelers can usually account for recurring congestion quite well in their trip planning, because they encounter it every day.

Nonrecurring congestion occurs when a crash or other unusual event (such as a construction zone or severe weather) causes a reduction in the traffic-carrying capacity of a road. Nonrecurring congestion represents a very significant proportion of all delays that drivers experience in an urban area. It has been estimated that between 52 and 58% of the total delay experienced by drivers in urban areas is due to incidents or other forms of nonrecurring congestion.[1] Because crashes could theoretically happen anywhere on the roadway network, nonrecurring congestion has the potential to develop any place where the road's capacity could be restricted to a level below the number of vehicles that want to use the road. By definition, nonrecurring congestion is unpredictable. This can create a lot of difficulty for drivers when planning trips, because they cannot account for how much time they might lose due to congestion caused by crashes.

The impact of crashes or other incidents on traffic-carrying capacity can be very significant. The *Highway Capacity Manual* defines the theoretical number of vehicles that can use a section of highway.[4] This is called the capacity of the road. The theoretical maximum capacity for a lane of a freeway under ideal conditions is 2400 vehicles per hour (vph). Based on data from field studies, the *Highway Capacity Manual* defines how much the capacity of a road may be reduced because of crashes or lane closures due to incidents. Table 1.1 shows how incidents can impact capacity.

Table 1.1 shows that incidents can have tremendous impacts on capacity. Even a crash on a shoulder that does not close a travel lane can reduce traffic flow by more than 1000 vph due to "rubbernecking" as people slow down to look at a crash as they travel past. Table 1.1 also shows that the impact of blocking lanes is not linear. If one of two lanes is blocked, then more than half of the capacity of the road is removed. As people merge into a single lane, additional turbulence is introduced during the merging maneuver, reducing the capacity beyond what would occur by removing a single lane.

TABLE 1.1 Impact of Incidents on Freeway Capacity

Number of Freeway Lanes by Direction	Capacity (Vehicles per Hour)				
	Theoretical Maximum	Crash on Shoulder	1 Lane Blocked	2 Lanes Blocked	3 Lanes Blocked
2	4800	3888	1680	N/A	N/A
3	7200	5976	3528	1224	N/A
4	9600	8160	5568	2400	1248

Since the late 1980s, there has been a growing recognition by state and federal governments that it will be impossible to build new road capacity at a rate fast enough to counteract increasing congestion. As a result, there has been an increasing focus on maximizing the operational efficiency of existing facilities. Agencies have been investing in operational improvements such as providing real-time traveler information, incident management programs, and improved traffic signal operations. Having timely and accurate data on traffic conditions is a basic requirement to develop and implement operational improvements, so many agencies have significantly expanded their traffic monitoring programs.

1.2 Traffic Monitoring Data

Traditional traffic monitoring programs are typically focused on collecting four types of data: traffic volume, vehicle classification, traffic speed, and traffic density. Travel time is also being monitored by more and more agencies as it provides a direct measurement of what travelers may experience. This section reviews common types of data that DOTs collect, and discusses the characteristics of each type of data.

1.2.1 Traffic Volume

Traffic volume refers to the number of vehicles that cross a specific point on the road in a particular amount of time. Volume represents the most basic data element in traffic engineering, and it is the most widely available piece of data. Volume data is typically summarized in increments ranging from 20 sec to an entire day. Data is most often available in hourly or daily increments, although more disaggregate data is available at locations where real-time operational decisions are made.

Traffic volume data is commonly reported as the annual average daily traffic (AADT) on a road. AADT represents the average number of vehicles that pass a particular location on a road in a day over an entire year. Although AADT is often used in traffic engineering, it does not reflect the variability in traffic volume that could be observed over the course of a year. Traffic volume typically varies by day of the week and month of the year. An average traffic volume representing an entire year may not fully account for these variations. For example, areas around major recreational or tourist destinations often see significant increases in traffic during the summer months. Likewise, the distribution of traffic on the weekend is different than during a weekday. AADT provides a good baseline estimate of the amount of traffic on a road, but engineers need to be aware that traffic volumes can vary significantly from this average due to these temporal changes in volume.

The amount of volume data available and its level of aggregation depend on a number of factors. In most major urban areas, Traffic Management Centers (TMCs) actively manage traffic. TMCs are typically focused on urban freeways and major arterials, and often collect data continuously on these facilities. Roads monitored by TMCs often have volume data available at spacings between 0.5 and 1 mi. The data at these sites is often available at a relatively disaggregate level, usually in the 20 sec to 3 min range.

In most states, actual volume data is collected continuously at a relatively small number of locations. On facilities not monitored by TMCs, continuous volume data is only available at permanent count stations. Permanent count stations collect data continuously at a limited number of predefined, strategic locations in order to help estimate seasonal and daily variations in traffic volumes. As an example, the state of Virginia has approximately 300 permanent count stations to cover over 55,000 miles of state-maintained roads—an average of one station per 183 mi. This data is typically summarized in increments of 15 min.

Volume data for roads without permanent count stations is generated through periodic coverage counts. For this, portable traffic counting equipment is placed on a road to collect volume data at regular intervals, such as every three to five years. The portable traffic counters collect data for some limited period, often only two to three days. Seasonal and daily adjustment factors are then applied to estimate AADTs on the road. For the vast majority of roads, coverage count data is the only type of volume data available. As a result, many roads in a state may only be counted once every several years.

An additional measure related to volume is vehicle miles of travel (VMT). VMT is the product of the volume of traffic using the road and the length of the road segment being examined. For example, a road with an AADT of 30,000 vehicles per day and a 5 mi length would have a daily VMT of 150,000. VMT is often used to normalize data to provide equitable comparisons of travel on roads with different characteristics.

1.2.2 Vehicle Classification

Vehicle classification data records traffic volume with respect to the type of vehicle that passes a point on the road (car, truck, and so on). The Federal Highway Administration (FHWA) has defined a set of 13 vehicle classes that are commonly used by most states.[5] Those vehicle classes include the following:

1. Motorcycles.
2. Passenger cars.
3. Other two-axle, four-tire single-unit vehicles other than passenger cars. This class mostly consists of pickup trucks, sports utility vehicles, and vans.
4. Buses.
5. Single-unit trucks, for which there are three separate classifications depending on the number of axles present.
6. Tractor trailers, for which there are five different classifications depending on the number of axles and the number of trailers present.

Vehicle classification data serves several important functions in transportation engineering. The design of bridges and pavements is heavily driven by the amount of truck traffic using the road. Likewise, trucks typically have much worse acceleration and deceleration characteristics than passenger vehicles. Analyses of traffic flow have to have accurate information on the number of trucks in the traffic stream to adequately take this into account. Detailed vehicle classification data is often available on a much more limited basis than traffic volume data. Often, the only information available on vehicle classification is an estimate of the percentage of traffic that consists of buses, single-unit trucks, and tractor trailers. There is often significant instability in these estimates, particularly at locations where coverage counts are made every two to three years.

1.2.3 Traffic Speed

Speed data is most often collected in major urban areas covered by a TMC. At those locations, the speed data is used for a variety of purposes, ranging from detecting congestion to tracking historic congestion trends on a road. The availability and level of aggregation is very similar to that of traffic volume data. Often, there is little data available in rural areas or on arterial streets where there are no permanent count stations. Speed data is therefore often only available when a special study has been conducted using portable equipment. Those studies are usually associated with requests to change speed limits.

In transportation engineering, two types of speeds are defined: time mean speed (TMS) and space mean speed (SMS).[6] Time mean speed is simply the arithmetic average of the speeds of all vehicles that pass a point on a road in a given amount of time. Time mean speed is calculated as

$$\bar{\mu}_{TMS} = \frac{\sum_{i=1}^{n} \mu_i}{n} \tag{1.1}$$

where μ is the speed of a vehicle and n is the number of vehicles.

An alternative way to define speed is based on average speeds over an extended segment of road. This is termed space mean speed. Space mean speed is defined as

$$\bar{\mu}_{SMS} = \frac{1}{\frac{1}{n}\sum_{i=1}^{n} t_i} \tag{1.2}$$

where t is the travel time expressed in hours per mile for the segment of road being examined.

A simple example can help illustrate the difference between these two methods of calculating average speed. Consider a 1 mi segment of road, where there is one vehicle traveling at 30 mi/h and another vehicle traveling at 60 mi/h. If the speeds were to be examined at a single point in the middle of the segment, the TMS would be 45 mi/h—the average of 30 and 60 mi/h. To calculate the SMS, we would first need to determine how long it would take the vehicles to travel over the 1 mi segment. To travel 1 mi, it would take the 30 mi/h vehicle 2 min and the 60 mi/h vehicle 1 min. Using the SMS formula:

$$\bar{\mu}_{SMS} = \frac{1}{\frac{1}{n}\sum_{i=1}^{n} t_i} = \frac{1}{\frac{1}{2}\left(2\frac{min}{mi} + 1\frac{min}{mi}\right)} = 0.67\frac{mi}{min} = 40 \text{ mi/h}$$

Thus, there is a 5 mi/h difference in the average speed attributable solely to how we determined the mean speed of the road. This is an important distinction, because different detection methods define average speeds differently.

Most of the speed data available are TMSs collected at a single point on a road. The location of speed data collection stations is usually at mid-block on an arterial road or between interchanges on freeways. Although these locations minimize the impact of vehicles that are turning, accelerating, or decelerating, there can be problems if these speeds are extrapolated to long roadway segments links. Congestion on freeways usually forms near interchange ramps, whereas most congestion on arterial roads is at intersections. If speeds midway between interchanges or intersections are assumed to represent average conditions, travel speeds are likely to be overestimated.

DOTs are increasingly interested in collecting SMS data directly. Probe-based traffic monitoring systems (discussed at the end of this chapter) measure speeds of vehicles over a segment of road. As a result, they are defining the SMS of the road, including any delays at intersections or interchanges. The relationship between TMS and SMS is defined as

$$\overline{\mu}_{TMS} = \overline{\mu}_{SMS} + \frac{s^2_{SMS}}{\overline{\mu}_{SMS}} \tag{1.3}$$

where s is the standard deviation of SMS on the link. As a result, it is theoretically possible to convert SMS values into TMS values. The fact that the two methods of determining speed produce different results is important, and engineers should be aware of the type of speed data that a particular sensor produces.

1.2.4 Traffic Density

Traffic density refers to the number of vehicles per mile on a section of road. Engineers are interested in this measure because it is often a more reliable indicator of the presence or onset of congestion than looking at traffic volume or speed. For example, a low number of vehicles passing a detector could mean that there is not much traffic on the road or it could mean that traffic is so congested that nobody can pass. Actual direct collection of the number of vehicles per mile is a difficult task using conventional sensors. Aerial photography has occasionally been used to gather data at known bottlenecks, but this is not cost effective in most cases.

Most often, detector occupancy is used as a surrogate measure for traffic density. Some detectors, like inductive loop detectors, sense the presence of a vehicle passing over the sensor. The percentage of time that the detector is active due to the presence of a vehicle has been found to be a good surrogate measure for the direct measurement of traffic density. The equation for the percent detector occupancy is given by

$$\text{percentage occupancy} = \frac{\sum_{n=1}^{N} t_{occ}}{T} \times 100 \tag{1.4}$$

where T is the selected time period, N is the number of vehicles detected in time period T, and t_{occ} is the time that the selected detector is turned on.

When density (or occupancy), speed, and volume are examined collectively, engineers can characterize the traffic flow state accurately using available macroscopic models.[6] These measures can be examined to assess traffic state, and also to assess whether the sensor collecting the data is producing valid data.

1.2.5 Travel Time

Transportation agencies are increasingly interested in directly collecting the travel time to go between two points. Historically, it has been a challenge to gather this data. Special studies have been required where drivers manually drive a route and record the travel time. Statistical validity is sometimes difficult to achieve in these studies, and it is often challenging to have enough drivers to capture sudden changes in traffic conditions (such as the onset of congestion).

As a result, travel time data is not widely available in most areas of the country, although a number of emerging methods (discussed at the end of this chapter) are starting to generate this data. One of the advantages of travel time data is that the public can readily understand how this piece of data impacts their lives. Traffic volume and occupancy have little meaning to the average driver, so they are often not good measures to use when communicating with politicians or the public. Travel time, although not commonly available yet, appears to be an emerging measure of effectiveness that will gain greater prominence in the future.

1.3 Commonly Used Sensor Technologies

The sensor technologies commonly used by DOTs can be broadly classified into two categories: intrusive and nonintrusive sensors. Intrusive sensors require intrusion into the pavement to perform installation or maintenance activities. An example of an intrusive detector would be an inductive loop detector (ILD) installed in the pavement surface. One of the problems with intrusive detectors is that travel lanes may have to be closed in order to perform maintenance on the sensor. This can be a significant barrier to timely maintenance, particularly in a congested urban area where it may be difficult to close lanes during the day due to the impact on traffic. Likewise, intrusive detectors need to be replaced every time a road is repaved. Nonintrusive detectors are mounted either above or beside the roadway. Installation and maintenance of these detectors does not typically require that a lane be closed to traffic. In this section, three of the most common sensors are introduced, and the advantages and disadvantages of each sensor are discussed.

1.3.1 Inductive Loop Detectors

ILDs are the most widely used form of sensor for collecting traffic data. ILDs have been used since the 1960s and are commonly used for detection at traffic signals and for freeway monitoring. Inductive loops are intrusive detectors, consisting of a coiled wire that is cut into the pavement. The ILD senses the presence of metal objects that pass over the coiled wire. When a vehicle passes over or stops on top of an ILD, it reduces the loop inductance of the wire. Currents are induced in the vehicle, reducing loop inductance. The reduction in loop inductance is then translated by a controller into a detection of the presence of a vehicle.

ILDs are termed "presence" detectors, meaning that they detect whether a vehicle is passing over them. Lengths of vehicles can be estimated, but the number of axles is not explicitly counted. ILDs can be installed as either a single- or double-loop configuration. Single-loop detectors comprise just one loop of coiled wire installed in the pavement, and are generally used to provide traffic volume and density information. With double loops, two loops are installed in a lane of road one after another, with a short space between the two ILDs. Double loops are necessary to generate speed estimates. As the two ILDs are placed a known distance apart, the speed of a vehicle can be estimated by examining the time that elapses between the activations of the two loops.

There are several advantages to using ILDs. The technology is mature, and there is a large experience base with the sensors. They can provide all of the fundamental traffic data commonly used by transportation agencies, including volume, occupancy, TMS, and vehicle classification. ILDs also generally have the highest accuracy of commonly used sensors, with accuracy reported to be between –1 and +2% across a variety of studies.[7]

Although ILDs can produce high quality data, there are several drawbacks to using them. First, ILDs are prone to failure, with data from Texas suggesting that an average of 2 to 4% of ILDs fail annually.[7] ILDs also require regular tuning to ensure that speeds and vehicle classification data are of high quality. Maintenance and installation of ILDs can be problematic, particularly in congested urban areas. If an ILD needs to be replaced, an entire lane of road would need to be closed so that the loop could be cut out and reinstalled. Closing a lane of traffic is often difficult in an urban area, due to the traffic congestion that could be created by removing a lane. This can act as a barrier to the timely maintenance and replacement of ILDs. As ILDs are installed in the pavement, they are usually milled out when a road is repaved and have to be replaced following completion of repaving.

Because of ILD failure rates and maintenance difficulties, large portions of an ILD network may not be returning quality data at any given time. For example, in 2005 the Virginia Department of Transportation Traffic Management Center in Hampton Roads estimated that about 40% of their ILDs were not returning quality data.[8] The difficulties in maintaining ILDs have been a significant reason why alternative detection technologies have been pursued.

1.3.2 Video Detection Systems

Video detection systems utilize cameras and image processing software to collect data on traffic flow. A camera is pointed at a roadway, and image processing software analyzes changes in pixels between successive frames. The processing software identifies when a vehicle has entered the frame, and then translates the movement of the vehicle on the video into traffic flow parameters. Video detection systems operate by essentially creating "virtual inductive loops." When a detection system is set up, a user will overlay detection zones on top of a video image of the road surface. These detection zones function much the same way as an inductive loop. When the image processing software determines that a vehicle has entered the detection zone, the virtual loop is activated. Detailed information on the operation and architecture of these systems can be found in the book by L.A. Klein.[9]

Video detection offers several advantages over ILDs. First, it can collect all of the same traffic flow parameters as an inductive loop. A single controller and camera combination can also be used to detect multiple lanes on an approach. Because video detection cameras are mounted above the road, system maintenance can often be accomplished without closing down traffic lanes. Video detection systems are nonintrusive detectors, so they also do not need to be replaced if a road is repaved.

The most common application of video detection systems is for traffic signal control. Cameras are typically mounted at least 30 ft above the roadway to get a good view point of the approaches to an intersection. For traffic signal detection applications, cameras are usually mounted on the mast arm of a traffic signal pole. Virtual loops are then drawn on the lanes approaching the intersection. Video detection has become widespread for traffic signal detection due to some of the advantages discussed earlier. A single camera can detect traffic on all lanes on an approach to an intersection. The nonintrusive nature of video detection often makes video detection a more cost-effective solution for traffic signal detection over the lifecycle of the equipment.

There are some limitations to using video detection, however. Some periodic maintenance, such as cleaning camera lenses, does require shutting down lanes. Fog and snow can also create problems with the video processing software because they reduce the contrast in the image between vehicles and the background. When heavy fog or snow events occur, the video detection system will sometimes constantly register that a vehicle is present on all approaches. This can cause the signal to effectively cease operating in a traffic demand responsive mode, increasing the delay experienced by vehicles. Video detection systems can also be affected by high winds. High winds can cause the field of view of the camera to move, which can create a situation where the detection zones are not aimed appropriately.

One of the biggest potential problems with video detection is occlusion. Occlusion occurs when a vehicle obscures another vehicle within the camera's field of view. Figure 1.1 shows

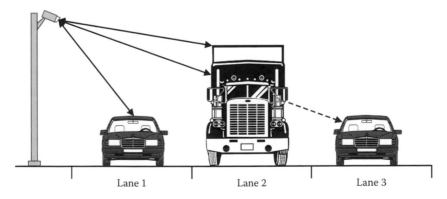

FIGURE 1.1   Impact of occlusion on video detection.

an example of the impact of occlusion on video detection. In Figure 1.1, the car in lane 3 is hidden behind the truck in lane 2. As a result, that car would not be detected by the video detection system. Occlusion can cause undercounting of traffic volumes or poor speed estimation. If video detection is being used with a demand-responsive traffic signal, occlusion could result in vehicles not being given a green phase to proceed. As a result, it is important that cameras be located so that occlusion effects are minimized. This typically involves mounting the camera so that the angle of the camera is as close as possible to straight down to minimize the impact of occlusion.

1.3.3 Microwave Radar Sensors

Microwave radar sensors provide another nonintrusive alternative to ILDs. Microwave sensors can also provide volume, occupancy, TMS, and vehicle classification data. Because microwave radar sensors are nonintrusive detectors, there is a great deal of flexibility in where they can be placed. The sensors are typically mounted on existing structures that pass over the highway (such as sign structures or bridges) or on posts adjacent to the roadway. They can be mounted to collect data for a single lane or to collect data across multiple lanes. The most common application of microwave sensors is to supplement data collected from ILDs on major freeway facilities.

The accuracy of microwave radar systems is generally good, although well-functioning ILDs generally produce superior results. A Texas Transportation Institute (TTI) study compared microwave sensors to loop detectors. The results showed that the tested microwave radar unit was consistent with ILDs, with differences in volume counts usually averaging around 5%.[7] Microwave sensors suffer from many of the same occlusion problems as video detection, but weather characteristics generally do not negatively influence the data produced by microwave radar, provided detection distances are short.[10] The occlusion impacts do mean that care should be taken in positioning microwave radar sensors so that consistent, accurate data is obtained.

1.4 Common Applications of Traffic Data

The data collected by transportation agencies is used to perform a variety of functions, ranging from tracking historic performance of roads to supporting real-time traffic control and traveler information. This section briefly describes some common functions of transportation agencies, noting the types of data that are often collected to support those tasks. Although this discussion is not exhaustive, it does provide an indication of areas where traffic monitoring data serves as an important input for agencies.

1.4.1 Link Characterization

The most basic reason to collect traffic data is to provide a historic record of travel characteristics along a roadway segment over time. This data is usually based on the permanent count stations and coverage count process described in the volume data section (Section 1.2.1). The vast majority of the road system in the United States does not have real-time traffic data available, so this historic roadway link information is often the only data that is available, especially in rural areas or away from freeways in urban areas. This data is used to support a variety of engineering decisions, such as:

1. Deciding how transportation funding dollars should be distributed to states, cities, and counties
2. Determining how bridges, roads, and pavements are designed
3. Determining when a road should be repaved
4. Examining whether a location has a disproportionate number of crashes
5. Determining where roads may need to be in the future
6. Performing air quality conformity analyses
7. Guiding where development should occur

This is just a small sample of the ways that traffic data is used in transportation engineering, but it gives an indication of why state DOTs spend a considerable amount of money collecting this type of data.

Another reason states invest in traffic monitoring programs is that the federal government requires states to regularly report data on the condition and performance of roads. Federal regulations give the authority for the FHWA to request any information necessary to administer the federal-aid highway program. Furthermore, Congress has mandated that FHWA produce a biennial estimate of future highway investment needs. Along with other factors, the traffic data reported by the states influences how much federal transportation funding is allocated to a state.

Table 1.2 shows some of the demographic and transportation characteristics of New York and Oklahoma and serves to illustrate some of the reasons why traffic data is needed to determine how funding should be allocated.[11,12] New York and Oklahoma have about the same number of lane miles of road, but New York has more than five times the population and almost three times the VMT. The New York apportionment is about three times that of Oklahoma, indicating that the VMT is a strong driver of the amount of money that a state receives. The VMT estimates could not be developed without comprehensive, statewide traffic monitoring programs in each of the states.

One of the major federal reporting programs is the Highway Performance Monitoring System (HPMS).[13] The HPMS is used to provide fundamental traffic data that the FHWA and the U.S. Congress use to define funding needs and the scope of the federal highway program. It is also used broadly as a data source for transportation planning functions, providing summary data for highway statistics, and long-term performance measurement. The HPMS includes some data on all public roads, with more detailed data available on a sample of collector and arterial routes. The data that populates the HPMS is

TABLE 1.2   Comparison of State Characteristics, 2005 Data

State	Population	Miles of Road	Vehicle Miles of Travel (Millions)	Estimated Annual Apportionment ($)
New York	19,254,630	113,341	137,521	1,683,966,455
Oklahoma	3,547,884	112,938	47,019	558,612,851

provided by the states, and includes data on issues ranging from pavement condition to congestion-related data.[14]

1.4.2 Real-Time Freeway Monitoring and Control

Although data collected to characterize conditions on road segments plays an important role in supporting many analyses, it is not intended to provide real-time support for engineers to make operational decisions. Real-time data is, in fact, not that necessary on many areas of the transportation system that experience little congestion. Characterizing the average amount of traffic that uses a low-volume, two-lane rural road is usually all that is needed. Real-time data on uncongested rural roads would not serve much purpose, because no active interventions are needed to keep traffic flowing well.

One area where real-time data has proven to be very valuable is on urban freeways. Urban freeways comprise only about 2.4% of the total road mileage in the United States, but they are estimated to carry about 20% of all traffic.[15] Given the high levels of traffic on urban freeways, recurring congestion and unexpected incidents can have significant impacts on what motorists experience during a particular journey. As a result, real-time datastreams can help engineers make informed decisions that could help mitigate the impact of congestion through actively managing traffic, responding to incidents, and informing the public about current conditions.

1.4.2.1 Role of Traffic Management Centers

Many urban areas have TMCs. The TMC serves as the focal point for the management of the freeway transportation system in an urban area. It integrates data from a variety of different sensor sources and provides a means for operators to manage traffic and inform the public from a centralized point. Personnel at the TMC examine data from sensors and closed circuit television (CCTV) to identify potential traffic problems, and then develop strategies to address the problems. Many TMCs are also colocated with emergency responders to help facilitate coordination when a crash or other emergency arises. TMCs often serve as the central media contact point for any freeway-related problems.

The centerpiece of the typical physical TMC building is a control room where operators sit behind workstations with a video wall on one side of the room. The video wall has a series of monitors that show images from the CCTV cameras looking at the freeway system. The workstations offer functionality to do tasks like changing messages on dynamic message signs (DMSs), viewing police dispatch reports, and controlling the pan/tilt/zoom of CCTV cameras.

Figure 1.2 shows some of the main functions that are typically located at the TMC. The ultimate goal of the TMC is to develop corrective actions to deal with problems. The major functions of the TMC are the following:

1. *Surveillance*. The surveillance function involves the collection of data on traffic flow conditions on the roadways being monitored.
2. *Traveler information*. In the traveler information function, the TMC provides information on current conditions to the public, enabling them to change routes or times of departure to avoid congestion.

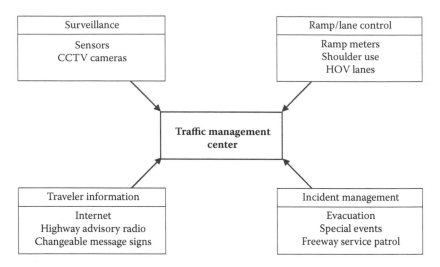

FIGURE 1.2   Functions of a traffic management center.

3. *Incident detection and management.* Incident detection and management includes the timely detection of sources of nonrecurring congestion and developing strategies to mitigate their impact.
4. *Ramp and lane control.* This involves dynamically changing traffic control devices on ramps and main freeway lanes to improve traffic flow.

In addition to these functions, the data from the TMCs is often used to support performance measurement activities that track congestion over time. Each of these functions is briefly discussed in the following sections.

1.4.2.2 Surveillance

The ability of a TMC to respond to traffic conditions is dictated largely by its ability to monitor roadways. As mentioned earlier, TMCs maintain a large network of sensors and cameras along the freeway system in their area of coverage. TMCs use ILDs, microwave radar, and other sensors to collect data on how traffic is flowing. This data is often reported back to the TMC at relatively short intervals, often around 20 sec. The data is often archived for future use in reporting and also serves as inputs to traffic management decisions made at the TMC. A number of TMCs use this data to provide real-time estimates of traffic conditions to the public on DMSs or the Internet.

TMCs also support CCTV cameras. The cameras are placed at elevated locations near important sections of road and interchanges. Their field of view can be moved so that TMC operators can see what is happening on the road in the vicinity of the camera. The CCTV cameras are used for several purposes. First, they are used to confirm whether the sensor data coming in to the TMC is accurate. For example, if an ILD was reporting abnormally slow speeds at a location, a CCTV camera could be focused on the location to see if traffic was in fact traveling slowly or if the sensor had malfunctioned. In the event that a severe crash occurs, the CCTV camera can also be used to remotely monitor the response at the scene. CCTV video feeds are also often shared with the public and media outlets as a way to provide information about current traffic conditions.

1.4.2.3 Traveler Information

One of the most visible functions of TMCs is providing information to the traveling public about current traffic conditions. The data from the sensors and CCTV cameras is used to provide valuable information on the location of congestion and potential travel times. The TMC synthesizes this information and presents it to the public by a variety of means. The goal is to provide travelers with as much information about conditions as possible so that they can avoid known problems. Some common ways that information is disseminated to the public include the following:

1. *Dynamic message signs (DMSs).* The DMSs installed on freeways are one of the most direct ways to get information to drivers while they are traveling on the road. The DMSs are controlled from the TMC, where operators decide which messages to post. Most TMCs have standard message libraries to ensure that drivers receive information in a consistent, easy-to-understand format. The types of information presented on DMSs vary, and may include estimates of traffic backups, locations of crashes, or travel time estimates between certain points on the road. The primary advantage of DMSs is that all traffic on the road can see the information, and it can easily reach travelers that are *en route*. The main disadvantages are that only a limited amount of information can be presented to drivers while they are traveling at freeway speeds, and the information is only available at the discrete points where the DMSs are installed.
2. *Highway advisory radio (HAR).* Some TMCs also operate HAR systems that provide traveler information over low-powered AM radio stations. One of the main advantages of HAR systems is that they can provide more detailed information than could be presented on a DMS. Access to HAR information is contingent on the driver actively changing a radio station to hear the HAR message, so all traffic on the road may not hear information broadcast on the HAR.
3. *511 systems.* Nationally, the 511 telephone number has been set aside as a way to provide traveler information by phone. As of 2007, 32 states have created 511 numbers to dispense traveler information.[16] Many TMCs are actively involved in providing information to 511 systems. Each state's 511 system is slightly different, but most use a voice-activated menu system to provide information on travel delays.
4. *Web sites.* Many TMCs also operate websites that display traffic condition information. Several pieces of information are typically contained on these websites. A color-coded traffic condition map that indicates the speeds on the freeway network is often shown. Many Web sites also provide access to the CCTV camera images. The main advantage of the Web site presence is that it allows travelers to plan their trip before they leave their home or office. Many travelers, however, may not have access to this information while on their way to a destination.
5. *Media reports.* TMCs also work with local media outlets to help disseminate information about traffic conditions. CCTV camera feeds are often shared with local television stations, and local traffic reporters often interact with TMC staff on a regular basis to gather information about traffic conditions.

1.4.2.4 Incident Detection and Management

As mentioned earlier in the chapter, nonrecurring delay is a major problem in urban areas. As a result, another core function of a TMC is to identify when incidents happen and then try to mitigate their effect. Studies in many urban areas have shown that formal incident management programs can result in significant delay reductions. Even if the time to clear the incident is reduced by only a small amount, it can have dramatic effects on the delay experienced in a congested urban environment. It has been estimated that incident management programs across 272 urban areas have reduced delay by 129.5 million hours annually.[1]

First, the TMC must quickly identify when an incident has occurred. Incidents can be identified in several ways:

1. Visual observation of CCTV cameras
2. Phone calls from the general public
3. Reports from DOT staff or police that are patrolling the freeways
4. Results of automated incident detection (AID) algorithms that examine changes in sensor data

The majority of incidents reported to a TMC are identified either by phone calls from drivers or reports from staff patrolling the road. If CCTV cameras cover the area where a crash was reported, they are then aimed at the crash so that the TMC can verify whether the incident has occurred and determine when it has been cleared.

One of the most direct uses of TMC sensor data is in the form of inputs into automated incident detection algorithms. These algorithms use sensor data to determine whether or not it is likely that an incident has occurred. AID algorithms have been the subject of a great deal of academic research, as researchers attempt to make algorithms that detect incidents quickly but also do not generate an excessive number of false alarms. Data from adjacent sensors are compared to one another and against historic trends to determine whether potential reductions in capacity are present. Two well-known approaches are called the California and McMaster algorithms. The California algorithm tries to identify whether an incident has occurred by comparing increases in the occupancy of an upstream sensor to decreasing occupancy in a downstream sensor.[17] This provides an indication that there is a bottleneck somewhere between the two sensors. The McMaster algorithm looks at volume, occupancy, and speed during congested and uncongested flow. Alarms are triggered when traffic flow characteristics change significantly.[18] Other methods incorporating artificial neural networks and fuzzy logic have also been proposed, but have not yet been widely adopted in practice.

Although researchers have developed automatic incident detection methodologies, implementation of these techniques has been problematic. A 1996 survey found that only 5 of 26 traffic management centers used some form of AID algorithm.[19] Four of these TMCs noted significant problems with their AIDs, including excessive false alarms, errors during low volume conditions, and excessively long detection times. The problem of excessive false alarms has proven to be a critical problem when using AIDs to identify whether an incident has occurred. Given the widespread ownership of cellular phones, many TMCs have found that drivers, DOT staff, and police on the road are often able to report incidents more quickly and reliably than AIDs can detect them.

Once an incident has been identified, the TMC is involved in coordinating efforts to remove the incident and return traffic to normal conditions. Many TMCs dispatch freeway service patrols (FSPs) to provide a relatively quick response to incidents. FSP vehicles patrol a set route looking for potential incidents. FSP vehicles are typically outfitted to handle a range of minor emergencies ranging from changing flat tires to providing fuel. They can also quickly call in tow trucks, police, and set up temporary traffic control to help move traffic around the incident. This relatively small service has been shown to produce significant benefits, with several studies showing benefits that outweigh costs by a ratio of at least 5:1.[20,21]

1.4.2.5 Ramp and Lane Control

Another key function of TMCs that relies heavily on sensor data is ramp and lane control. This function involves dynamically controlling lane usage or entry onto the freeway based on current conditions. By actively managing lane usage, traffic flow on highways can be improved. One of the most common types of control is ramp metering. The goal of the ramp meters is to prevent the freeway from breaking down into congestion by restricting entries onto the freeway from the ramps. In ramp metering, traffic signals are placed on entrance ramps to control the amount of traffic that accesses the highway. Capacity bottlenecks tend to occur near freeway ramp junctions. Ramp meters seek to improve freeway flow by reducing the number of vehicles that enter the highway, effectively shifting potential delays from the freeway to the ramps.

Ramp metering systems can either operate in a simple, pretimed manner or in a traffic responsive manner.[22] In simple systems, a preset number of vehicles are allowed to access the highway from the ramp during a certain time of day. Traffic responsive systems look at the vehicle occupancy on the mainline freeway, and dynamically change how many vehicles are allowed to access the highway based on that occupancy. As the road becomes more congested, the rate at which vehicles are allowed to enter the highway is reduced.

Ramp meters have been shown to produce positive effects on freeway operations when they have been used. In Washington State, a reduction in braking maneuvers of up to 79% has been observed at locations near the ramp meters.[23] One of the most comprehensive evaluations of ramp meters occurred in Minneapolis/St. Paul.[24] In 2000, the Minnesota legislature mandated that the state DOT conduct a complete shutdown of ramp meters in the Twin Cities and evaluate whether the ramp meters improved traffic operations. At the time, 430 ramp meters were installed over 210 mi of freeway. Following the shutdown of ramp meters, the DOT observed a 9% reduction in freeway volumes with no corresponding increase on parallel arterial routes. During the peak hours, throughput declined by 14% on the freeway. This showed conclusively that the ramp meters were having a positive impact on traffic flow.

Another example of lane control is dynamically allowing traffic to use the shoulder as a travel lane during peak periods. Several cities in the United States allow travel on predesignated shoulder lanes during peak travel times. Lighted signals indicating whether or not the shoulder is open to through traffic are placed over the shoulder. In Figure 1.3, lane 4 represents the shoulder lane. When the green arrow is illuminated, cars are free to travel in the shoulder lane. A red "X" is displayed when the shoulder is closed to travel. In most cases, shoulders are opened at predefined times based on historic data indicating

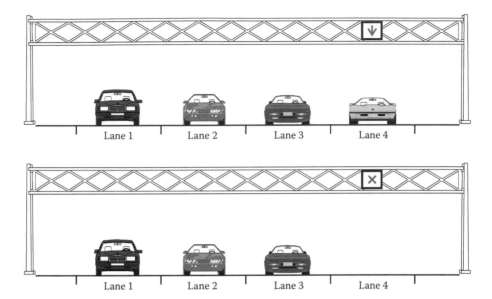

FIGURE 1.3   Dynamic use of the shoulder as a travel lane.

when peak travel occurs. The activation of these signals is typically done from the TMC. One of the key considerations in allowing travel on the shoulder is the impact on safety. Disabled vehicles will no longer be able to stop on the shoulder if it is used as a travel lane. As a result, the TMC must closely monitor traffic when the shoulder is in use to ensure that there are no conflicts between disabled vehicles and through traffic.

In several European countries, shoulders are dynamically opened to travel based on observed traffic conditions in the main travel lanes.[25] As traffic in the main lanes becomes more congested, the TMC will open the shoulders for travel. Sensor data from the main lanes are used to determine when traffic conditions meet thresholds to trigger this shift. Additional sensors on the shoulder are used to identify whether any vehicles have broken down or stopped on the shoulder. This allows the TMC to know whether it is safe to open the shoulder for travel at any given time. This sort of active traffic management is gaining increasing interest in the United States, and several DOTs are investigating the feasibility of transferring this technique to the United States.

1.4.2.6 Performance Measurement

State DOTs are under increasing pressure to show that transportation funds are being used wisely. Politicians often want to see quantitative data that tracks how well certain transportation facilities are performing. As a result, state DOTs have begun to regularly report the performance of key facilities over time. Archived data from TMCs is a key input into many of these activities, because the TMC data is usually the most robust dataset available. One of the leaders in performance measurement is the Washington State DOT. They produce a quarterly report, *Measures, Markers, and Mileposts*, commonly called the Gray Notebook.[23] The Gray Notebook covers a broad range of issues, including safety, congestion, project delivery, and infrastructure condition. Some of the congestion measures tracked by Washington include the following:

1. The number of severe incidents lasting more than 90 min
2. The average peak travel time
3. Vehicle throughput
4. Delay
5. The percent of days that speed falls below 35 mi/h
6. The duration of congestion

Obviously, the data generated by the TMCs plays an important role in producing the Gray Notebook. The real-time data is archived, and provides a rich data source that can be used to provide performance measurement and other historic analysis for freeway facilities. Many TMCs are increasingly viewing performance measurements as one of their core functions.

1.4.3 Traffic Control on Surface Streets

The term surface streets refers to roads that do not control how adjacent properties access the road (nonfreeways, in other words). The most common reason for installing sensors on surface streets is to support traffic signal operations. Sensors are a critical component of any traffic signal system that is responsive to the traffic demand on different approaches. There are three basic ways that traffic signals are controlled:

1. *Pretimed control.* In pretimed control, the amount of time that a traffic signal shows a green indication to each movement at an intersection is set to a fixed value. This type of control is not responsive to traffic at all, and no sensors are used to determine traffic conditions. Pretimed traffic control is used only when traffic volumes remain relatively consistent or where there are closely spaced, interconnected signals (such as in a downtown area).
2. *Semiactuated control.* In semiactuated control, sensors are present on the minor road at an intersection. When vehicles are detected on minor road approaches, the signal controller is notified that a green indication should be shown to that traffic. This type of control is often used when there is a definite higher volume route, and the minor road experiences intermittent traffic demand.
3. *Fully actuated control.* Fully actuated control uses sensors to detect traffic on all approaches to an intersection. This a demand responsive system that can accommodate changing traffic flow patterns on different approaches.

Thus, sensors are a critical component of semiactuated and actuated control. Actual vehicular demand at the intersection dictates which movements are shown a green signal, and how long the green signal is shown to each movement.

Traffic signals can operate as either isolated intersections or as part of a coordinated system. In isolated intersections, semiactuated and actuated controllers are free to allocate green time to different movements as traffic dictates, within certain parameters set by an engineer for a specific intersection. Coordinated systems seek to move traffic efficiently along a route by ensuring the vehicles traveling on a major road do not have to stop at every intersection. Signal timings are set so that vehicles arrive at adjacent signals during a green indication.

As mentioned earlier, ILDs and video detection are the primary means used to perform intersection detection. Although sensor data is a critical input into demand responsive signals, this data is often not retained. The sensor data is fed to a controller placed on the roadside that operates the signal. Once the signal cycle is completed, the data is deleted. A number of major urban areas operate traffic operations centers that allow signals to be monitored and operated remotely at a central point. This may or may not be a separate entity from the TMC. In these cases, signal timing plans can be dynamically changed by the traffic operations center based on observed traffic conditions. Some traffic operations centers also archive the sensor data at the signals so that they can use the volume counts at a future time for other purposes. This archived signal system data, however, is not commonly available.

1.5 Probe-Based Detection Methods

DOTs have traditionally relied on networks of point detectors to collect information. ILDs, video detection, and microwave radar all collect very detailed information at a single location. Although these methods may be very successful at collecting data for that specific point, difficulties can arise when that data is extrapolated to cover extended lengths of road. On freeways, congestion tends to start at locations where there are large volumes of traffic entering or leaving the freeway, as well as at locations where the number of lanes on the freeway mainline is reduced. Point detectors placed between interchanges may not do a good job of characterizing what travelers actually experience as they drive along a road because the congestion at interchanges may not be captured. This problem is even more pronounced on arterial roads, where the influence of traffic signals and vehicles turning in and out of driveways can create significant reductions in average travel speed between midblock sensors.

DOTs have shown a growing interest in capturing performance data across entire roadway links in order to gain more knowledge about issues relating to travel time and travel time variability. A number of new monitoring methods that rely on the use of probe vehicle data have been deployed as a way to gather this information. Probe vehicle-based systems track the movements of a subset of the vehicle population in order to estimate the travel characteristics of all vehicles on the road. Probe vehicle systems can generate estimates of speed and travel time on a section of road, but they generally do not produce estimates of the volume or density of traffic. By using this approach, estimates of SMS (rather than TMS) are generated for the road. Thus, the speed estimates more fully characterize what drivers actually experience.

This section discusses four methods that rely on the use of intelligent transportation systems technology as a way to generate this information. Although they are all probe vehicle-based systems, the level of technical maturity of these systems varies. Automatic vehicle identification (AVI) and automatic vehicle location (AVL) have both been shown to be technically viable though large-scale field deployments, but have not achieved widespread implementation in the field. Several pilot tests of wireless location technology have been carried out, but data quality has not been shown to be sufficient for most traffic engineering applications. The Vehicle Infrastructure Initiative is still in early development, with no production-level deployments under way. The potential benefits and disadvantages of each of these systems are discussed in the following sections.

1.5.1 Automatic Vehicle Identification

AVI systems rely on the toll transponders used in electronic toll collection to determine travel times on roads. In electronic toll collection, drivers acquire toll transponders that they mount on their car. Toll transponder tags reflect encoded radio signals transmitted from roadside antennas or readers. The reflected signals are modified by the toll tag identification code so that the tag's information can be read by the system. In an electronic tolling environment, that information is used to debit a customer's account to pay a toll. By using the transponder, drivers can travel through toll booths at high speeds and are not forced to slow down or stop to manually pay a toll.

This type of system has also been adapted to collect traffic monitoring data. In a monitoring application, roadside antennae are installed along major highways where the DOT wants to collect information on travel times or speeds. The unique toll tag identification numbers are logged each time a vehicle passes by an antenna, although the customer's account is not charged. The travel time of the vehicle can then be explicitly calculated by examining when a vehicle passes known antenna locations on the highway. This provides true point-to-point travel times for all vehicles with toll transponders. Figure 1.4 shows an example of how this would work. The antennae are mounted on sign structures that span the freeway, and the vehicle's transponder is logged each time it passes an antenna. Figure 1.4 is not to scale, and these antennae would be at least 1 mi apart from each other.

There are several issues related to the deployment of AVI-based monitoring systems. First, the ability of AVI-based systems to provide useful data is directly linked to the number of potential probes on the road. A sufficient number of probe vehicles must travel a route for the travel time estimate to have statistical validity. Past research has estimated

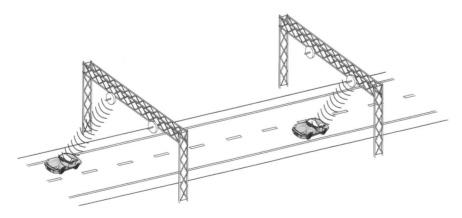

FIGURE 1.4   Automatic vehicle identification for traffic monitoring (not to scale).

that between one and four vehicles are required to traverse a link every 5 min to generate travel time estimates at a 95% level of confidence.[26] In other words, this system is only viable when there are enough toll roads and electronic toll collection users to justify its installation and many urban areas do not have a sufficient number of transponders to make this system viable. Likewise, these systems require the installation of significant roadside infrastructure in the form of AVI tag readers and communications in order to gather data. Capital costs for a single detector site on a six-lane highway range from $18,000 to $38,000, with annual operating costs of $4000 to $6000 per site.[27,28] If sites are to be spaced every 1 to 2 mi, this can be a significant cost.

One of the most robust AVI-based traffic monitoring systems in the United States is located in Houston, TX.[29] Antennas are installed on major freeways in the Houston area at an average spacing of 3 mi. The system covers 227 mi of freeway and 100 mi of high-occupancy vehicle lanes. Houston has several major toll facilities, and over one million toll tags were in use in 2003 to produce travel time estimates.[30] In this sort of operating environment, AVI-based monitoring systems can produce quality speed and travel time estimates.

1.5.2 Automatic Vehicle Location

AVL refers to a suite of technologies that track the location of vehicles traveling through the roadway network. AVL systems are commonly used by transit companies to track the location of buses on their routes, and by trucking and package delivery companies to assist in fleet management and routing activities. Much like AVI systems, AVL location data can be mined to develop speed and travel time estimates.

There are several ways that AVL systems can work. The first method is called a sign-post system, and is most commonly used by transit agencies that have fixed routes.[31] In this system, roadside transmitters at fixed locations emit unique identification codes. Approaching vehicles receive this identification code, and store this information with a corresponding time stamp, vehicle identification number, and odometer reading. This information is then transmitted back to a central control facility at predefined intervals. The more commonly used method relies on global positioning system (GPS) data. GPS data is collected continuously by the vehicle, and then periodically transmitted back to a central control facility over a radio backbone, cellular service, or satellite communications network.

The location data generated by AVL systems could be mined to generate traffic data. With GPS-based AVL systems, fixed infrastructure requirements are much less than for AVI-based systems. Vehicles could also be monitored anywhere on the network, because there is no roadside hardware for GPS-based systems. There are several instances where AVL-based systems have been used to provide estimates of speed and travel times on roads where no point sensors were available.[32]

There are several limitations to using AVL-based systems. First, only a small subset of the vehicle population is outfitted with AVL equipment. Thus, AVL systems could encounter many of the same sample size limitations as AVI systems. Transit vehicle AVL systems also may not provide data that is representative of all traffic, because transit vehicles have to stop to load and discharge passengers. Trucking companies have been reluctant to share their AVL data with others due to concerns about losing competitive advantages in the marketplace, so that data source is also not widely available. The amount of data generated by transit agencies and individual trucking companies may also not be sufficient to generate high-quality estimates of speeds and travel times for an entire region.

1.5.3 Wireless Location Technology-Based Monitoring Systems

Although AVL- and AVI-based monitoring systems can generate speed and travel time information, they both focus on monitoring a relatively small population of vehicles. Anonymous tracking of cellular phones, termed wireless location technology (WLT)-based traffic monitoring, has the potential to expand both the number of vehicles being monitored and the size of the roadway network where data could be obtained. WLT-based monitoring involves anonymously tracking the location of cellular phones in vehicles as they travel through the network. By looking at a series of positions, the travel speed for a particular phone (and by extension the vehicle it is riding in) can be estimated.

The technology used to perform this location estimation has not yet fully matured. Early-generation WLT systems developed in the 1990s estimated phone locations by examining wireless signal data as seen from multiple towers. This required installing specialized equipment at actual cellular towers to collect the data. Since about 2000, WLT-based systems have focused on mining existing data collected by cellular companies in order to estimate locations.[33] The cellular communications network knows the approximate location of a phone at any point in time in order to route calls efficiently to that phone. As a phone travels through the roadway network, it checks the strength of the signal with the nearest cellular tower. When the signal strength drops below a preset threshold, a handoff occurs whereby the call is transferred to an adjacent cellular tower. This creates seamless voice communications for the caller. Many current WLT systems mine this handoff data to try to determine vehicle speeds. The approximate geographic locations where handoffs occur can be determined, and algorithms have been developed that assign a path and determine travel times based on a series of handoffs.

There are a number of obvious advantages to using WLT-based approaches for generating speed and travel time information. First, there is an extremely large pool of potential probes because any cellular phone could act as a probe. In September 2006 it was estimated that over 72% of households in the United States had wireless service, so probe penetration would not be a problem with this technology.[34] Likewise, any road with cellular service could theoretically be monitored without installing any infrastructure on the road.

The two major barriers to widespread use of WLT-based monitoring relate to the maturity of the technology and institutional issues. A number of deployments have occurred in the United States, and none has been able to produce data of sufficient accuracy to support traveler information or real-time operations.[33] The spatial accuracy of the location estimates used by WLT systems is not as precise as GPS data. For example, existing systems cannot distinguish between different phones in the same vehicle or even determine differences in travel speeds between adjacent lanes of traffic. Because of these difficulties in providing precise location estimates, many systems have had difficulty producing precise estimates of speed and travel times, particularly on closely spaced urban roads. Some recent deployments have shown some improvements in accuracy, but some real concerns still remain about data latency and accuracy in some situations.[35]

The second barrier to the use of WLT is institutional. In WLT-based monitoring systems, the data is generated by a third party vendor that sells the data as a service to a DOT. Questions about who owns the data and what rights a DOT has to distribute the data still remain. As a production-level system has not been deployed, the final costs to deploy a large system remain unknown. As a result, it is unclear what cost savings might be achieved by using a WLT-based monitoring system. The public is also often concerned about the privacy implications of these systems. Federal communications law specifically prohibits cellular companies from releasing individually identifiable information, but aggregate information on link speeds does not appear to be protected.[33] As a result, it appears that there is no legal barrier to using the technology for most traffic management applications. Despite this, public perception of the privacy issues related to this technology will likely need to be addressed.

1.5.4 Vehicle Infrastructure Integration

The vehicle infrastructure integration (VII) program is developing a platform for the exchange of real-time data between vehicles and roadside infrastructure elements.[36] The VII program is a cooperative effort involving the U.S. DOT, several state DOTs, and the automobile industry. The program's primary focus is on improving safety by providing timely alerts of hazardous conditions, but there is also the potential to collect real-time traffic data using this technology. The VII program is built on using dedicated short-range communications (DRSC) technology using the 5.9 GHz band. The VII program is still in the research and development stage, and no market-ready systems have been deployed.

The principal components of the VII architecture are on board units (OBUs) in vehicles and roadside units (RSUs). OBUs collect information about vehicles and their surroundings and store that information for some period of time. The information is collected both periodically and also when certain events occur (like skidding or sudden deceleration). A total of 3 to 15 min of data would be stored on the OBU. Whenever an OBU passes an RSU, the OBU passes its data to the RSU using DSRC. The RSU can also pass information that it has gathered from other vehicles to the OBU at that time. Through this mechanism, vehicles could be alerted of conditions experienced by other vehicles that pass by the RSU. As an example, data could be exchanged between vehicles approaching an intersection and roadside hardware. The hardware would know what the other vehicles approaching the intersection are doing, and would be able to alert oncoming vehicles if it is not safe to enter the intersection because someone was going to run a red light. Numerous applications are in development that are exploring the ability of VII to improve safety at intersections and in other situations.

There are also significant implications in the VII on traffic operations and data collection. Real-time information on traffic speeds and congestion could be used to dynamically change traffic signal timings or provide traveler information to avoid congestion. Early simulation work indicates that reasonably good estimates of speed (with 3 to 4 mi/h) would be possible with VII OBU penetrations as low as 1%.[37] These results, however, have not yet been corroborated with actual field data.

The VII program is still in its infancy, and a number of obstacles need to be overcome before this is a viable system. First, the technology must be thoroughly tested and evaluated to ensure that it is reliable and accurate. Then, VII must be deployed in enough vehicles and at enough roadside locations to be able to generate quality information. If only a few vehicles have VII technology, then little valuable information will be generated. Likewise, sufficient roadside locations must be equipped for the VII concept to produce demonstrable safety and operational benefits. Another unresolved issue for the VII program is cost. Although the cost for in-vehicle systems will be borne by the purchaser of the vehicle, states will have to bear the cost of roadside equipment. As noted earlier, many states have difficulty maintaining the existing sensor infrastructure. Adding a large additional network of communications and roadside hardware will represent a significant capital, operating, and maintenance expense. Agencies will have to determine if they have the resources to operate a VII system effectively, and if the benefits of running a system outweigh the costs.

1.6 Chapter Summary and Future Trends

This chapter summarizes the current state-of-the-practice in traffic monitoring. Some of the key points of this chapter include the following:

1. Expansion of roadway capacity has not kept up with the growth in travel in the United States, causing significant increases in congestion in major urban areas. Nonrecurring congestion created by crashes or other incidents is a major component of delays experienced by drivers in urban areas.
2. Most traffic data is currently collected using a network of point detectors, such as inductive loops, microwave radar sensors, or video detection systems. Although these systems can collect very accurate data, they only gather this information at a single point. The data cannot always be easily extrapolated to long road segments.
3. DOTs use data for a wide variety of purposes, including characterizing travel on links, providing real-time monitoring and control, and controlling traffic signals on urban streets.
4. The type and quantity of traffic data available varies depending on where the data is collected. Major urban freeways tend to have closely spaced sensors that provide real-time data for traffic management. Rural roads often only have data collected at wide intervals, often with years elapsing between data collection efforts. Those roads often only have data collected to track how traffic volumes change over time.
5. DOTs are increasingly interested in probe-based traffic monitoring methods. AVI and AVL systems have been implemented in the field to generate travel time and speed information. WLT and VII approaches are still under development, but offer additional capabilities to collect data if technical issues can be resolved.

Given the discussion in this chapter, it is worthwhile to briefly discuss several trends that are influencing how transportation agencies might go about collecting and using traffic data in the future. As mentioned earlier, politicians and the general public are demanding greater transparency and accountability in how transportation agencies conduct their business. Agencies need to show that they are being responsible stewards of the public's money, and that projects are creating demonstrable improvements in traffic flow. Second, transportation agencies are being asked to provide services as efficiently as possible. The motor vehicle fuel tax is the primary revenue stream used to fund transportation improvements. Politicians have been reluctant to increase the fuel tax, especially with oil prices being at record high levels. Simultaneously, construction costs have increased substantially due to rising costs for materials such as structural steel and asphalt. The net result is that the real purchasing power of the fuel tax has declined and agencies have been asked to stretch funding dollars further. Finally, agencies have realized that they cannot build their way out of congestion. There is a new focus on maximizing the efficiency of existing facilities through better operations and management.

One of the impacts of these trends is that states are increasingly focused on providing performance measurement data that politicians and the general public can easily relate to, with travel time being the most obvious. Agencies are definitely focusing more on probe-based systems to generate this information, and this trend is likely to continue. The reduced buying power of the transportation dollar has also created several impacts in the traffic monitoring world. There is a growing interest in outsourcing traffic data collection to private vendors. The costs of operating and maintaining sensor systems have caused many agencies to look to the private sector to install and maintain sensors. These vendors then sell the data as a service to DOTs, television stations, and in-vehicle navigation companies. Another impact of the reduced funding stream is the increased interest in non-intrusive detectors. Nonintrusive detectors can be installed and maintained more easily than ILDs, and replacement costs and traffic impacts are mitigated.

Based on these trends in sensor usage and data acquisition, it seems likely that the following trends will occur in the future:

1. DOTs will try to increase the size of the network they actively monitor to support performance measurement activities and real-time management of facilities. This will include renewed efforts to gather real-time data on urban arterial roads.
2. Agencies will increasingly move from point detectors to probe-vehicle based systems. Travel time data will become a critical piece of data for many agencies.
3. Intrusive detectors will be replaced by nonintrusive sensors wherever feasible, in an attempt to reduce costs and impact on traffic.
4. DOTs will be looking to the private sector to augment sensor facilities, rather than expanding the size of the network that the DOT must operate and maintain.

Given the current landscape, it seems likely that transportation agencies will be undergoing some significant changes in how they collect traffic data. Vehicular networks offer the potential to provide a new data source that could be used in future years as the technology matures and becomes widespread.

References

1. Schrank, D. and Lomax T., *The 2007 Urban Mobility Report*, Texas Transportation Institute, College Station, TX, 2007.
2. U.S. Department of Transportation, National Household Transportation Survey, available at http://www.bts.gov/programs/national_houshold_transportation_survey/, 2007.
3. Pisarski, A., *Commuting in America III: National Cooperative Highway Research Program Report 550*, Transportation Research Board, National Research Council, Washington, DC, 2006.

4. Transportation Research Board, *Highway Capacity Manual*, Transportation Research Board, Washington, DC, 2000.

5. Federal Highway Administration, *Traffic Monitoring Guide*, Federal Highway Administration, Washington, DC, 2001.

6. May, A.D., *Traffic Flow Fundamentals*, Prentice-Hall, Englewood Cliffs, NJ, 1990.

7. Middleton, D., Jasek, D., and Parker, R., *Evaluation of Some Existing Technologies for Vehicle Detection*, Texas Transportation Institute, College Station, TX, 1999.

8. Hanshaw, S., "New" Data Sources, Paper presented at the North American Traffic Monitoring Exhibition and Conference, Minneapolis, MN, 2006.

9. Klein, L.A., *Sensor Technologies and Data Requirements for ITS*, Artech House, Boston, 2001.

10. Klein, L., Mills, M., and Gibson, D., *Traffic Detector Handbook*, 3rd ed., vol. I, Federal Highway Administration, Washington, DC, 2006.

11. Federal Highway Administration, Estimated Average Annual Highway Apportionment, available at http://www.fhwa.dot.gov/reauthorization/rta-000-1664ar.xls, 2005.

12. Research and Innovative Technology Administration, *State Transportation Statistics*, U.S. Department of Transportation, Washington, DC, 2006.

13. Federal Highway Administration, About Highway Performance Monitoring System (HPMS), available at http://www.fhwa.dot.gov/policy/ohpi/hpms/abouthpms.htm, 2002.

14. Federal Highway Administration, Overview of Highway Performance Monitoring System (HPMS) for FHWA Field Offices, available at http://www.fhwa.dot.gov/policy/ohpi/hpms/hpmsprimer.htm, 2006.

15. Meyer, M.D., *A Toolbox for Alleviating Traffic Congestion and Enhancing Mobility*, Institute of Transportation Engineers, Washington, DC, 1997.

16. Federal Highway Administration, 511 Deployment, available at http://www.fhwa.dot.gov/trafficinfo/511.htm, 2007.

17. Payne, H. and Tignor, S.C., Freeway incident detection algorithms based on decision trees with states, *Transportation Research Record*, 682, 30–37, 1978.

18. Persaud, B.N., Hall, F.L., and Hall, L.M., Congestion identification aspects of the McMaster incident detection algorithm, *Transportation Research Record*, 1287, 167–175, 1990.

19. Parkany, E. and Shiffer, G., *Survey of Advanced Technology Deployment in Traffic Management Centers with an Emphasis on New Sensor Technologies and Incident Detection*, University of California—Irvine, Irvine, 1996.

20. Dougald, L. and Demetsky, M., *Performance Analysis of Virginia's Safety Service Patrols: A Case Study Approach*, Virginia Transportation Research Council, Charlottesville, VA, 2006.

21. Dougald, L., *A Return on Investment Study of the Hampton Roads Safety Service Patrols*, Virginia Transportation Research Council, Charlottesville, VA, 2007.

22. Roess, R.P., McShane, W.R., and Prassas, E.S., *Traffic Engineering*, Prentice-Hall, Upper Saddle River, NJ, 1998.

23. Washington State Department of Transportation, *Measures, Markers, and Mileposts*, Washington State Department of Transportation, Olympia, WA, 2007.

24. Cambridge Systematics, *Twin Cities Ramp Meter Evaluation*, Cambridge Systematics, Oakland, CA, 2001.

25. Federal Highway Administration, *Active Traffic Management: The Next Step in Congestion Management*, U.S. Department of Transportation, Washington, DC, 2007.

26. Holdener, D.J. and Turner, S.M., Probe Vehicle Sample Size for Real-Time Information: The Houston Experience, in *Intelligent Transportation: Realizing the Benefits*, ITSAmerica, Houston, 1996.

27. Ullman, G.L., Balke, K.N., McCasland, W.R., and Dudek, C.L., Benefits of Real-Time Travel Time Information in Houston, in *Intelligent Transportation: Realizing the Benefits*, ITS America, Houston, TX, 1996.

28. Wright, J. and Dahlgren, J., *Using Vehicles Equipped with Toll Tags as Probes for Providing Travel Times*, University of California, Berkeley, CA, 2001.

29. Smalley, D.G., Hickman, D.R., and McCasland, W.R., *Design and Implementation of Automatic Vehicle Identification Technologies for Traffic Monitoring in Houston, Texas*, Texas Transportation Institute, College Station, TX, 1996.

30. Houston TranStar, Houston TranStar fact sheet, available at http://www.houstontranstar.org/about_transtar/, 2003.

31. Turner, S.M., Eisele, W.L., Benz, R.J., and Holdener, D.J., *Travel Time Data Collection Handbook*, Federal Highway Administration, Washington, DC, 1998.

32. Dailey, D.J. and Cathey, F.W., *AVL Equipped Vehicles as Speed Probes (Final Phase)*, Washington State Transportation Center, Seattle, WA, 2005.

33. Smith, B.L. and Fontaine, M.D., *Private Sector Provision of Congestion Data*, National Cooperative Highway Research Program, Washington, DC, 2007.

34. Cellular Telephone and Internet Association, Wireless Statistics, available at http://www.ctia.org/research_statistics/index.cfm/AID/10202, 2006.

35. Telvent Farradyne, Inc., *Kansas City Scout Traffic Management Center Special Report: Evaluation of Cellular Probe Data Final Report*, Telvent Farradyne, Kansas City, KS, 2007.

36. ITS Joint Program Office, *VII Architecture and Functional Requirements, Version 1.1*, U.S. Department of Transportation, Washington, DC, 2005.

37. Smith, B.L., Park, B., Tanikella, H., and Zhang G., *Preparing to Use Vehicle Infrastructure Integration in Traffic Operations, Phase I*, Virginia Transportation Research Council, Charlottesville, VA, 2007.

2

Models for Traffic Flow and Vehicle Motion

Martin Treiber and
Arne Kesting
*Institute for Transport
and Economics
Technical University
of Dresden*

2.1 Introduction

Transportation is a central economic factor in industrialized countries. The associated mobility is also an integral part of our quality of life, self-fulfillment, and personal freedom. Today's traffic demand is predominantly served by individual motor vehicle travel, which is the primary means of transportation. In Germany, for example, motorized vehicles constitute 77% of the individual transport (measured in terms of passenger-kilometers) and 70% of the total freight traffic (measured in tonne-km).[1] This leads to a growing socioeconomic burden. According to a study by the European Commission, the external costs of congestion already amount to 0.5% of the gross national product (GNP) in the European Union and will increase to 1% by the year 2010.[2] However, in view of the restricted space and societal concerns about the environmental impact and energy consumption, building new infrastructure is rarely a viable option. So, the efficient management of traffic, particularly individual motorized traffic, is one of the greatest challenges, and optimizing the capacity and efficiency of the vehicular networks is of prime importance. In the presence of increasing computing power, realistic microscopic traffic models and the associated simulations are becoming a more and more important tool.

Traffic models have been successful in reproducing the observed *collective, self-organized traffic dynamics*, including phenomena such as breakdowns of traffic flow, the propagation of stop-and-go waves (with a characteristic propagation velocity), capacity drop, and different spatiotemporal patterns of congested traffic due to instabilities and nonlinear interactions.[3–7] For an overview of experimental studies and the development of miscellaneous traffic models, we refer to the recently published extensive review literature.[3,8–12]

The primary means by which traffic flow models may help in optimizing the vehicular network is the evaluation and simulation-based optimization of various control measures. This includes "classic" measures such as permanent speed limits, or the optimal guidance of traffic past locations of roadworks, and more modern and flexible concepts and services that often are summarized by the term *intelligent transportation systems* (ITS).

Road-based ITS strategies that are accessible to traffic flow modeling include advanced traffic control systems such as variable message signs, adaptive speed limits, dynamic route guidance, incident management, entrance ramp metering, temporary or local overtaking bans for trucks, and many more. Although it is often problematic or even impossible to test/optimize some aspects of these measures in real traffic, traffic-flow models allow these tests for virtual traffic.

Vehicle-based ITS strategies that can be optimized by the models discussed here include driver assistance systems with two dimensions: (i) "driver information," such as detailed information about the traffic situation, including an estimation of delays due to congestion, and (ii) automation of some driving tasks such as acceleration and deceleration by means of adaptive cruise control (ACC). With respect to traffic information, models of traffic flow can be used to enhance the performance of traffic-state recognition and short-term prediction, either directly by running real-time simulations,[13,14] or indirectly by using the deeper understanding ("stylized facts"; see Section 2.2.3) of traffic flow dynamics obtained from the models.[15,16] When including actual "vehicle control," models for traffic flow can also be used for simulating and assessing the efficiency of vehicle-based control strategies. In contrast to pure information systems, ACC provides feedback on the traffic flow. In view of the low percentage of vehicles equipped with ACC, simulations are the only means to assess the future impact that such assistance systems have on traffic flow. Currently, it is not even clear whether this effect is positive[17–19] or negative.[4,20] Finally, traffic models are used to generate surrounding traffic, both in computer games and in driving simulators used for behavioral and psychological studies of drivers.

Generally, the various dynamic aspects of vehicular transportation can be sorted by the associated typical timescales as shown in Table 2.1. Models of traffic flow are useful for all phenomena with timescales between ca. 1 sec and ca. 1 h. Dynamic aspects on a shorter timescale, including some assistance systems such as antilock braking systems (ABS), electronic stability control (ESC), lane-departure warning, precrash-braking are purely in the domain of vehicle rather than traffic dynamics. Timescales of 1 h and longer are the domain of transportation planning.

TABLE 2.1 Subjects in Transportation Systems Sorted by Typical Timescales Involved

Timescale	Subject	Models	Aspects
0.1 sec	Vehicle dynamics	Submicroscopic	Drive-train, brake, ESP[a]
1 sec			Reaction time, time gap
10 sec			
1 min	Traffic dynamics	Car-following models	Accelerating and braking
10 min		Fluid-dynamic models	Traffic light period
1 h			Period of stop-and-go wave
			Peak hour
1 day	Transportation planning	Traffic assignment models	Day-to-day human behavior
1 year		Traffic demand model	Building measures
5 years		Statistics	Changes in spatial structure
50 years		Prognosis	Changes in demography

[a] Electronic stability program, a driver assistance system detecting and preventing lateral skidding.

2.2 Models for Longitudinal Vehicle Movement and Traffic Instabilities

The mathematical description of the dynamics of traffic flow already has a long history. The scientific activity had its beginnings in the 1930s with the pioneering studies on the fundamental relations of traffic flow, velocity, and density conducted by Greenshields.[21] By the 1950s, scientists had started to describe the physical propagation of traffic flow by means of dynamic macroscopic and microscopic models. During the 1990s, the number of scientists engaged in traffic modeling grew rapidly because of the availability of better traffic data and higher computational power for numerical analysis. At present, traffic-flow models are widely used for traffic-state estimation and the growing field of vehicle-based ITS systems.

2.2.1 Macroscopic vs. Microscopic Approaches

There are two major approaches to describing the spatio-temporal propagation of traffic flow. *Macroscopic traffic flow models* make use of the picture of traffic flow as a physical flow of a continuous fluid (Figure 2.1). They describe the traffic dynamics in terms of aggregated macroscopic quantities such as the traffic density $\rho(x,t)$, traffic flow $Q(x,t)$, or the average velocity $V(x,t)$ as a function of space x and time t corresponding to partial differential equations. The underlying assumption of all macroscopic models is the conservation of vehicles, expressed by the continuity equation

$$\frac{\partial \rho}{\partial t} + \frac{\partial (\rho V)}{\partial x} = \nu_{rmp}(x,t),\tag{2.1}$$

where the source term ν_{rmp} represents inflows from and outflows to ramps or side roads.[3] In the simplest case initially considered by Lighthill, Whitham and Richards,[22,23] the continuity equation is supplemented by a static velocity–density relation

$$V(x,t) = V_e(\rho(x,t)).\tag{2.2}$$

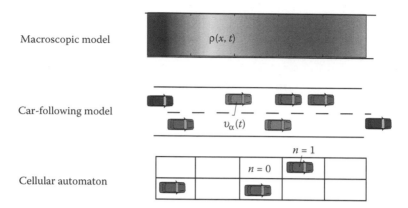

FIGURE 2.1 Illustration of different traffic modeling approaches. A snapshot of a road section at time t_0 is either characterized by *macroscopic traffic quantities* such as traffic density $\rho(x, t_0)$, flow $Q(x, t_0)$, or average velocity $V(x, t_0)$, or, *microscopically*, by the positions $x_\alpha(t_0)$ of single driver-vehicle agent α. For cellular automata, the road is divided into cells, which can be either occupied by a vehicle or empty.

Many functional forms have been proposed for the equilibrium velocity function $V_e(\rho)$. This approach does not allow for a dynamic generation of traffic instabilities or stop-and-go traffic, which is completely sufficient for city traffic where the traffic dynamics are essentially controlled externally by the traffic lights. In contrast, more advanced so-called "second-order" models are necessary for a realistic description of the traffic instabilities arising in freeway traffic due to the finite acceleration capability of vehicles.[24,25] Although the continuity Equation 2.1 remains unchanged, the static velocity relation 2.2 is replaced by a dynamic acceleration equation, which takes the general form

$$\frac{\partial V}{\partial t} + V\frac{\partial V}{\partial x} + \frac{1}{\rho}\frac{dP(\rho, V)}{dx} = A\left(\rho, V, \rho_a, V_a, \frac{\partial \rho}{\partial x}, \frac{\partial V}{\partial x}, \dots\right). \qquad (2.3)$$

Here, the "traffic pressure" results from the microscopic velocity variance, and the function $A(\cdot)$ represents the macroscopic equivalent of the vehicle acceleration that may depend on the state variables $\rho(x,t)$, $V(x,t)$, and their spatial derivatives. In the case of nonlocal models such as the gas-kinetic-based traffic model described in Ref. 25, the acceleration function also depends on the state variables ρ_a and V_a taken at anticipated positions $x_a > x$.

By way of contrast, *microscopic traffic models* describe the motion of each individual vehicle. They model actions such as accelerations, decelerations, and lane changes of each driver as a response to the surrounding traffic. The result is individual trajectories $x_\alpha(t)$ of all vehicles α, and, in the case of multilane models, the lateral movement $y_\alpha(t)$ or the lane number $l_\alpha(t)$ as well. Microscopic traffic models are especially suited to the study of heterogeneous traffic streams consisting of different and individual types of *driver–vehicle units*. As all vehicle-based ITS concepts lead to heterogeneous traffic (namely, equipped and nonequipped vehicles), microscopic models are best suited for these applications.

Specifically, one can distinguish the following major subclasses of microscopic traffic models (cf. Figure 2.1):

1. *Time-continuous models* are formulated as ordinary or delay-differential equations and, consequently, space and time are treated as continuous variables. *Car-following models* are the most prominent examples of this approach.[26–29] In general, these models are deterministic, but stochasticity can be added in a natural way.[30] For example, a modified version of the Wiedemann model[31] is used in the commercial traffic simulation software PTV-VISSIM™.

2. *Cellular automata* (CA) use integer variables to describe the dynamic state of the system. The time is discretized and the road is divided into cells that can be either occupied by a vehicle or empty. Besides rules for accelerating and braking, most CA models require additional stochasticity. The first CA for describing traffic was proposed by Nagel and Schreckenberg.[32] Although CA lack the accuracy of time-continuous models, they are able to reproduce some traffic phenomena.[33–35] Owing to their simplicity, they can be implemented very efficiently and are suited to simulating large road networks.[13]

3. *Iterated coupled maps* lie between CA and time-continuous models. In this class of model, the update time is considered as an explicit model parameter rather than an auxiliary parameter needed for numerical integration.[36] Consequently, the time is discretized while the spatial coordinate remains continuous. Popular examples are the Gipps model[37] and the Newell model.[38] However, these models are typically associated with car-following models as well.

2.2.2 Car-Following Models

Time-continuous car-following models and the related lane-changing models are arguably the most intuitive class of traffic models because their dynamic variables have a direct relation to the actual driving experience. In the rest of this contribution, we will focus on this class of models. For each vehicle α, the following are the basic dynamic state variables (cf. Figure 2.2):

- Location x_α
- Velocity $v_\alpha = \mathrm{d}x_\alpha/\mathrm{d}t$
- Lane l_α.

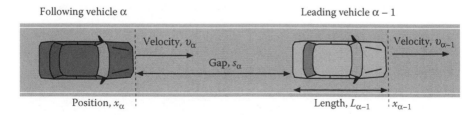

FIGURE 2.2 The dynamic state variables of car-following models. Notice that the vehicle indices α are ordered such that $(\alpha - 1)$ denotes the leading vehicle.

Derived quantities such as the spacial gap $s_\alpha = x_{\alpha-1} - x_\alpha - L_{\alpha-1}$ with the vehicle length $L_{\alpha-1}$ of the leading vehicle or the acceleration dv_α/dt have an intuitive meaning as well. When considering ACC systems, even the "jerk" (the time derivative of the acceleration) becomes essential as a measure of driving comfort. Most car-following models can be written in the form

$$\frac{dx_\alpha}{dt} = v_\alpha, \tag{2.4}$$

$$\frac{dv_\alpha}{dt} = a_{mic}(s_\alpha(t), v_\alpha(t), \Delta v_\alpha(t)). \tag{2.5}$$

This class of car-following models is completely defined by the microscopic acceleration function $a_{mic}(\cdot)$, which is the equivalent of the acceleration function $A(\cdot)$ of the macroscopic models. The acceleration function must encode at least the following aspects of driving:

- An *acceleration strategy* towards a desired speed in the free-flow regime
- A *braking strategy* for approaching other vehicles or obstacles, including emergency braking situations
- A *car-following strategy* for maintaining a safe distance when driving behind another vehicle

Depending on the purpose, the argument list of the acceleration function $a_{mic}(\cdot)$ can be generalized to include further leading and trailing vehicles (multianticipation of experienced human drivers, "pushing" effects of aggressive drivers), vehicles on the neighboring lanes (anticipation/preparation of lane changes), and accelerations ("braking lights"). Furthermore, the state variables can be taken at delayed times (modeling finite reaction times), which leads to delay-differential equations, or stochastic terms can be added (nonperfect driving and estimation errors), resulting in stochastic differential equations. It is important to model these additional influencing factors when modeling the human driver in a more detailed way.[39] It is interesting to note that the simpler car-following models described by acceleration functions of the type of Equation 2.5 correspond more closely to a controller for adaptive cruise control rather than to a human driver.

2.2.3 The Intelligent Driver Model

By way of example, we will consider the Intelligent Driver Model (IDM)[27] in this section. The IDM acceleration function is of the type of Equation 2.5. It is given by

$$\frac{dv_\alpha}{dt} = a_{mic}(s_\alpha, v_\alpha, \Delta v_\alpha) = a\left[1 - \left(\frac{v_\alpha}{v_0}\right)^\delta - \left(\frac{s^*(v_\alpha, \Delta v_\alpha)}{s_\alpha}\right)^2\right]. \tag{2.6}$$

This expression combines the acceleration strategy $\dot{v}_{free}(v) = a[1 - (v/v_0)^\delta]$ towards a *desired speed* v_0 on a free road with the parameter a for the *maximum acceleration* with a braking strategy $\dot{v}_{brake}(s, v, \Delta v) = -a(s^*/s)^2$ serving as a repulsive interaction when vehicle α comes too close to the vehicle ahead. If the distance to the leading vehicle, s_α, is large, the interaction term $\dot{v}_{\text{brake}}$ is negligible and the IDM equation is reduced to the free-road acceleration $\dot{v}_{\text{free}}(v)$, which is a decreasing function of velocity with a maximum value $\dot{v}(0) = a$ and a minimum value $\dot{v}(v_0) = 0$ at the desired speed v_0. For more dense traffic, the deceleration term becomes relevant. It depends on the ratio between the effective "desired minimum gap"

$$s^*(v, \Delta v) = s_0 + vT + \frac{v\,\Delta v}{2\sqrt{ab}}, \tag{2.7}$$

and the actual gap s_α. The *minimum distance* s_0 in congested traffic is significant for low velocities only. The main contribution in stationary traffic is the term vT, which corresponds to following the leading vehicle with a constant *desired time gap* T. The last term is only active in nonstationary traffic corresponding to situations in which $\Delta v \neq 0$ and implements an "intelligent" driving behavior including a braking strategy that, in nearly all situations, limits braking decelerations to the *comfortable deceleration b*. Note, however, that the IDM brakes more strongly than b if the gap becomes too small. This braking strategy makes the IDM collision-free.[27] All IDM parameters v_0, T, s_0, a, and b are defined by positive values. These parameters have a reasonable interpretation, are known to be relevant, are empirically measurable, and have realistic values.[40] Moreover, the IDM acceleration $a_{\text{mic}}(s, v, \Delta v)$ is a smooth function of the three controlling variables with plausible properties, in particular $\partial a_{\text{mic}}/\partial s \geq 0$, $\partial a_{\text{mic}}/\partial v \leq 0$, $\partial a_{\text{mic}}/\partial \Delta v \leq 0$, and $a_{\text{mic}}(\infty, v_0, 0) = 0$ (Figure 2.3).

The *stationary properties* of the IDM are influenced by the parameters for the desired time gap T, the desired speed v_0, and the minimum distance between vehicles at a standstill s_0. A stationary traffic situation is characterized by constant velocities, $v_\alpha(t) = v_e$ for all time instants and vehicles, that is, $\Delta v = dv_\alpha/dt = 0$ for all vehicles and times. In the case of *heterogeneous traffic*, that is, every driver–vehicle unit α has individual model parameters, this corresponds to individual *equilibrium gaps* $s_e^{(\alpha)}(v_e)$. In the following, we

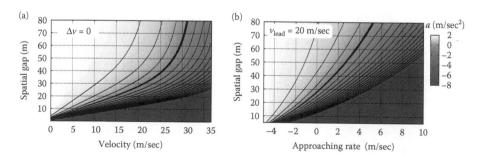

FIGURE 2.3 Acceleration function (2.6) of the IDM (a) as a function of the velocity and the gap for zero-velocity difference to the leader, (b) as a function of the velocity difference and the gap for a constant velocity v_{lead} of the leading vehicle. The IDM model parameters are given by $v_0 = 120\,\text{km/h}$, $T = 1.2\,\text{sec}$, $s_0 = 2\,\text{m}$, and $a = b = 2\,\text{m/sec}^2$.

consider a homogeneous ensemble of identical driver–vehicle agents corresponding to identical parameter settings. Then, the IDM acceleration Equation 2.6 with the constant setting $\delta = 4$ simplifies to

$$s_e(v) = \frac{s_0 + vT}{\sqrt{1 - (v/v_0)^4}}. \tag{2.8}$$

The equilibrium distance depends only on the minimum jam distance s_0, the safety time gap T and the desired speed v_0. The diagrams in Figure 2.4 show the equilibrium distance as a function of the velocity $s_e(v)$ for different settings of the relevant parameters v_0, T, and s_0. In particular, the equilibrium gap of homogeneous *congested* traffic (with $v \ll v_0$) is essentially equal to the desired gap, $s_e(v) \approx s^*(v, 0) = s_0 + vT$. It is therefore composed of the minimum bumper-to-bumper distance s_0, kept in stationary traffic at $v = 0$, and an additional velocity-dependent contribution vT corresponding to a constant safety time gap T, shown in the diagrams by straight lines. For $v \to 0$, the equilibrium distance approaches the minimum distance s_0. If the velocity is close to the desired speed, $v \approx v_0$, the equilibrium distance s_e is clearly larger than the distance vT according to the safety time gap parameter. For $v \to v_0$, the equilibrium distance diverges due to the vanishing denominator in Equation 2.8. That is, the free speed is reached *exactly* only on a free road.

In the literature, the equilibrium state of homogeneous and stationary traffic is often formulated in the macroscopic quantities traffic flow Q (vehicles per time unit and lane), (local) average velocity V, and traffic density ρ (vehicles per km and lane). The translation from the microscopic net distance s into the density is given by the *micro–macro relation*

$$s = \frac{1}{\rho} - l, \tag{2.9}$$

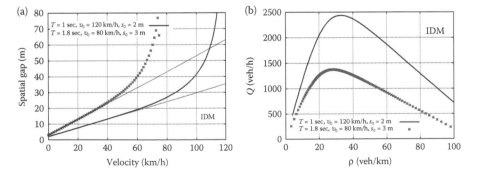

FIGURE 2.4 Equilibrium relations for the velocity and the traffic flow as a function of the traffic density for the IDM and two sets of the relevant parameters v_0, s_0, and T. The vehicle length has been set to $L_\alpha = 6$ m.

where l is the vehicle length. Furthermore, the mean velocity is simply $V = v_e$ and the traffic flow follows from the hydrodynamic relation

$$Q = \rho V. \tag{2.10}$$

This leads to the flow-density diagram of Figure 2.4b, which is sometimes called a *fundamental diagram*. In the low-density limit $\rho \ll 1/(v_0 T)$, the equilibrium flow can be approximated by $Q \approx v_0 \rho$. In the high-density regime, one has a linear decrease of the flow,

$$Q(\rho) \approx \frac{1 - \rho(l + s_0)}{T}, \tag{2.11}$$

which can be used to determine the effective length $l + s_0$, and the time gap T from aggregated detector data.

The *dynamic properties* of the IDM are mainly controlled by the maximum acceleration a, and the parameter for the comfortable braking deceleration, b. Let us now consider the following scenario: If the distance s is large (corresponding to the situation of a nearly empty road), the interaction $\dot{v}_{\text{brake}}$ is negligible and the IDM Equation 2.6 is reduced to the free-road acceleration $\dot{v}_{\text{free}}(v)$. The driver accelerates to his or her desired speed v_0 with the maximum acceleration $\dot{v}(0) = a$. The acceleration exponent δ specifies how the acceleration decreases when approaching the desired speed. The limiting case $\delta \to \infty$ corresponds to approaching v_0 with a constant acceleration a, while $\delta = 1$ corresponds to an exponential relaxation to the desired speed with the relaxation time $\tau = v_0/a$. In the latter case, the free-traffic acceleration is equivalent to that of the Optimal Velocity Model.[26] However, the most realistic behavior is expected between the two limiting cases of exponential acceleration (for $\delta = 1$) and constant acceleration (for $\delta \to \infty$). Therefore, we set the acceleration exponent constant to $\delta = 4$.

Figure 2.5 shows an example that is typical for city traffic controlled by traffic lights: At the beginning (a traffic light switches to green), a standing queue of vehicles accelerates to the desired velocity $v_0 = 50\,\text{km/h}$. When the resulting platoon of vehicles approaches the (red) traffic light at the next intersection 1 km downstream, the platoon decelerates to a standstill, again. The typical accelerations and decelerations are consistent with the parameter settings $a = 1\,\text{m/sec}^2$ and $b = 2\,\text{m/sec}^2$. Notice that the term $\propto \Delta v$ in the dynamic desired distance (2.7) includes an anticipative braking reaction such that the actual deceleration is essentially restricted to the comfortable deceleration parameter b. This means that smaller values of b automatically lead to a higher sensitivity with respect to velocity differences, and to a more anticipatory driving style. We require that, in this situation, the restriction of decelerations to values not much above b is not only true for the first vehicle of the platoon, but for the subsequent vehicles as well. This constitutes a critical test for car-following models, which most models of the type of Equation 2.5 fail.

2.2.4 Simulating Traffic Instabilities

Figure 2.6 shows the reaction to a perturbation in a freeway traffic situation with dense (but not congested) traffic. In contrast to city traffic, the higher velocities driven on

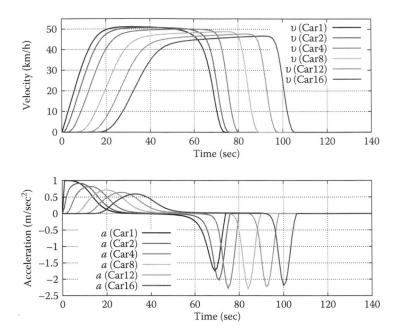

FIGURE 2.5 IDM simulation in a city environment. Shown is a situation where a traffic light with waiting vehicles turns green at time $t = 0$, and the vehicles drive to the (red) traffic light at the next intersection located 1 km downstream.

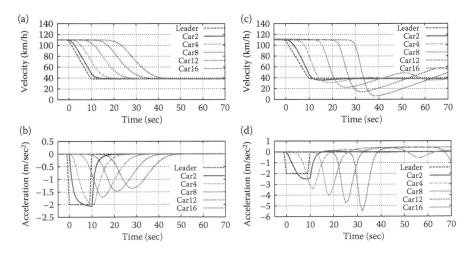

FIGURE 2.6 Simulated reaction of a vehicle platoon (desired velocity 144 km/h) driving behind a slower vehicle with no overtaking possibility after this vehicle decelerates. Plots (a) and (b) show the velocity and acceleration for stable acceleration parameters ($a = b = 2$ m/sec^2), while plots (c) and (d) show the same situation for parameters leading to a traffic instability ($a = 0.5$ m/sec^2, $b = 5$ m/sec^2). The values for T and s_0 are the same as in Figure 2.3.

freeways will lead to collective instabilities of traffic flow if the parameter settings correspond to little anticipation (high values of b) and a low agility (low values of a). In agreement with observations, collective instabilities are rarely observed in city traffic.

Figure 2.7 shows collective traffic instabilities in form of trajectory data. The plot on the left-hand side shows one of the first sets of empirical trajectory data from the pioneering work of Treiterer.[41] The IDM simulation[42] on the right-hand side reproduces the essential properties of the observations, particularly the growth of the amplitude of the perturbation until saturation (temporarily standstill traffic), and the propagation velocity of the stop-and-go wave traveling *against* the direction of traffic flow with a velocity between 15 km/h and 20 km/h.[3]

One of the nicer aspects of time-continuous car-following models is the possibility for interactive simulation both for educational and research purposes. Figure 2.8 shows screenshots of Java applets implementing the IDM and the lane-changing model to be discussed in Section 2.3.1 that are publicly available on the Internet.[43] The left-hand screenshot of Figure 2.8 shows stop-and-go waves (cluster of vehicles) in a "pure" homogeneous road context that is most easily implemented by a ring road. Notice that, recently, stop-and-go waves on a ring road have been observed in real traffic, at least in an experimental setting.[44] The on-ramp situation on the right-hand side shows the realistic situation of an open inhomogeneous system where an on-ramp serves as a bottleneck. Both in the simulation and in real traffic observations, one finds that the stop-and-go waves are emanating from the bottleneck region. As they propagate against the driving direction and can reach locations as far as ten or more kilometers upstream of the generating bottleneck, drivers might get the illusion of a "phantom traffic jam."

Are there genuine "phantom traffic jams" as suggested by the Japanese driving experiment?[44] An extensive empirical survey[7] of the breakdown phenomena on the

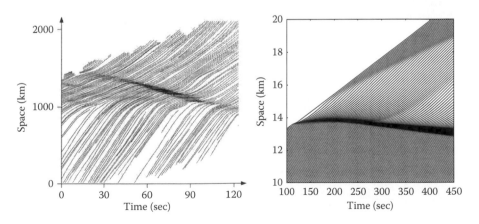

FIGURE 2.7 Trajectory data showing a collective instability leading to a stop-and-go wave in observation and in an IDM simulation.

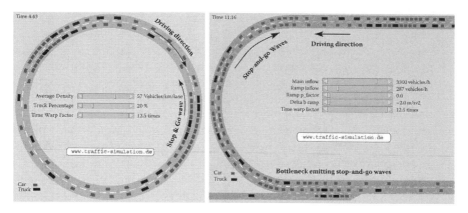

FIGURE 2.8 Screenshots of propagating stop-and-go waves on a ring road (left), and a two-lane freeway with an on-ramp (right) from the website. The source code of the simulator implements the IDM and is publicly available as an open source. Further scenarios are available that show the effects of speed limits, uphill gradients, and traffic lights.

German freeway A5 shows that the vast majority of all traffic breakdowns are caused by bottlenecks such as on-ramps and off-ramps at intersections and junctions, obstructions by accidents or road construction sites, uphill or downhill gradients, and even by accidents on the *opposite* driving direction causing a "behaviorally induced" bottleneck. As an example, Figure 2.9b shows a composite congestion on the German freeway A8-East caused by uphill and downhill gradients ("Irschenberg") in the region $38\,km \leq x \leq 41\,km$, and an additional obstruction by an accident at $x = 43.5\,km$ that is active in the time period between 17:40 h and 18:15 h. Finally, the congestion patterns of Figure 2.9a are caused by combinations of off-ramps and on-ramps (junctions, intersections).

Besides the bottleneck as a universal source of traffic breakdowns, the empirical results can be summarized by the following stylized facts for the spatiotemporal evolution of congested traffic states:

1. There are three "ingredients" that are nearly always present in real-world (in contrast to academic) situations: (i) a high traffic demand, (ii) a stationary road inhomogeneity in the form of a bottleneck, and (iii) dynamic perturbations of the traffic flow caused, for example, by an unexpected braking or lane-changing maneuver, or by trucks overtaking each other.
2. The resulting region of congested traffic is either localized with an almost constant spatial extension of typically less than 1 km, or extended, with a time-dependent spatial extension.
3. The downstream front of congested traffic is either fixed at the bottleneck, or moves upstream with a characteristic velocity of about 15 km/h. Both cases can occur within one congestion pattern, as shown in Figure 2.9a, where the stationary downstream front at $x = 510\,km$ detaches at about 09:45 h. In the congested pattern displayed in Figure 2.9b, the stationary downstream front at $x = 43.5\,km$ starts moving upstream at about 18:20 h.

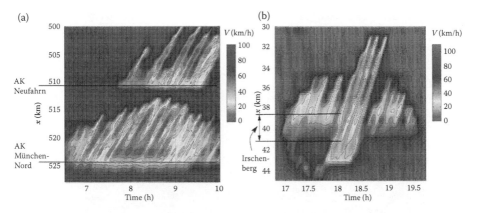

FIGURE 2.9 Spatiotemporal dynamics of the average velocity. (a) German freeway A9 South in the region north of Munich. We show two intersections (labelled "AK") playing the role of bottlenecks. Traffic flows in the direction of increasing values of x. (b) German freeway A8-East about 40 km to the east of Munich. Here, the bottlenecks are caused by the uphill and downhill gradients of the hill "Irschenberg" and by an accident at $x = 43.5$ km in the time interval between 17:40 h and 18:20 h.

4. It follows from this and the previous fact that localized congested patterns are either pinned at the bottleneck (called "pinned localized traffic,"[27,45] or "localized synchronized traffic"[4]) or they propagate upstream in the form of isolated stop-and-go waves ("moving localized clusters"[45]) at a velocity of about 15–18 km/h.
5. The upstream front of extended congestions has no characteristic velocity. Depending on the traffic demand and the bottleneck capacity, it can propagate upstream or downstream; this is essentially governed by the continuity equation. This can be seen in all the congestion patterns of Figure 2.9 (see also Refs. 4 and 46).

The states of *extended congested traffic* (ECT) are particularly interesting because they occur quite frequently. About 200 out of 400 breakdowns observed in the research of Ref. 7 were ECT states. From observations, we can obtain the following empirical properties of ECT states:

1. In most cases, ECT states show some "internal structure" (e.g., oscillations, stop-and-go traffic, or small jams) that can be distinguished from statistical fluctuations. Generally, these internal structures increase in amplitude while propagating upstream, resulting in a region of essentially stationary traffic near the bottleneck and oscillating congested traffic further upstream. At the upstream congestion end, some of the growing perturbations may eventually become isolated "wide jams"; see, for example, Figure 20 in Ref. 7. In other situations, no such transition is observed. The congestion caused by the bottleneck at $x \approx 510$ km of Figure 2.9a may be considered a borderline case.
2. The propagation velocity of the internal structures seems to correspond to that of propagating downstream fronts of congestions; that is, the structures propagate

upstream with a velocity of about 15 km/h. From the previous facts, it follows that, with the exception of the upstream fronts, *all* moving structures propagate upstream with the same velocity. Consequently, all the spatiotemporal structures in Figure 2.9 are essentially parallel.

3. In contrast to the propagation velocities, the typical periods (or, equivalently, wavelengths) of the internal oscillations vary widely between about 4 min and 60 min, corresponding to wavelengths of between 1 km and 15 km. In general, the periods and wavelengths decrease with the severity of the congested state (measured in terms of the average velocity), that is, with the bottleneck strength. In Figure 2.9a, the greater bottleneck is located at the Intersection München-Nord, whereas in Figure 2.9b, it corresponds to an accident at $x = 43.5$ km.

2.3 Lane Changes and Other Discrete-Choice Situations

Besides longitudinal control, which requires an essentially continuous reaction, drivers encounter many situations on the road network, where a discrete decision between two or more alternatives is required. This relates not only to lane-changing decisions but also to considerations as to whether or not it is safe to enter the priority road at an unsignalized junction, to cross such a junction, or to start an overtaking maneuver on a rural road. Another question concerns whether or not to stop at an amber-phase traffic light. All of the above problems belong to the class of *discrete-choice problems*; since the pioneering work of McFadden,[47] these have been extensively investigated in an economic context as well as in the context of transportation planning. In spite of the relevance to everyday driving situations, there are fewer investigations attempting to incorporate the aforementioned discrete-choice tasks into microscopic models of traffic flow, and most of them are restricted to modeling lane changes.[48] Only very recently have acceleration and discrete-choice tasks been treated more systematically.[49,50]

In the following, we will present a recently formulated general framework for modeling traffic-related discrete-choice situations in terms of the acceleration function of a longitudinal model.[50] For the purpose of illustration, we will apply the concept to model mandatory and discretionary lane changes (Section 2.3.1). Furthermore, we will consider the decision process in determining whether or not to brake when approaching a traffic light turning from green to amber (Section 2.3.3).

2.3.1 Modeling Lane Changes

Lane-changing models are a required component in many situations because a realistic description of heterogeneous driver–vehicle units and of most types of bottlenecks is only possible within a multilane modeling framework, allowing the realization of the necessary ("mandatory") lane changes at on-ramps or lane-closing bottlenecks as well as the "discretionary" lane changes in preparation for passing slower vehicles. We emphasize that the basic mechanisms of traffic instabilities that are summarized by the stylized facts of the previous section are essentially unaffected by the lane-changing dynamics.

Modeling of lane changes is typically considered as a multistep process. On a strategic level, the driver knows about his or her route on the network, which influences the lane choice, for example, with regard to lane blockages, on-ramps, off-ramps, or other mandatory merges.[51] In the *tactical* stage, an intended lane change is prepared and initiated by advance accelerations or decelerations of the driver, and possibly by cooperation of drivers in the target lane.[52] Finally, in the *operational* stage, one determines if an immediate lane change is both safe and desired.[48] Although mandatory changes are performed for strategic reasons, the driver's motivation for discretionary lane changes is a perceived improvement of the driving conditions in the target lane compared with the current situation. In the following, we will describe the last step more closely.

When considering a lane change, a driver typically makes a trade-off between the expected own advantage and the disadvantage imposed on other drivers. For a driver considering a lane change, the subjective utility of a change increases with the gap to the new leader in the target lane. However, if the speed of this leader is lower, it may be favorable to stay in the present lane despite the smaller gap. A criterion for the utility including *both* situations is the difference between the accelerations after and before the lane change. This is the core idea of the lane-changing algorithm MOBIL,[50] which is based on the expected (dis)advantage in the new lane in terms of the difference in the acceleration, which is calculated with an underlying microscopic longitudinal traffic model such as the Intelligent Driver Model (Section 2.2.3). Generally, any microscopic acceleration model can be used as the underlying model, as long as the acceleration can be defined in some way. This even includes most cellular automata.[53]

For the lane-changing decision, we first consider a *safety constraint*. In order to avoid accidents by the follower *n* in the prospective target lane (see Figure 2.10), the safety criterion

$$\dot{v}_n \geq -b_{safe} \qquad (2.12)$$

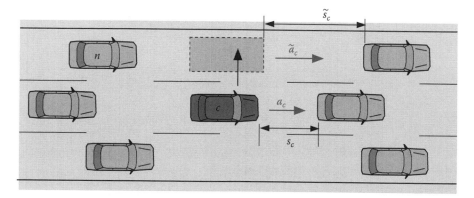

FIGURE 2.10 Definition of the basic variables necessary for the mathematical modeling of lane changes.

guarantees that the deceleration of the successor $\dot{v}_{follow}$ in the target lane does not exceed a safe limit $b_{safe} \simeq 4\,\text{m/sec}^2$ after the lane change. In other words, the safety criterion essentially restricts the deceleration of the lag vehicle on the target lane to values below b_{safe}. Although formulated as a simple inequality, this condition contains all the information provided by the longitudinal model via the acceleration $\dot{v}_{follow}$. In particular, if the longitudinal model has a built-in sensitivity with respect to *velocity differences* (such as the IDM) this dependence is transferred to the lane-changing decisions, which is illustrated in Figure 2.11. In this way, larger gaps between the following vehicle in the target lane and the vehicle's own position are required to satisfy the safety constraint if the speed of the following vehicle is higher than the vehicle's own speed. In contrast, lower values for the gap are allowed if the back vehicle is slower. Moreover, by formulating the criterion in terms of safe braking decelerations of the longitudinal model, crashes due to lane changes are *automatically* excluded as long as the longitudinal model itself guarantees crash-free dynamics.

For discretionary lane changes, an additional *incentive criterion* favors lane changes whenever the acceleration in one of the target lanes is higher. The incentive criterion for a lane change is also formulated in terms of accelerations. Neglecting specific legislations such as a ban to overtake on the right-hand side or the rule to keep to the right lanes, an incentive to change lanes is given if (cf. Figure 2.11)

$$\underbrace{\tilde{a}_c - a_c}_{\text{driver}} + p(\underbrace{\tilde{a}_n - a_n}_{\text{new follower}} + \underbrace{\tilde{a}_o - a_o}_{\text{old follower}}) > \Delta a_{\text{th}}. \qquad (2.13)$$

FIGURE 2.11 Illustration of the safety criterion (2.12) when applied to car-following models that are sensitive to velocity differences. In the situation illustrated on the right, the lag vehicle in the target lane is insignificantly faster and lane changing is allowed. In the illustration on the left, a fast vehicle approaches and, at decision time, the distance is the same as in the former illustration. Lane changing may be forbidden for safety reasons because the velocity difference leads to a higher deceleration $-da_n/dt$ of the vehicle in the target lane.

A lane change is executed if the sum of the own acceleration and those of the affected neighboring vehicle–driver agent is higher in the prospective situation than in the current local traffic state (and if the safety criterion 2.12 is satisfied, of course). The innovation of the MOBIL framework[50] is that the immediately affected neighbors are considered by the "politeness factor" p. For an egoistic driver corresponding to $p = 0$, this incentive criterion simplifies to $\dot{v}_{new} > \dot{v}_{old}$. However, for $p = 1$, lane changes are only carried out if this increases the combined accelerations of the lane-changing driver and all affected neighbors. This strategy can be paraphrased by the acronym "Minimizing Overall Braking Induced by Lane changes" (MOBIL). We observed[50] realistic lane-changing behavior for politeness parameters in the range $0.2 < p < 1$. Additional restrictions can easily be included. A "keep-lane" behavior is modeled by an additional constant threshold Δa_{th} when considering a lane change. The "keep-right" directive valid in most European countries can be implemented by adding an additional bias term on the right-hand side of Equation 2.13 that is positive (negative) when changing to a lane to the right (left).

2.3.2 Turning Decisions at Unsignalized Junctions

The reasoning in formulating the lane-change model of the previous section can be adapted to the turning decision at unsignalized intersections in a natural way. We assume that the considered driver c waits at an unsignalized junction for a sufficiently large gap to enter the priority road as depicted in Figure 2.12 left. At a given moment, there are two options: (i) to wait or (ii) to enter the road and start accelerating.

The safety criterion 2.12 can be adopted literally if vehicle n denotes the next vehicle approaching on the priority road. In most cases, the velocity of the considered driver F at decision time is zero, so expression 2.12 becomes

$$a_n(s_n, v_n, v_n) > -b_{safe}. \tag{2.14}$$

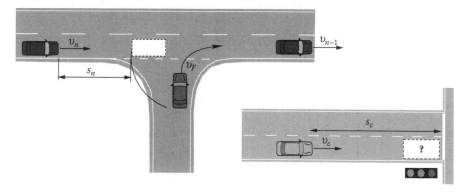

FIGURE 2.12 Two typical discrete-choice situations encountered by drivers in a vehicular network: entering a priority road (left), and approaching a traffic light that changes from "green" to "amber."

To apply the incentive criterion 2.13 for this situation, we observe that the acceleration of driver c for the option "wait" is given by $a_c^{\text{wait}} = 0$. Furthermore, it is obvious to set $p = \Delta a_{\text{thr}} = 0$, so a decision to turn according to Equation 2.13 is made as soon as its own acceleration as calculated with the longitudinal model becomes positive:

$$a_F(s_{n-1}, 0, -v_{n-1}) > 0. \tag{2.15}$$

Here, v_{n-1} is the velocity of the last vehicle on the main road that has already crossed the junction, and s_{n-1} denotes the distance of this vehicle to the junction. Notice that, for this case, Equation 2.15 is, in fact, another safety criterion preventing the acceleration if vehicle $n - 1$ has not yet cleared the junction area.

2.3.3 Approaching a Traffic Light

When approaching a traffic light that switches from green to amber, a decision has to be made whether to stop just at the traffic light or to pass the amber-phase light with unchanged speed. For an empirical study on the stopping/running decision at the onset of an amber phase we refer to Ref. 54. If the first option is selected, the traffic light will be modeled by a standing "virtual" vehicle at the position of the light. Otherwise, the traffic light will be ignored. The criterion is satisfied for the "stop at the light" option if the own braking deceleration at the time of the decision does not exceed the safe deceleration b_{safe}. The situation is illustrated in Figure 2.12, right. Denoting the distance to the traffic light by s_c and the velocity at decision time by v_c and assuming a longitudinal model of the form of Equation 2.6, the safety criterion 2.12 can be written as

$$\dot{v}(s_c, v_c, v_c) \geq -b_{\text{safe}}. \tag{2.16}$$

Notice that the approaching rate and the velocity are equal ($\Delta v_c = v_c$) in this case. The incentive criterion is governed by the bias towards the stopping decision because legislation requires that one stop at an amber-phase traffic light if it is safe to do so. As a consequence, the incentive criterion is always fulfilled, and Equation 2.16 is the only decision criterion in this situation.

Similarly to the lane-changing rules, the "stopping criterion" (2.16) will inherit all the sophistication of the underlying car-following model. In particular, when using realistic longitudinal models, one obtains a realistic stopping criterion with only one additional parameter b_{safe}. Conversely, unrealistic microscopic models such as the Optimal Velocity Model[26] or the Nagel–Schreckenberg cellular automaton[32] will lead to unrealistic stopping decisions. In the case of the Optimal Velocity Model, it is not even guaranteed that drivers deciding to stop will be able to stop at the lights.

2.4 Simulating Vehicle-to-Vehicle and Infrastructure-to-Vehicle Communication

With the increasing availability of ad hoc wireless networks, concepts of a local noncentralized communication between vehicles, also known as intervehicle communication

(IVC), become more and more important for motorized road traffic.[17,55–59] This new technology has important and promising applications in traffic safety, information, and control. On a small scale, a vehicle involved in an accident or suffering from a breakdown might constantly send out messages stating that there is an emergency situation, thus warning oncoming vehicles in advance. Furthermore, with the emergence of ACC systems and associated sensors for the distance to the preceding vehicle, equipped vehicles can inform their environment about the local traffic state, and past actions (e.g., a hard braking maneuver). The next ACC and IVC equipped vehicles in the upstream direction can process this information and automatically adapt their driving style to the situation or at least display suggestions and warnings to the driver, which may serve as a means to homogenize traffic and thus increase the overall traffic throughput in a self-organized way.[17] On a larger scale, the information passed along the IVC network can be aggregated to produce live floating car data. In addition to the conventional cross-sectional data from loop detectors, this information can be used by traffic-state recognition and traffic-forecast models, thereby increasing their performance and reliability.

In most applications, it is necessary to carry messages over distances that are significantly longer than the device's broadcast range. As illustrated in Figure 2.13, this can be achieved in two ways. First, a message can be passed to the following vehicle, which then passes it to the following vehicle, and so on. We call this *longitudinal hopping*, because the message always travels parallel to the desired travel direction. Or, a vehicle may transfer a message to a vehicle driving in the opposite direction. This vehicle can

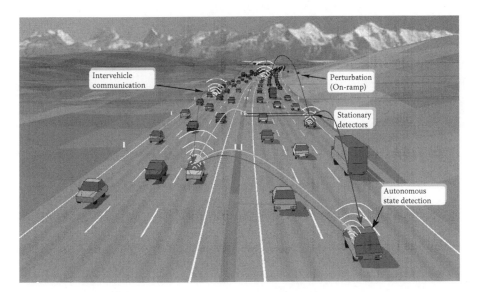

FIGURE 2.13 Illustration of the the two principal concepts for intervehicle communication. Transversal hopping, indicated by the gray arrows, uses equipped vehicles traveling in the opposite direction as information carriers, while longitudinal hopping requires a chain of communicating vehicles traveling in the same direction. Also indicated is infrastructure–vehicle communication (white arrows).

store the message and continuously broadcast it for a certain period of time while mechanically transporting the message upstream. Although this message is of no use for this relay vehicle, it might eventually jump back to the original driving direction by a second transversal hop. We therefore call this method *transversal hopping* or "store and forward."

The longitudinal hopping mode has the advantage of virtually instantaneous message transmission, so the transmitted information is always up to date. However, its reliability may be low because a fully connected chain of IVC equipped vehicles is needed. This restriction is overcome by the transversal hopping mode where the successful transmission is only a matter of time, but the information may be obsolete when it finally arrives. There is, in fact, a third possibility of intervehicle communication involving road-side units that may act as repeaters between two vehicles that are too far apart for direct communication. However, this concept requires an extensive amount of stationary hardware.

2.4.1 Simulating the Longitudinal Hopping Mode

In this subsection, we test the assumptions of an analytical model[60] by means of simulations with a microscopic traffic model. The model assumptions are (i) longitudinal hopping along a linear chain, (ii) a deterministic and instantaneous transmission mechanism in which a message is available for receiving within a certain radius r from the sender with certainty, but unavailable further away, (iii) randomly distributed nodes (IVC vehicles) with a linear node density λ, that is, the distances Y between two nodes are i.i.d. exponentially distributed stochastic variables whose density is given by

$$f_Y(y) = \lambda e^{-\lambda y}. \tag{2.17}$$

From these assumptions, following analytical expression for the distribution $P_{\text{chain}}(x)$ of the length of the communication chain, and the distribution $P_c(x)$ of the maximum distance of availability of a message can be derived[60,61]:

$$P_{\text{chain}}(x) = P_c(x + r), \tag{2.18}$$

$$P_c(x) = 1 - \sum_{k=1}^{m} e^{-k\lambda r} \left[\frac{(-\lambda(x - kr))^{k-1}}{(k-1)!} \left(1 + \frac{\lambda(x - kr)}{k} \right) \right], \tag{2.19}$$

where m denotes the integer part of x/r. In order to test the assumptions of the analytical model we have performed several simulations of longitudinal message hopping with the following two-stage approach. In a first step, vehicle trajectories are obtained from a microscopic simulation with the Gipps model.[37] Similar results would be obtained when simulating the intelligent-driver model discussed above. Then, the communication is simulated by analyzing snapshots of the vehicle trajectories for several times t.

The top graphics of Figure 2.14 shows the resulting simulated connectivities $P_{\text{chain}}(r)$ for free traffic assuming a range $r = 200$ m and various penetration rates α. For $\alpha < 40\%$,

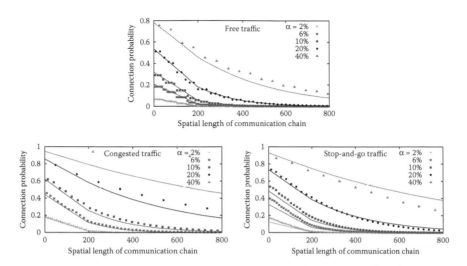

FIGURE 2.14 Probability of message reception at a certain distance from the initial sender in free traffic for $r = 200$ m and several values of the penetration rate α. Shown are the analytical values $P_{chain}(x)$ (solid lines), and simulation generated with the Gipps model for free traffic (top), homogeneous congested traffic (left), and stop-and-go traffic (right).

we found good agreement with the analytical model, but, for larger penetration rates, the observed connectivity was higher than the analytical result. In congested traffic (Figure 2.14, left), the situation is similar, with good agreement for small α, while the theory underestimates the observed connectivity for larger penetration rates. In quantitative terms, the deviation reaches the order of 0.05 for $\alpha \approx 15\%$ while, in free traffic, such discrepancies are only observed for $\alpha \geq 40\%$. Remarkably, the situation is drastically different for stop-and-go traffic (Figure 2.14, right) where, for sufficiently high penetration rates, we observed *lower* reception probabilities compared to theory, while, for penetration rates below 20%, they were *higher*.

All findings can be understood in terms of correlations of the distances between the vehicles that are observed in real (and simulated) traffic, but are absent in the analytical communication model. Obviously, one source of correlation is the repulsive interaction that is necessary to keep a certain minimum "safe" distance to the leading vehicle. This leads to short-ranged correlations of the distances between the vehicles decaying rapidly over a few vehicle distances.[62] With respect to the theoretical exponential distribution of Equation 2.17, this effect compresses the actual distance distribution towards the mean value $1/\rho$. If the IVC penetration rate α is sufficiently high such that the mean distance $1/\lambda = 1/\alpha\rho$ of the *equipped* vehicles does not significantly exceed the correlation length, the correlation carries over to that of the equipped vehicles. If, in addition, the mean distance satisfies $1/\lambda < r$, the probability of distances exceeding r, thereby breaking the chain, is reduced. As the correlation length and the assumed communication range $r = 200$ m are of the same order, both conditions are satisfied simultaneously for sufficiently high penetration rates, explaining the higher connectivities observed in this regime. If, however, the average node distance $1/\lambda$ exceeds the correlation length, the positions of the equipped vehicles are nearly independent (even if that of all vehicles are not), and the analytical model is approximatively valid. Because the repulsive force acts only between vehicles on the same lane, the effect described above is expected to be weaker for multilane traffic.

2.4.2 Simulating the Transversal Hopping Mode

As in the longitudinal communication path, we will compare an analytical model with simulations of a microscopic traffic model. The analytic model is derived under the same general assumptions as in the previous subsection, that is, an exponential gap distribution of the equipped vehicles, and communication between equipped vehicles within a limited broadcast range r. Inside this range, the communication is assumed to be error free and instantaneous. For a description of information propagation via the opposite driving direction, we consider a bidirectional freeway or arterial road (cf. Figure 2.15). In both directions, we assume essentially homogeneous traffic flows that are characterized by the lane-averaged velocities v_1 and v_2, and the total densities ρ_1 and ρ_2, respectively. As in the longitudinal hopping model, we assume a global market penetration α such that the relevant density variables are the partial densities of equipped vehicles, $\lambda_1 = \alpha\rho_1$ and $\lambda_2 = \alpha\rho_2$, respectively. By virtue of the assumed constant velocities, the first relay vehicle having received the message will also be the first vehicle reaching the target region, so we will ignore message reception and transport by subsequent relay vehicles.

Let us consider a message that is generated by an equipped vehicle driving in direction 1 at position $x = 0$ and time $t = 0$. As illustrated in Figure 2.15, we use the same coordinate system for both directions such that driving direction 1 is parallel and direction 2 antiparallel to the x-axis. Together with the assumptions above, this eventually leads to following cumulative distribution function $P_3(\tau)$ for the total communication

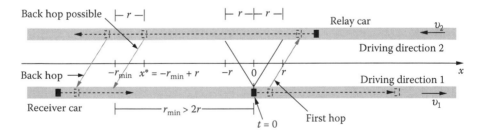

FIGURE 2.15 Illustration of message transport via the opposite driving direction. At time $t = 0$, a car traveling in direction 1 generates a message and starts broadcasting it. The message is transported forward at speed v_1 until it will be received by an equipped car moving at speed v_2 in direction 2. The relay car transports the message at least up to the position $-r_{\min} + r$ (the target region) before broadcasting it again. Finally, the message is received by an equipped car in direction 1. Transversal distances are neglected when determining the communication range r.

time τ between the sending of a message and successful reception by an equipped vehicle in the target region[63]:

$$P_3(\tau) = \Theta(\tau - \tau_{min})\left(1 - \frac{\tilde{\lambda}_1}{\tilde{\lambda}_1 - \lambda_2} e^{-\lambda_2 x_e(\tau)} + \frac{\lambda_2}{\tilde{\lambda}_1 - \lambda_2} e^{-\tilde{\lambda}_1 x_e(\tau)}\right), \qquad (2.20)$$

where the minimum transport time for a successful complete transmission is given by $\tau_{min} = (r_{min} - 2r)/v_2$, and $x_e(\tau) = v_2(\tau - \tau_{min}) = v_2\tau - r_{min} + 2r$ denotes the part of the target region that intersects (or has been intersected by) the range of the relay vehicle for the "best case." Furthermore, $\tilde{\lambda}_1 = \lambda_1[(v_1 + v_2)/v_2]$, and the Heavyside function $\Theta(x)$ is equal to 0 for $x < 0$ and 1 for $x \geq 0$.

In the case of identical traffic conditions in both driving directions (i.e., $v_1 = v_2 = v$ and $\lambda_1 = \lambda_2 = \lambda$), we have $\tilde{\lambda}_1 = 2\lambda$, resulting in the more intuitive expression[63]

$$P_3(\tau) = \Theta\left(\tau - \frac{r_{min} - 2r}{v}\right)\left[1 - e^{-\lambda(2r + v\tau - r_{min})}\right]^2 \qquad (2.21)$$

which is shown as solid lines in Figure 2.16. Note that the quadratic term in Equation 2.21 reflects the fact that two encounters of equipped vehicles are needed for propagating a message by means of transverse message hopping.

From Figure 2.16, we observe that the analytical model 2.20 agrees well with the simulations unless (i) there is only one lane and (ii) the percentage of equipped vehicles is 5% or higher. We conclude that, for most practical situations, the analytic considerations are remarkably accurate although the assumption of an exponential gap distribution is clearly violated in vehicular traffic.

Finally, Figure 2.17 provides an illustrative example of the working of the transversal hopping mechanism. A simulation is shown where a bottleneck causing the vehicles to stop is introduced at $x = 3$ km between $t = 100$ sec and 200 sec.[64] The solid lines indicate equipped vehicles in the considered direction, and the dotted lines equipped vehicles in the opposite direction serving as information carriers. Messages are sent on an event-oriented basis whenever an equipped vehicle is forced to drive very slowly (triangles). The arrival of messages is shown for the equipped vehicle, whose trajectory is drawn as a thick solid line. One notices that even for percentages as low as 1% this vehicle receives enough messages warning about the stopped traffic ahead.

2.5 Discussion and Outlook

In this contribution, we have given an overview of traffic flow modeling and simulation with the focus on microscopic models; that is, the dynamics of individual vehicles is considered. The applications of such models range from testing classic and modern concepts for improving the operations of the vehicular network, providing better information to the drivers, to using simulators for visualization and behavioral studies.

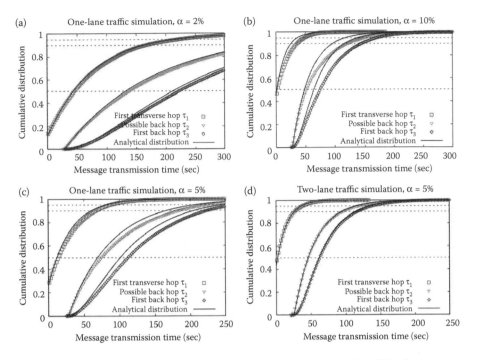

FIGURE 2.16 Simulated vs. analytical results for the statistical properties of the characteristic communication times for different market penetrations and numbers of lanes. The broadcast range r is 200 m. The average speed v (density ρ) of both driving directions was determined from the simulations as approximately 78.8 km/h (15.2/km).

As an example application, we applied simulations of microscopic models for traffic flow to assess the potential of IVC, which is a promising and scalable concept for exchanging traffic-related information among vehicles over relatively short distances. However, like all technologies relying on local communication, IVC faces the "penetration threshold problem." Thus, the system is effective only if there is a sufficient number of communication partners to propagate the message between equipped cars. Therefore, it is crucial to assess the feasibility of different communication variants in terms of the necessary critical market penetration. Because, to date, only prototype realizations exist, simulations cannot be replaced by other means of testing. As a main result, we have found that, for both the longitudinal and transversal communication paths, the analytical model agrees remarkably well with the simulated results of the more realistic model variants. Although the main assumptions made in deriving these models—homogeneous traffic and exponentially distributed intervehicle distances—are normally not perfectly met in real traffic flow, it turns out that the theoretical results are remarkably robust with respect to violations of the models' assumptions. In particular, for multilane traffic and penetration levels below 5%, the errors are typically a few percent only.

With respect to the applicability for different traffic situations, we have found that the considered mechanisms supplement each other. The longitudinal hopping mechanism

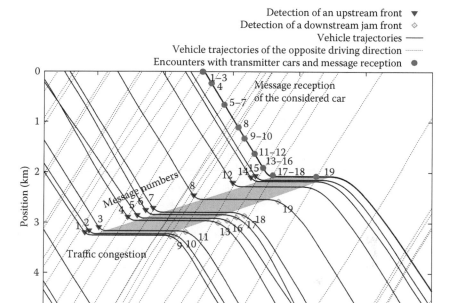

FIGURE 2.17 Space–time diagram of a scenario with traffic congestion. The contour plot shows the average speed of all vehicles in driving direction 1. Trajectories of equipped vehicles (only 1% of all vehicles) are indicated by solid and dashed lines, depending on the driving direction. Dots indicate the generation of messages when detecting the shock fronts of the jam. Messages are transmitted between vehicles whenever two trajectories cross each other.

typically fails for too low percentages of equipped vehicles whereas the transversal mechanism will always succeed but the communication delay inherent in this method will increases for low equipment rates. Furthermore, the transverse hopping mechanism is disturbed if both traffic directions are congested, because the message transport by relay vehicles becomes slower. However, in such situations, the longitudinal coupling becomes more effective, because the partial density of equipped vehicles increases with traffic density. Therefore, a hybrid system is advisable for the intermediate stages of the deployment of IVC systems.

Finally, we note that IVC is only one building block of a future integrated traffic communication system. As a straightforward next step, including police cars and emergency vehicles into the IVC fleet will lead to a timely production of event-related messages. Furthermore, adding infrastructure–vehicle communication to the system may help to overcome the penetration barrier. This will be particularly economic and efficient when placing the infrastructural communication units near to known bottlenecks, where the necessary sensors for producing event-based traffic messages are already in place. Microscopic models will become an essential means to assess the performance of all these future communication systems.

References

1. Statistisches Bundesamt, Verkehr in Deutschland 2006, 2006.
2. European Commission, Energy & Transport, White Paper European Transport Policy for 2010: "Time to Decide," COM (2001) 370 final, 2001.
3. Helbing, D., Traffic and related self-driven many-particle systems, *Reviews of Modern Physics*, 73, 1067–1141, 2001.
4. Kerner, B.S., *The Physics of Traffic*, Springer, Heidelberg, 2004.
5. Kerner, B.S. and Rehborn, H., Experimental features and characteristics of traffic jams, *Physical Review E*, 53, R1297–R1300, 1996.
6. Cassidy, M.J. and Bertini, R.L., Some traffic features at freeway bottlenecks, *Transportation Research Part B: Methodological*, 33, 25–42, 1999.
7. Schönhof, M. and Helbing, D., Empirical features of congested traffic states and their implications for traffic modeling, *Transportation Science*, 41, 1–32, 2007.
8. Chowdhury, D., Santen, D., and Schadschneider, A., Statistical physics of vehicular traffic and some related systems, *Physics Reports*, 329, 199–329, 2000.
9. Nagatani, T., The physics of traffic jams, *Reports of Progress in Physics*, 65, 1331–1386, 2002.
10. Maerivoet, S. and DeMoor, B., Cellular automata models of road traffic, *Physics Reports*, 419, 1–64, 2005.
11. Hoogendoorn, S.P. and Bovy, P.H.L., State-of-the-art of vehicular traffic flow modelling, *Proceedings of the Institution of Mechanical Engineers, Part I: Journal of Systems and Control Engineering*, 215(4), 283–303, 2001.
12. Leutzbach, W., *Introduction to the Theory of Traffic Flow*, Springer, Berlin, 1988.
13. Ministry of Transport, Energy and Spatial Planning of Nordrhein-Westfalen, Traffic State Prediction for the Freeway Network, 2007, available at http://autobahn.nrw.de, accessed May 2007.
14. Treiber, M., Modellgestützte Verkehrs-Zustandsschätzung unter Berücksichtigung verschiedener Datenquellen, Technical Report, Technische Universität Dresden, 2006.
15. Treiber, M. and Helbing, D., Reconstructing the spatio-temporal traffic dynamics from stationary detector data, *Cooperative Transportation Dynamics*, 1, 3.1–3.24, 2002, available at www.TrafficForum.org/journal.
16. Kerner, B.S., Rehborn, H., Aleksic, M., and Haug, A., Recognition and tracking of spatio-temporal congested traffic patterns on freeways, *Transportation Research Part C: Emerging Technology*, 12, 369–400, 2004.
17. Kesting, A., Treiber, M., Schönhof, M., and Helbing, D., Adaptive cruise control design for active congestion avoidance, *Transportation Research Part C: Emerging Technologies*, 16, 668–683, 2008.
18. Kesting, A., Treiber, M., Schönhof, M., and Helbing, D., Extending adaptive cruise control to adaptive driving strategies, *Transportation Research Record*, 2000, 16–24, 2007.
19. Davis, L.C., Effect of adaptive cruise control systems on traffic flow, *Physical Review E*, 69, 066110, 2004.

20. Marsden, G., McDonald, M., and Brackstone, M., Towards an understanding of adaptive cruise control, *Transportation Research Part C: Emerging Technologies*, 9, 33–51, 2001.
21. Greenshields, B.D., A Study of Traffic Capacity, in Proceedings of the Highway Research Board, Vol. 14, pp. 228–477, Highway Research Board, Washington, DC, 1959.
22. Lighthill, M.J. and Whitham, G.B., On kinematic waves: II. A theory of traffic on long crowded roads, *Proc. Roy. Soc. London A*, 229, 317–345, 1955.
23. Richards, P.I., Shock waves on the highway, *Operations Research*, 4, 42–51, 1956.
24. Kerner, B.S. and Konhäuser, P., Structure and parameters of clusters in traffic flow, *Physical Review E*, 50, 54–83, 1994.
25. Treiber, M., Hennecke, A., and Helbing, D., Derivation, properties, and simulation of a gas-kinetic-based, non-local traffic model, *Physical Review E*, 59, 239–253, 1999.
26. Bando, M., Hasebe, K., Nakayama, A., Shibata, A., and Sugiyama, Y., Dynamical model of traffic congestion and numerical simulation, *Physical Review E*, 51, 1035–1042, 1995.
27. Treiber, M., Hennecke, A., and Helbing, D., Congested traffic states in empirical observations and microscopic simulations, *Physical Review E*, 62, 1805–1824, 2000.
28. Rui Jiang, Qingsong Wu, and Zuojin Zhu, Full velocity difference model for a car-following theory, *Physical Review E*, 64, 017101, 2001.
29. Tilch, B. and Helbing, D., Generalized force model of traffic dynamics, *Physical Review E*, 58, 133–138, 1998.
30. Treiber, M., Kesting, A., and Helbing, D., Understanding widely scattered traffic flows, the capacity drop, and platoons as effects of variance-driven time gaps, *Physical Review E*, 74, 016123, 2006.
31. Wiedemann, R., Simulation des Straßenverkehrsflusses, in *Heft 8 der Schriftenreihe des IfV*, Institut für Verkehrswesen, Universität Karlsruhe, 1974.
32. Nagel, K. and Schreckenberg, M., A cellular automaton model for freeway traffic, *J. Phys. I France*, 2, 2221–2229, 1992.
33. Lee, H.K., Barlovic, R., Schreckenberg, M., and Kim, D., Mechanical restriction versus human overreaction triggering congested traffic states, *Physical Review Letters*, 92, 238702, 2004.
34. Helbing, D. and Schreckenberg, M., Cellular automata simulating experimental properties of traffic flows, *Physical Review E*, 59, R2505–R2508, 1999.
35. Knospe, W., Santen, L., Schadschneider, A., and Schreckenberg, M., Human behaviour as origin of traffic phases, *Physical Review E*, 65, 015101, 2001.
36. Kesting, A. and Treiber, M., How reaction time, update time and adaptation time influence the stability of traffic flow, *Computer-Aided Civil and Infrastructure Engineering*, 23, 125–137, 2008.
37. Gipps, P.G., A behavioural car-following model for computer simulation, *Transportation Research Part B: Methodological*, 15, 105–111, 1981.
38. Newell, G.F., Nonlinear effects in the dynamics of car following, *Operations Research*, 9, 209, 1961.

39. Treiber, M., Kesting, A., and Helbing, D., Delays, inaccuracies and anticipation in microscopic traffic models, *Physica A*, 360, 71–88, 2006.
40. Kesting, A. and Treiber, M., Calibrating car-following models using trajectory data: Methodological study, *Transportation Research Record*, 2008, preprint http://arxiv.org/pdf/0803.4063v1.
41. Treiterer, J. and Myers, J.A., The hysteresis phenomenon in traffic flow, in Proc. 6th Int. Symp. on Transportation and Traffic Theory, Buckley, D.J., Ed., Elsevier, New York, 1974, p. 13.
42. Kesting, A., Microscopic Modeling of Human and Automated Driving: Towards Traffic-Adaptive Cruise Control, doctoral thesis, Technische Universität Dresden, 2008.
43. Treiber, M., available at http://www.traffic-simulation.de (an interactive simulation of the Intelligent Driver Model in several scenarios including open-source access), 2007.
44. Sugiyama, Y., Fukui, M., Kikuchi, M., Hasebe, K., Nakayama, A., Nishinari, K., Tadaki, S., and Yukawa, S., Traffic jams without bottlenecks—experimental evidence for the physical mechanism of the formation of a jam, *New Journal of Physics*, 10, 033001, 2008.
45. Helbing, D., Hennecke, A., and Treiber, M., Phase diagram of traffic states in the presence of inhomogeneities, *Physical Review Letters*, 82, 4360–4363, 1999.
46. Schönhof, M. and Helbing, D., Criticism of three phase traffic theory, *Transporation Research Part B: Methodological*, 2008 (submitted).
47. Hausman, J. and McFadden, D., Specification tests for the multinomial logit model, *Econometrica*, 52(5), 1219–1240, 1984.
48. Gipps, P.G., A model for the structure of lane-changing decisions, *Transportation Research Part B: Methodological*, 20, 403–414, 1986.
49. Toledo, T., Koutsopoulos, H.N., and Ben-Akiva, M., Integrated driving behavior modeling, *Transportation Research Part C: Emerging Technologies*, 15(2), 96–112, 2007.
50. Kesting, A., Treiber, M., and Helbing, D., General lane-changing model MOBIL for car-following models, *Transportation Research Record*, 1999, 86–94, 2007.
51. Toledo, T., Choudhury, C.F., and Ben-Akiva, M.E., Lane-changing model with explicit target lane choice, *Transportation Research Record*, 1934, 157–165, 2005.
52. Hidas, P., Modelling vehicle interactions in microscopic traffic simulation of merging and weaving, *Transportation Research Part C: Emerging Technologies*, 13, 37–62, 2005.
53. Treiber, M., Kesting, A., and Helbing, D., Alternative models and interpretations of empirically observed traffic patterns, *Transportation Research Part B: Methodology*, 2008 (submitted).
54. Rakha, H., El-Shawarby, I., and Setti, J.R., Characterizing driver behavior on signalized intersection approaches at the onset of a yellow-phase trigger, *IEEE Transactions on Intelligent Transportation Systems*, 8(4), 630–640, 2007.
55. Shladover, S.E., Polatkan, G., Sengupta, R., VanderWerf, J., Ergen, M., and Bougler, B., Dependence of cooperative vehicle system performance on market penetration, *Transportation Research Record*, 2000, 121–127, 2007.

56. Wischhof, L., Ebner, A., and Rohling, H., Information dissemination in self-organizing intervehicle networks, *IEEE Transactions on Intelligent Transportation Systems*, 6, 90–101, 2005.
57. Zhang, J., Ziliaskopoulos, A.K., Wen, N., and Berry, R.A., Design and implementation of a vehicle-to-vehicle based traffic information system, *IEEE Intelligent Transportation Systems*, 473–477, 2005. Proceedings.
58. Mahmassani, H.S., Dynamic network traffic assignment and simulation methodology for advanced system management applications, *Networks and Spatial Economics*, 1(3), 267–292, 2001.
59. Kato, S., Tsugawa, S., Tokuda, K., Matsui, T., and Fujii, H., Vehicle control algorithms for cooperative driving with automated vehicles and intervehicle communications, *IEEE Transactions on Intelligent Transportation Systems*, 3(3), 155–161, 2002.
60. Dousse, O., Thiran, P., and Hasler, M., Connectivity in Ad-Hoc and Hybrid Networks, in Proc. IEEE Infocom. Twenty-First Annual Joint Conference of the IEEE Computer and Communications Societies, New York, 2002.
61. Thiemann, C., Treiber, M., and Kesting, A., Longitudinal hopping in inter-vehicle communication: Theory and simulations on modeled and empirical trajectory data, *Physical Review E*, 78, 036102, 2008.
62. Knospe, W., Santen, L., Schadschneider, A., and Schreckenberg, M., Single-vehicle data of highway traffic: Microscopic description of traffic phases, *Physical Review E*, 65, 056133, 2002.
63. Kesting, A., Treiber, M., Thiemann, C., Schönhof, M., and Helbing, D., Alternative modes of message transmission by inter-vehicle communication: Analytic and simulated connectivity statistics, *Transportation Research Part C: Emerging Technologies*, 2008 (submitted).
64. Schönhof, M., Treiber, M., Kesting, A., and Helbing, D., Autonomous detection and anticipation of jam fronts from messages propagated by intervehicle communication, *Transportation Research Record*, 1999, 3–12, 2007.

II

U.S. and European Initiatives

3

Vehicle-Infrastructure Cooperation

James A. Misener,
Susan Dickey, and
Joel VanderWerf
*Partners for Advanced
Transit and Highways
University of California*

Raja Sengupta
*Department of Civil and
Environmental Engineering
University of California*

3.1 Introduction

The inclusion of the roadside element is a logical extension of Vehicular Ad-hoc Networks (VANET) from the perspective of those with vehicle-centric and also those with infrastructure-centric network applications backgrounds. Consider that stationary motes (or roadside equipment, RSE) intersect moving cars and represent the edge of, and therefore access to, an extensive landside network, replete with existing or (with vehicle-infrastructure cooperation, VIC) *new* user services. The idea of connecting onboard equipment (OBE) and RSE suggests a wide range of information-based applications and, important to the traveler, a plethora of information-based services. What is this plethora of services? Mobility applications should have the basic attribute of delivering dynamic traveler information and may include the following:

1. Traffic and travel conditions, including route-specific travel times and delays
2. Route assistance and route diversion

3. Map database assistance
4. Adverse weather information
5. Multimodal trip planning; such as transit connections, fare information, schedules, and real-time bus/train arrival information, and airport and port authority information
6. Parking information, to include transit and commercial vehicle parking
7. Signal phase timing information (for signal coordination)
8. Road surface conditions

Moreover, safety services may include the following:

1. Vehicle-based sensor and communication systems, to include vehicle–vehicle communications
2. Cooperative intersection collision avoidance systems, to potentially include signal violation warnings
3. Merge assistance systems
4. Rear-end collision warning systems
5. In-vehicle signing for both static and dynamic advisories
6. Transit applications such as precision docking and automation

These applications lists are not comprehensive; they are merely a list, and given VIC as an enabler, transportation services that depend on telematics would have the attribute of real-time or on-demand information, and the list may grow as large as the marketplace of useful service and application providers and willing subscribers (and their road authorities) will allow. Another dimension to this marketplace is the provision of on-demand infotainment, as the delivery of these products to vehicles may provide an economic incentive to build an infrastructure that provides additional mobility services and to some, the safety imperative.

The real question, then, is how do these services begin to be delivered? A key to the answer will be the existence of RSE that are in two-way communication with cars, or VIC. Given VIC and the existence of a backhaul network to landside operations there can be delivery to and from *the first equipped vehicle* of existent telematics and infotainment services, such as travel time information. Subsequently, as OBE-equipped vehicles proliferate, market-penetration-based services available by information exchange between vehicles and the stationary RSE become more real. Take for example, the VII California test bed along the North–South corridor in the San Francisco Bay Area, c. 2007: a "mere" presence of 12 RSE units intersect an average of 400,000 vehicles a day, traveling three major routes, two limited access freeways and one major signalized arterial. Assuming that applications delivered from these RSE provide some benefit to travelers, there is no need to await market penetration to allow acceptable peer-to-peer connectivity and VANET. In fact, a compelling aspect of having some installation of RSE is that the roadside component and services it may provide could lead to OBE being installed in vehicles. What may make this even more compelling is that these OBEs brought into vehicles may not only be automotive company installations; they could also be handheld consumer electronic devices that have short- and medium-range connectivity; such as mobile phones, perhaps equipped with WiFi or even Dedicated Short Range Communication (DSRC). Imagine dynamic route advisories given to drivers to route themselves out of traffic jams; imagine windshield wiper data giving roadway authorities insight into a moving storm front; imagine the wealth of services that could enhance travel, traffic management, and indeed transportation efficiency and quality of life.

However, the main application of DSRC is safety, and it is necessary to establish how to deliver the quality of service—that is, low latency and high availability—necessary to enact an OBE-to-OBE or OBE-to-RSE/RSE-to-OBE wirelessly delivered safety message. In this case, the entire network need not be accessible; rather, the edge of the network car-to-road or road-to-car can exchange detailed in-vehicle information that can be displayed in a nearly immediate and indeed urgent manner to the driver. Imagine an intersection that warns the inattentive driver that the yellow phase will imminently turn to red—and prevent a red light violation; imagine a car that brakes hard or conversely one that brakes moderately, with that safety information communicated via DSRC to the car or string of cars behind such that onboard processors and displays could provide or not provide a warning; imagine again the wealth of services that could enhance safety and along the way also increase transportation efficiency.[1]

The arguments above, one a mobility and efficiency argument and another a safety argument, have led institutions in the United States to mobilize around the 75 MHz of free unlicensed bandwidth from 5.85 to 5.925 GHz and to begin to standardize and institutionalize DSRC. This, until recently, was the thrust of the United States efforts in VIC, and this is the starting point of this chapter. However, in order to impart a better understanding of the state of VIC, we will begin a little more broadly by briefly covering the worldwide context, then describe more pointedly the case in the United States (and probably by *de facto* precedent all of North America). We will describe some of the emergent research with DSRC, some of the conceptual applications, and then we will point toward a possible future that transcends a "just DSRC" wireless paradigm, in work enabled by the United States Department of Transportation (U.S. DOT), California Department of Transportation (Caltrans), and conducted by the authors. This potential future comes full circle, placing work across many regions of the world into the same trajectory, by recognizing the proliferation of consumer handheld devices in the world and therefore taken by travelers on most trips, and then by leveraging multiband attributes beginning to appear on these devices and finally transitioning into a world where applications, safety, and mobility alike may be ubiquitously applied across the heterogeneous base of portable consumer devices.

In this chapter we first discuss the development of VIC in Europe, Japan, and the United States We discuss the current VIC communication technology in the United States and present results on its performance derived using our VII California Testbed. Finally we conclude with our reading of the future of VIC.

3.2 Developments in VIC: 2004–2008

3.2.1 World Context

The idea of VIC has captured the imagination worldwide, and it is interesting to observe the regional and in some cases national ideas and means to instantiate them into a true VIC deployment and in all dimensions: the wireless communication frequency, the degree of infrastructure at the roadside, the complexity of onboard equipment, the timeline to deployment, and the applications or services considered. Clearly, transportation needs, legacy infrastructure, and existing policy and other institutional arrangements dictate the degree and ubiquity of ideas.

Although there are several world-regional ideas, in this section we describe the European Commission's Continuous Air interface for Long and Medium distance (CALM) and the Japanese SmartWay efforts. The description is topical and brief, as the primary message is that, despite the focus on this chapter on VIC in the United States, there are indeed other ideas, efforts and flavors under consideration that will affect VIC. These ideas—and CALM and SmartWay in particular—have recently influenced thinking and policy in the United States, namely the work therein has posed the questions, "Is there a better way?" and "Isn't it better defined as deployment sooner, not in the distant future?" These are excellent questions and, continually asked, they will always point toward new and more relevant research.

3.2.1.1 CALM (Europe)

The Continuous Air interface for Long and Medium distance (CALM)[2] is integral to the European Commission's Cooperative Vehicle-Infrastructure System (CVIS) effort, as it provides both the architecture and a means to interweave existing protocols such that multiband communications may occur simultaneously, and in principle seamlessly. Therefore, in essence, transportation applications may be delivered to a consumer device that has, for example, a 3G communication channel. As market forces drive the consumer device to more and more connectivity, such as 4G (e.g., WiMax), WiFi, or even IEEE 802.11p DSRC, the handheld device operating under the CALM architecture will still be able to enact two-way communication.

Therefore, CALM aims to develop into an intelligent agent that arbitrates between and provides security between nearly 25 different standards, many still in progress, and if and when those standards roll out, CALM-compliant handheld platforms may be a communications link-neutral or -independent applications environment. Via CALM and potentially as applied by CIVIS, the transportation system manager will receive traffic probe data, and the traveler will in turn receive local map data and that hoped-for plethora of various other services.

3.2.1.2 SmartWay (Japan)

In October 2007, several years' work in leveraging existing infrastructure and programs, "SmartWay 2007," was demonstrated by the Ministry of Land, Infrastructure, and Transport and the Transport National Institute for Land and Infrastructure Management.[3] SmartWay was hosted on the Tokyo Metropolitan Expressway and was the culmination in delivering a prototype of what the Japanese call an "on-board ITS experience." It delivered in demonstration fashion an ensemble of mobile services, either via an integrated center console driver interface or through an aftermarket audio-only version. Central to SmartWay is VIC, as services were all delivered by wireless means. Services delivered fall under three categories:

1. Near real-time driver assistance
2. In-vehicle messaging
3. Two-way communication, which enables e-payment or tolling

In Japan, there are unique technological and institutional underpinnings and years of public- and private-sector investment in the Japanese surface transportation system. Thus, virtually every new car sold in Japan is equipped with an in-vehicle navigation system and additionally equipped with a Vehicle Information and Communication System (VICS) component. Therefore, as a significant SmartWay enabler, road and congestion state information for major arterials and limited access highways is aggregated by the Japan Road Traffic Information Center, transported to the VICS Center, then delivered by optical (IR) beam or 2.4 GHz radio beacon back to cars. The widespread adoption of VICS has allowed cars to serve as probes, which multiplies the efficacy of this system.

As a second significant SmartWay enabler, at the time of the demonstration, approximately 75% of all cars using toll roads in Japan were equipped with 5.8 GHz DSRC-based Electronic Toll Collection (ETC), developed to Japanese standards and implemented by most expressway authorities in Japan. This particular DSRC range is 30 m, sufficient to accomplish its primary purpose of providing the short-range communications link for an automated tolling transaction.

Therefore combining the two enablers, or fusing VICS and ETC onto one OBE, allows probe information already available from VICS to be sent via DSRC from the infrastructure to the vehicle and vice versa. This new OBE serves as a gateway for telematics applications to be delivered from the roadside to the vehicle and then displayed in the VICS-equipped navigation unit. Importantly for VIC deployment, hardware development with such a system based on this unique Japanese legacy was not difficult. Essentially, the primary tasks are to develop applications delivered through the landside portion of the network and deliver by existing communication means to the SmartWay OBE and to the customized user interfaces.

Thus, a total of eight SmartWay applications have been demonstrated:

1. *Milepost ("positional") information.* The delivery of changeable message sign- and overhead-sign-panel-equivalent in-vehicle signage of distance to exits or significant destinations.
2. *Audio messaging.* Where link times are provided through in-vehicle auditory means, for example, "10 minutes to Exit A."
3. *Merging assistance.* Visual and audio information is provided at a merge point when other vehicles, which may be occluded, merge onto the mainline or vice versa.
4. *Information on conditions ahead.* Visual and audio information is supplied— including a still-frame video detection camera output—regarding congestion on the upcoming roadway. In Japan, a particular siting might be at tunnel entrances or other known bottlenecks, for example stating, "Current traffic flow ahead of Gaien entrance about 1 km ahead" as the caption to a surveillance camera image.
5. *Parking lot payment.* E-transactions conducted via a "credit card through DSRC" transaction at a parking lot adjoining the Metropolitan Expressway.
6. *Internet connection.* Delivery of HTTP via 5.8 GHz DSRC brought to the vehicle within the aforementioned parking lot. Internet content is carefully controlled in the demonstration, but the attendee can envision the freedom of the Internet through this demonstration.

7. *Road alignment.* This involves the use of an onboard map database, which when combined with vehicle speed, enables delivery of curve overspeed warnings, for example, "Sharp curve ahead! Drive carefully."
8. *Obstacle warning.* This is particularly important with upcoming blind curves. Visual and audio information on stopped vehicles around the bend of the curve is provided, for example, "Congestion ahead! Drive carefully."

3.2.2 VIC in the United States

The roots of DSRC in the United States may be formally traced to 2003, when the Federal Communications Commission (FCC) adopted what is termed a Report and Order that provided licensing and service rules for DSRC in the ITS Radio Service. This enabled free, licensed use of the 5.850–5.925 GHz frequency range, primarily for use in safety but also for other transportation and commerce applications. It was originally conceived as a general purpose Radio Frequency Identification (RFID) technology. In making implementation decisions, however, it was quickly recognized that the Orthogonal Frequency Division Multiplexing (OFDM) was the best protocol at the time, as it allows for closely spaced orthogonal subcarriers to carry separate streams of data in parallel. There existed an emergent consumer base for wireless office LAN (Local Area Network) under IEEE 802.11a standard, based on the Atheros 5141 chipset and using OFDM. Because of the similar operating frequencies of 802.11a and the DSRC spectrum granted by the FCC, the economies of scale in using chipsets already commercially available made this case compelling during the standards discussions at the time.

For highly mobile transportation applications, a major problem to be solved in the normal operation of chipsets designed for wireless office LAN is the time, measured in seconds, required for a mobile wireless station (e.g., a laptop) to associate with a wireless access point. The nature of the channel scanning and security mechanisms in the IEEE 802.11a devices necessitates this delay, which made the existing standard association protocols unsuitable for use with safety applications that require a vehicle moving at high speed to quickly exchange information with another vehicle or RSE while in range. Changes were required at both the basic channel connection level and at the level where decisions are made to accept messages and route to their destination in order to solve this problem.

There emerged three sets of standards development and associated application ideas. First, an amendment was required to the IEEE 802.11 Wireless LAN standard, to define the DSRC spectrum and band plan and to allow fast association, comprising the physical layer IEEE 802.11p. The changes at the IEEE 802.11 layer require providing new security and channel selection services at a higher layer and specifically in the control channel and multiple service channels.[4]

These are enabled, in addition to other network services, in the IEEE 1609 standards for Wireless Access in Vehicular Environments (WAVE): the resource manager, or 1609.1; security services and management, or 1609.2; networking services, or 1609.3; and finally, multichannel operations, or 1609.4. These standards enable a system in which vehicles can connect immediately on a common channel in the DSRC band for safety applications, while other channels are available for less urgent communications.

The final or top layer standard would be the application layer, SAE J2735, under development by the Society of Automotive Engineers, DSRC Technical Committee.

Because the focus of this chapter is on the potential of transportation applications from VIC, the emergent SAE J2735 Standard is important as it transforms the antecedent and additional IEEE standards to this, the end-use, transportation applications domain. The standard, once adopted, will assure interoperability between manufacturers and allow the lower IEEE-defined protocol stack to work for proprietary as well as common applications, thus enabling vehicle-to-vehicle and vehicle-to-infrastructure safety communications.[5]

These enabling standards have been regarded as an essential and parallel complement to the U.S. DOT's Vehicle-Infrastructure Integration (VII) Program. Their adoption was considered integral to an interoperable system of vehicles and the roadside and holds significant promise for roadway operations and safety. This promise captured the interest of local, regional, and state stakeholders in tandem with interest from the U.S. DOT and, in particular, the automotive industry. They fed a vision of a roadside-based network delivering low-latency, highly reliable data communications to support safety and mobility services to users. Indeed, the integrated transportation data network would be of unprecedented scope and complexity. It was envisioned that the fully deployed system would include OBE installed on every new vehicle manufactured after a specified date and RSE installed at all signalized intersections in major urban areas, at primary intersections in other areas, at all highway interchanges, and along major intercity and many rural highways. The coverage would be extensive, as the proposed VII system would cover all urban roads within 2 min travel times, 70% of all signalized intersections in 454 urban areas, and as a vehicle OBE complement to this network of RSE, up to 15 million new vehicles per year would be DSRC- and therefore VII-equipped.[6] This has been likened to an unprecedented marriage between industries and the public sector to a scale hitherto not imagined, especially in the transportation sector.

In addition to the across-the-board recognition that high quality of service (QoS) safety applications would be an important development, a temptation to the automotive industry would be that an essentially "free" wireless network (if one discounts that tax dollars may ultimately pay for it) would be available to deliver telematics and infotainment content to drivers. An additional allure to roadway operators and planners is that VII would make it possible to collect transportation operations data of great breadth and depth, to implement sophisticated traffic safety systems, and to manage traffic flows with previously unthinkable precision.

From this vision, the process of developing the answers to such an ambitious endeavor would consume considerable time and effort at the U.S. DOT level, that is, the VII program. In addition, lessons for deployment might be learned from regional efforts, notably in Florida,[7] Michigan,[8] and California[9] and most certainly in the forthcoming U.S. DOT VII Proof of Concept experiments.[6]

3.3 Making the U.S. Technology Work: The VII California Testbed

It is essential to note that within the push–pull between national and more local interests, a literal and figurative divide emerged: differences between deployment models and their applications became increasingly obvious. Would the U.S. DOT "big bang" prevail?

Or would local or regional needs grow separate regional areas where some type of VIC would locally manifest?

Focusing on California, state and regional stakeholders have specific transportation infrastructure, operating policies, and needs different than what may be universally addressed with a national VII program. These needs led to a partnership between the California Department of Transportation (Caltrans) and the San Francisco Bay Area Metropolitan Transportation Commission (MTC). Caltrans and MTC began addressing these needs with a multiyear effort to develop, demonstrate, and deploy a VII tested in a key corridor in Northern California with a formal program testbed design, development, installation, and associated engineering.[10] This corridor is comprised of roughly ten-mile segments of two routes North of Palo Alto and South of San Francisco Airport. It encompasses two highways: State Route 82 and US 101. This selection addresses both a high-volume freeway (US 101) and a major arterial complete with signalized intersections (SR 82), as shown in Figure 3.1.

Overall, the Caltrans and MTC aims in VII California are the following:

1. To evaluate exemplar public use cases from which we can generalize VII feasibility
2. To evaluate institutional, policy, and public benefit issues
3. To explore wireless communication deployment issues and options
4. To resolve key technical issues involving implementation and operation
5. To assess implementations of the VII infrastructure, architecture, and operations
6. To support private sector evaluation interests

The basis and requirements for the testbed are the VII use cases and in particular the public sector use cases. This list was developed to corroborate with distinct Caltrans and MTC state and regional interests, which in turn address the specific San Francisco Bay Area; moreover, private use cases (item 6) round out the following set of applications:

1. Traveler information
2. Ramps
3. Electronic payment (tolling)
4. Intersection safety
5. Curve overspeed warnings
6. OEM-specific applications

The VII California testbed, described below, was essentially the on-the-road, large-scale laboratory built expressly to address these goals and applications.

3.3.1 The Elemental Building Block within the Network: Roadside Equipment

The hardware experience accrued in developing the VII California testbed focuses on the RSE and its connection to the existing Caltrans roadside appurtenances and their function. The RSE is the basic infrastructure building block for VII. Each RSE serves as an in-the-field gateway between locally transmitted vehicle data and the roadside communications infrastructure. At the top level, an RSE is a computer with a radio transceiver and an antenna. The computer must have sufficient processing and storage

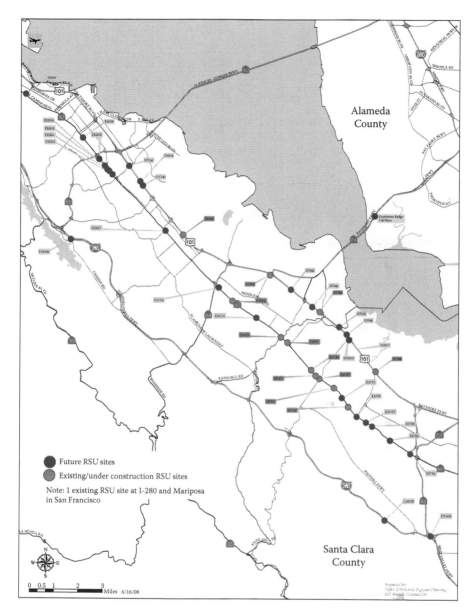

FIGURE 3.1 A map of the VII California testbed (November 2007). Installed testbed elements are shown in gray dots, planned expansion in black dots.

capability to run a gateway application between its two network interfaces, DSRC and backhaul (to the Traffic Management Center and other servers), and to run additional local safety processor software, with data filtering, buffering, aggregation, and formatting, as needed (Figure 3.2).

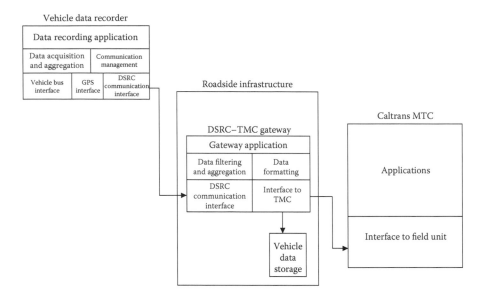

FIGURE 3.2 Roadside and backhaul communications architecture. This allows data to move from the vehicle, through the roadside infrastructure, and to the VII California network.

The RSE is therefore the link between the vehicle and the roadside backhaul network (which might already exist), providing the necessary interface. Physically, it could sit in a Type 332 cabinet, and consist of a wireless transceiver (with antenna) atop the cabinet and a computer within the cabinet, with connections.

The initial VII California computer was designed to explore a variety of applications and therefore consists of a PC-104 stack with processor and a PCMCIA card with connection to the WAVE radio and antenna. To fit within the Type 332 cabinet, form factor is low: approximately 6 × 6 × 12 in. Power is standard 120 V AC, with low current draw. A schematic of the VII California connected RSE, residing within a traffic controller cabinet, is shown in Figure 3.3.

The hardware is split into two main clusters. One resides in the roadside cabinet and the second in a separate weatherproof enclosure. This design minimizes the length of the 5.9 GHz antenna cable to maximize signal strength. Some installations require a separation of up to 200 ft between the existing roadside cabinet and the location where the 5.9 GHz antenna needs to be in order to provide coverage of the approaching roadways. Figure 3.4 illustrates a typical RSE installation.

3.3.2 The RS-UDP Transport Layer Protocol

While the aforementioned description of the hardware sets the stage, the primary objective of the testbed is to conduct experiments, or in other words, to operate the testbed. The network architecture and software, therefore, is in many senses more interesting—and at the very least—quite important.

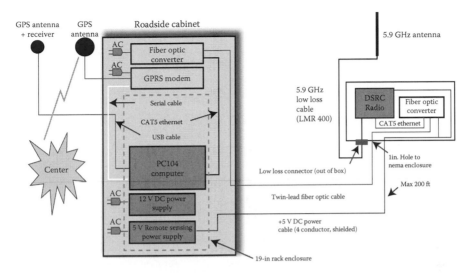

FIGURE 3.3 Schematic of the VII California RSE components.

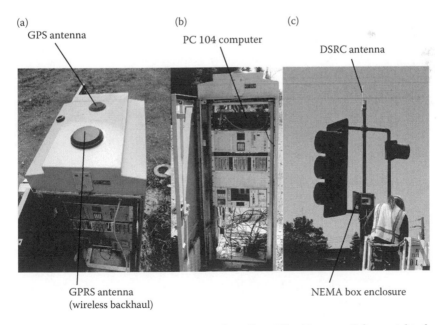

FIGURE 3.4 (a,b) VII California RSE mounted in a Type 332 cabinet on a Caltrans right of way. (c) DSRC radio enclosure and antenna mounted on a ramp meter signal pole at the freeway on-ramp.

A major goal of the VII California testbed is to support an initial set of applications, including probe vehicles, public information providers, and bidirectional nonpublic services (such as navigation and tolling). The testbed has been used as a development laboratory, with experiments and small-scale demonstrations of these applications conducted with public and vehicle manufacturer partners, starting in 2005. Heavy use of this testbed has exposed a number of issues for future VII implementations and has motivated development of a message transport layer built on top of Internet Protocol (IP) networking, a set of VII application servers, in-vehicle libraries, and administration tools. The characteristics of this software are best examined against the background of the constraints, many of which exist with any transportation and many wireless network implementations and are therefore important to note from a "lessons learned" perspective[11]:

1. The compressed development schedule has required the use of prototype-quality hardware. The Denso DSRC WAVE[4] Radio Modules used initially in the RSE were particularly limited; because they cannot transmit packets to more than four hosts on their wired side and they cannot assign dynamic IP addresses to vehicles. Firmware for several components (including DSRC radios and cellular modems) evolved to new generations during the project.
2. The backhaul network is heterogeneous, ranging from cellular modems to dedicated T1 landlines. Hence, not all sites can support all services. Furthermore, this diversity requires built-in limits on roadside-to-server communication, such as buffering, prioritizing, and possibly probe data aggregation.
3. The inherent unreliability of mobile wireless networks requires a degree of robustness. Packet loss is likely in an environment with nodes traveling at high speeds with limited range, line-of-sight occlusions, and radio frequency (RF) interference. No application can expect to have long-lasting, consistently available connections.
4. Application messages can be long. Probe messages can be up to 64 kB with multiple snapshots.
5. The transport layer should present VII California application code with an interface for communication between vehicles and servers that is message oriented (like User Datagram Protocol, UDP) but has inherent quality of service controls (to some extent like Transmission Control Protocol, TCP). It must be somewhat reliable, but not at the cost of excessive use of the channel. We call this new protocol roadside UDP (RS-UDP).
6. Information from the vehicle must be kept as private as possible.

Within these constraints, the VII California transport layer (RS-UDP) succeeds in several ways:

1. The first packet from the vehicle to the server is a data packet; there is no delay for dynamic IP address assignment (DHCP) association, Internet TCP session initialization, or other handshaking. The reduction in delay increases the chance that a return message (if any) will be received before the vehicle goes out of range.
2. Message reception probability is robust in the face of short-term failures. The transport layer can detect the loss of a fraction of the packets in a message and

retransmit the missing fragments. If packet loss is not extensive, this will quickly and efficiently complete the message transmission. Unlike TCP, if no fragments arrive, no bandwidth is wasted resending them, and no bandwidth is wasted on acknowledgment packets.

3. The suite includes a proxy that runs on the vehicle side to insulate the vehicle code from the complexity of the wireless user datagram protocol (UDP protocol) and provide it with a TCP socket interface for sending and receiving messages.

4. On the server side, similarly, the interface to the transport layer is a TCP socket, with multiple vehicle sessions multiplexed in one TCP session using transaction IDs.

5. Message addressing is encapsulated across the wireless link to overcome the Denso limitations.

6. No identifying information from the vehicle (outside of the application payload) propagates farther than the roadside or is stored anywhere.

Certain aspects of a complete VII architecture are not fully addressed by this prototype software, including security, scalability, and conformance to emerging national VII standards.[6]

3.3.3 Performance of Current Technology

The adoption of the IEEE 1609 DSRC/WAVE family of standards for trial use in 2007[4] has given the public sector and the auto industry a well worked out framework for conducting proof-of-concept testing. However, performance under congested traffic conditions remains an open issue. The density of vehicle communications on crowded freeways and busy intersections is high compared to the typical office WiFi hotspot, and furthermore, highway communications are expected to include an unusually large number of broadcasts. Analysis and simulations cannot accommodate the full complexity of practical systems with large numbers of vehicles in close proximity. Overall performance depends not only on physical layer channel effects and on the performance of the 802.11 MAC layer, but on operating system and application structuring and scheduling of messages. More research is needed to characterize the magnitude of congestion problems and the total amount of communication possible using real hardware in the field and to investigate methods of congestion mitigation.

In addition to the problem with the density of communications, safety applications have critical latency requirements. Communication measurement tools are often only concerned with the aggregate data transfer rates and error percentages, and do not measure the latency of individual messages. However, in DSRC systems the latency may be the limiting parameter on the data transfer rate, because the system load cannot be allowed to rise to a rate that will delay crucial safety communications.

The U.S. DSRC allocation is a 75 MHz channel. In replication of the WiFi model, the 75 MHz is divided into several non-overlapping channels to facilitate the separation of colocated network service providers by having them operate on different channels. However, because the different service providers may offer different services, a vehicle interested in consuming two services from two different providers would have to switch its radio from one channel to another. For example, a vehicle interested in listening to an intersection safety message provided by the intersection operator while also paying a toll due to a road operator may have to switch its radio back and forth between two channels. The standard IEEE 802.11 protocol model does not require this switching. Thus 802.11 hardware currently on the market has not been designed for fast channel switching.

The work described in this chapter represents the first steps in using inexpensive computer hardware with prototype radios to explore performance issues for multiple OBEs approaching a RSE. A major motivation for this research is the scenario described in the proceedings of the December 2004 DSRC workshop[12] in which many vehicles come in range of a transaction service advertised by a RSE simultaneously and immediately switch to a service channel and begin consuming the transaction service.

In the subsequent sections, we characterize the performance of the first generation of DSRC hardware, that is, the DENSO Wave Radio Module (WRM). We focus on the following:

1. Round-trip time from OBE to RSE
2. The impact of channel switching on communication performance

3.3.3.1 Mobile Transaction Time

The equipment used for our tests takes advantage of the availability of inexpensive miniature microprocessor portable assemblies. Four assemblies were constructed, each containing three 400 MHz gumstix expansion boards with two Ethernet ports[13] (Figure 3.5). Funding for the equipment was provided by ARINC Corporation, as part of their work on DSRC for U.S. DOT. ARINC also supplied 12 Mobile Mark magnetic mount antennas (Figure 3.6) specially constructed for the 5.9 GHz frequency.

Each of the 12 individual computers runs an embedded Linux system and has two Ethernet interfaces. On one Ethernet interface it is paired with its own DENSO WRM and rooftop antenna; on the other it can be connected to a host computer for data

FIGURE 3.5   Assembly of three gumstix netduo expansion boards.

FIGURE 3.6   Antenna placements with eight active OBUs on one vehicle.

storage. The assemblies were designed for flexible configuration. All the assemblies can be connected via a router and deployed in one vehicle for tests to an RSE, or individual assemblies can be placed in different vehicles for vehicle-to-vehicle research or to test multiple approach paths to an RSE simultaneously.

For most of the tests described in this section, the gumstix processors are connected through the router to a laptop, which provides a networked file system where trace data from the test run can be stored. In some tests, the laptop is also connected through a USB port to a GPS unit, so that location can be saved as well. All processors are synchronized through the router.

3.3.3.2 OBE-RSE Round-Trip Time

Figure 3.7 shows that the basic round-trip time (RTT) between two systems in good communication, with no competition from other stations for the airwaves, is typically 2 to 3 msec. Although the actual time transferring the symbols on the radio medium is much less than this, this includes the time buffering and copying of the message on transmission and reception, and scheduling the application processes that are the source and destination of the message. Figure 3.8 shows the situation when two different stations are sending messages pairwise at a moderate rate. Competition for the airwaves between the two pairwise streams causes the RTT to rise to between 3 and 5 msec. Figure 3.9 shows how a high rate of transfer between two computers can cause greatly increased delay on a pairwise stream between two other computers communicating at a more moderate rate. Note that the RTT scale on this graph is in seconds, not in milliseconds as for the previous

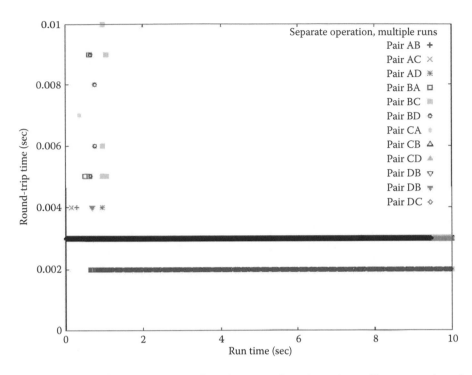

FIGURE 3.7 Baseline operation, 1480 byte data per packet, 10 msec interval between sends, and different pairwise runs of the four stations used for desk checking.

graphs. It can be seen on the lower graph in Figure 3.9 that on the higher rate stream packets were actually dropped, due to exceeding the retry limit, between sequence numbers 1454 and 1746.

3.3.3.3 Effects of Channel Switching

The prototype DENSO WRMS that we used for this preliminary testing were not designed to switch channels fast enough to allow a 50 msec control channel period and a 50 msec service channel period, as specified in the IEEE 1609.4 standard. However, we were interested in seeing how well communication could be maintained across channel switches even under adverse circumstances. We also wanted to test how well the gumstix processors were synchronized across the router. We found that, whereas for infrequent channel switching it was possible to continue communications with only a small percentage of lost packets, when channel switching was more frequent it was impossible with our current equipment to maintain synchronization well enough to avoid significant losses.

We began by performing a set of tests, with stationary equipment in close range (Table 3.1). Five gumstixs acted as "vehicles" sending to a sixth used as the "road side unit" or receiver. Channels were switched between 172 and 178 by a background process running on the host laptop that sends a message through the router telling each gumstix processor when it is time to switch channels. The gumstix processor writes a record to

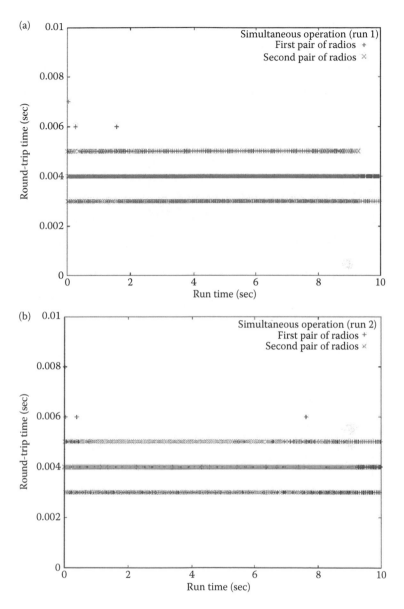

FIGURE 3.8 (a,b) Two runs showing interference between two pairwise communication streams causing slightly increased round-trip time (1480 byte data per packet, 10 msec interval between sends on both streams).

the shared file system of the time of the switch. Clocks on all gumstix processors are synchronized before the start of testing.

Figure 3.10 shows the baseline case with no channel switching. Note that in these graphs the sequence numbers for the different processors have had a constant added so

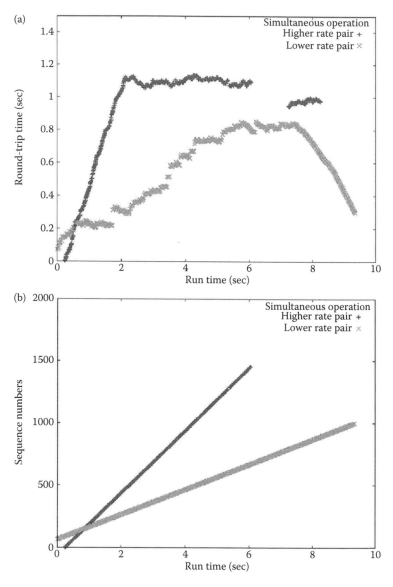

FIGURE 3.9   (a) Large increase in round-trip time due to interference between a higher rate pairwise communications stream (1480 byte packets at 4 msec interval between sends) and a lower rate stream (1480 byte packets at 10 msec interval between sends). Note that the scale for round trip time is in seconds. (b) Sequence numbers.

that they appear in different Y ranges; in fact the sequence numbers begin at one for every invocation of the program.

Figure 3.11, with case sw1, indicates that infrequent channel switching causes little or no degradation in performance. Figure 3.12 shows that for this run the channel changes

TABLE 3.1 Parameters Used for Tests of Frequent and Infrequent Channel Switching

Testname	Switch Interval (between Channel Changes)	Packet Interval (between Sends) (msec)	Packet Size (Data Bytes)	Load at Receiver (Bytes/ sec from All Senders) (K)
sw0	No change	100	200	10
sw1	5 sec	100	200	10
sw2	5 sec	200	200	5
sw3	5 sec	50	200	20
sw4	100 ms	25	200	40
sw5	100 ms	10	200	100

on the different gumstix processors are clustered within ~10 to 15 msec of each other, with delays through the router from the host laptop where the channel switch directive is initiated and scheduling delay on the individual processors adding up to this amount. Cases sw2 and sw3 show no appreciable difference from case sw1. However, for case sw4, as shown in Figure 3.13 the increased overall load and increased channel switching have caused some dropped packets (evidenced by missing sequence numbers) and some RTTs >5 msec. Notice in the close-ups in Figure 3.14 how for some processors, the time they are on a channel is skewed enough away from the time of the receiver that they may successfully send only one or two messages while they are connected on the same channel. Overall system load on the router may also be contributing to the greater skew on channel switch times, when sw5 is compared to sw4.

3.3.3.4 Mobile Urban Intersection Testing—Transaction Times

Tests were performed in San Francisco intersections with ARINC support, leveraging on the RSE installations that were made as part of the ITS America 2005 Innovative Mobility Showcase (ITS America 2005). A short paper on these tests was presented at the First IEEE International Symposium on Wireless Vehicular Communications in 2007.[14]

The data in this section were taken with the RSE that was installed at 6th and Brannan in San Francisco. The DENSO WRMs for these RSEs were mounted on the signal arm (see the location in Figure 3.15). The wired interface of the WRM was connected by fiber to a PC104 in the traffic signal controller cabinet, running Linux. The software in the PC104 during this test was source identical with the data acquisition software running on the gumstix computers in the vehicle. These tests were conducted on September 24, 2006.

For all runs, data gathering started at the intersection with 5th and Brannan and continued for two minutes, while driving straight along Brannan to 7th street, turning right, coming back along Bryant, and then turning right at 5th. See the illustration with the dotted line in Figure 3.15 of the active part of the run.

There were three types of experiments run with low, medium, and high overall load. For each set, the intervals between message sends were adjusted to maintain a constant load of messages at the RSE. In the first case the load was 40 kb/sec, in the second it was 100 kb/sec, and in the third 200 kb/sec. For all runs on this day of testing the data rate setting in the DENSO WRMs was set at 6 Mb/sec.

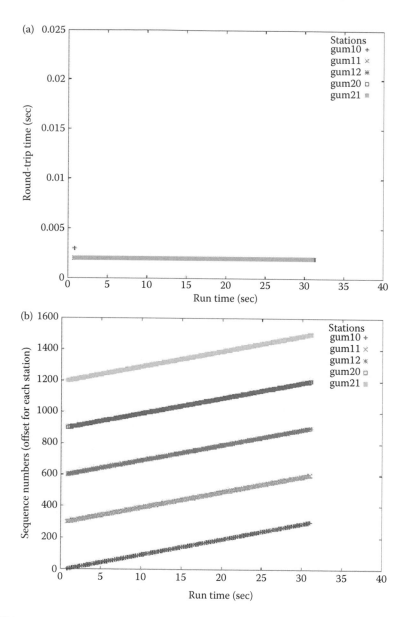

FIGURE 3.10 Baseline case with no channel switching, light load: (a) round-trip time versus seconds in run; (b) sequence numbers.

This data included two fields for local and remote Received Signal Strength Indication (RSSI), the radio energy per channel, measured in dBm. RSSI values for all the runs are plotted together against latitude and longitude in Figure 3.16.

Variable parameters for each run were DENSO WRM rate setting, in Mb/sec; the number of senders in the run; the packet size in bytes for this run; the number of

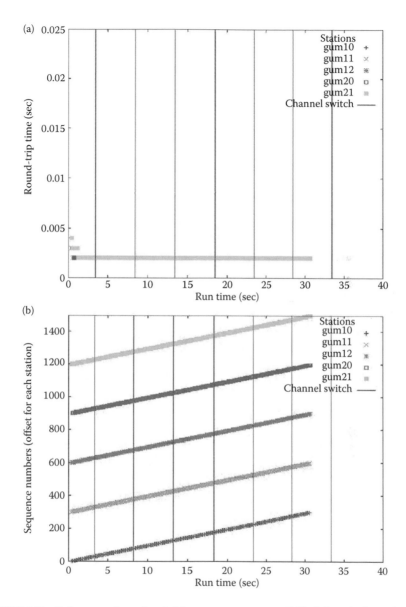

FIGURE 3.11 Infrequent channel switching, messages sent at 10 Hz (t1); each channel switch is indicated by a vertical line, and at that point the channel changes: (a) round-trip time; (b) sequence numbers.

milliseconds in the interval between packets; and the number of sends in the test (in some cases continuous and halted by the user when out of range).

To approximate a model of transaction processing in which the traffic is predominantly generated on a service channel in response to short broadcasts received on a control channel, post-processing created the following summary data for each run: run

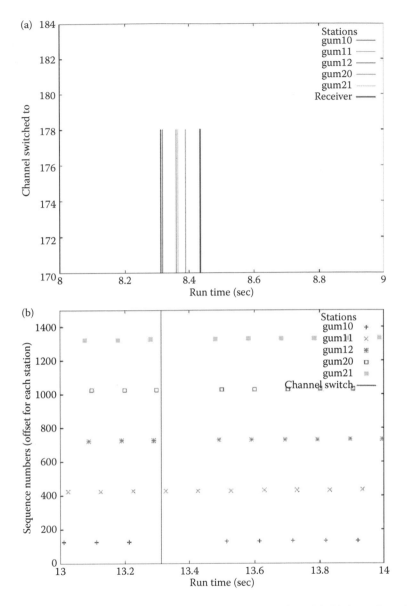

FIGURE 3.12   Close-ups, for the case of infrequent channel switching (t1): (a) channel switching; (b) sequence numbers.

type identifier string; B total load for run (in kilobytes); R, rate setting of DENSO WRM in Mb/sec; S number of senders; M, bytes per packet; I, time interval between packet sends; N, number of sends in test; successful sends in test; average RSSI (over round-trip); start time (time of first reception); message time (time of reception for

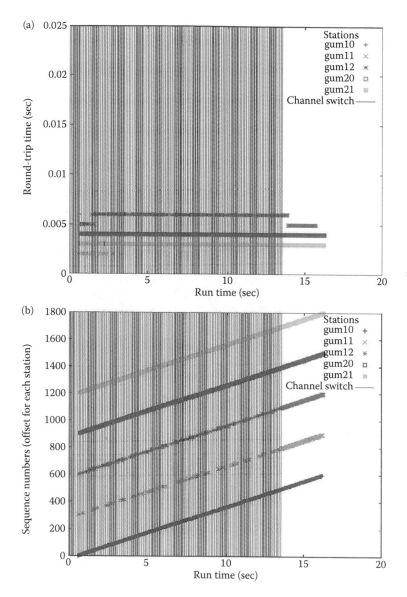

FIGURE 3.13 Case sw4, with increased overall load and increased channel switching: (a) round-trip time; (b) sequence numbers.

8000 bytes—line 8000/P); end time (time of last reception); message period (message time–start time); connected period (end time–start time); minimum distance from RSE; maximum distance from RSE; average distance from RSE; success rate (line count/possible transmissions; between start time and end time, i.e., while in range); and gumstix processor ID number.

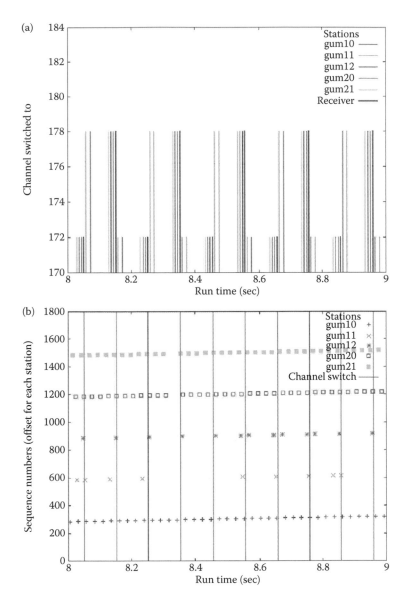

FIGURE 3.14   Close-up of switching and sequence number loss for case sw4: (a) channel switching;
(b) sequence numbers.

In our figures we have produced graphs of two measures of communications performance with respect to the interval between sends. In Figure 3.17, the success rates for
each run are used as a measure of probabilities of successful round-trip transmissions.
In Figure 3.18, the time required for successful reception of 8000 bytes is used as a
measure of the transaction time. Reading the graphs, for each line with a fixed packet

FIGURE 3.15   Location of RSE at 6th and Brannan in San Francisco.

size, the values on the left represent higher overall loads than the ones on the right. Due to testing time limitations, the set of run types with the highest load was completed only for 2 and 8 nodes.

Looking at the highest load values for 2 and 8 nodes, in these tests, for the same overall load, the 400 and 800 byte packet size tests (with smaller send intervals) have better performance in terms of the success rate of packet transmission than the 200 byte packet size. While this is interesting, clearly much more testing is required to characterize the regions of operation for this effect.

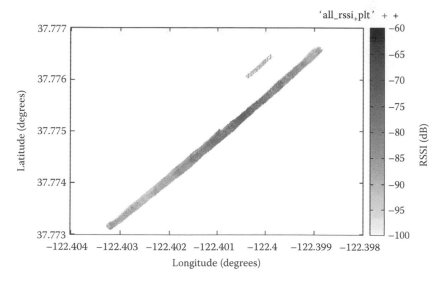

FIGURE 3.16   Plot of the average of intersection and vehicle received signal strength indicator (RSSI) by vehicle GPS location for all round-trip packets received in runs at 6th and Brennan.

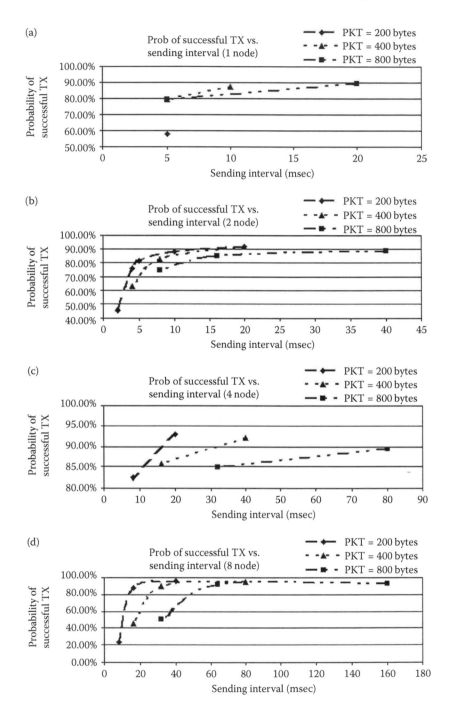

FIGURE 3.17 Probability of successful round trip transmission, as a function of the interval between packet sends, for (a) 1, (b) 2, (c) 4, and (d) 8 nodes, with varying packet sizes of 200, 400, and 800.

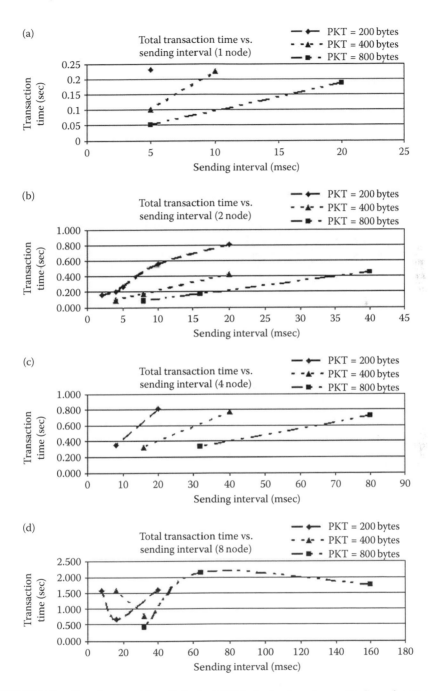

FIGURE 3.18 Time (in seconds) until the first 8000 bytes has been transmitted, as a function of the interval between packet sends, for (a) 1, (b) 2, (c) 4, and (d) 8 nodes, with varying packet sizes of 200, 400, and 800 for tests at 6th and Brannan.

Note that even with eight nodes, although the probability of individual packet success is low for smaller sending intervals, because of collisions, at the overall loading used in these tests the time until an 8000 byte transaction is completed, from the time of first connection, stays under 2.5 sec.

3.3.4 Performance Discussion

The differing performance measurements we have obtained so far indicate the wide variations in performance that can occur with the same installed equipment, depending on variations in sending interval, packet size, and detailed geographical characteristics of the approach. In many of these runs, we see evidence of two regions of performance with respect to the sending interval (see Figure 3.19 for an illustration of probability of successful transmission, and Figure 3.20 for transaction time). In region R1, the probability of a successful round-trip transmission increases as the sending interval increases, because there are fewer collisions. In region R2, the probability of a successful

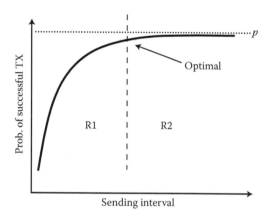

FIGURE 3.19 Illustration of regions of performance for successful transmission.

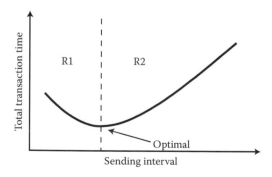

FIGURE 3.20 Illustration of regions of performance for transaction times.

round-trip transmission appears to approach a limit. Because the sending interval is already long enough for contention resolution, the performance approaches a bound of p, which is decided by physical layer performance. With respect to transaction times, too low a probability of success will cause delays if the sending interval is too short, but over the optimal sending interval, the transaction time will increase because of wasted time between packets. The points of optimality in Figures 3.19 and 3.20 need not be the same, and a great deal of testing, load characterization, and site characterization will be required to determine these intervals in practice. It is necessary to be able to integrate data acquired at all layers of the network stack for complete understanding. We hope to use this data acquisition software and equipment to carry out further vehicle-to-roadside and vehicle-to-vehicle tests at VII California RSEs, in conjunction with more physical layer testing, and to develop a methodology for site characterization that will allow us to predict the performance of communication with RSEs.

3.4 The Future of VIC

3.4.1 Transformation into Multiband, Wireless Communication

Lessons have indeed been learned from both the research and initial deployments of VIC. Internationally, SmartWay and CALM in particular are leaders in their respective regions of the world. In the United States, the idea of a network and deployment strategy envisaged in Section 3.2.2, while alluring, is recognized as quite expensive. To this end, the U.S. DOT has in 2008 embarked on an initiative to transform VII: Safe and Efficient Travel through Innovation and Partnerships for the 21st Century (SAFE TRIP-21).[15]

3.4.1.1 SAFE TRIP-21

The paradigm of SAFE TRIP-21 is probably a reaction to two fundamental forces:

1. The rapid emergence of multiband communications links and services available to surface transportation users, spurred by the ubiquity of mobile consumer devices.
2. The contrastingly tepid progress of the U.S. DOT's major VII program.

The U.S. DOT had indeed witnessed the worldwide developments and wanted to include those developments within North America and seek to, in effect, transform VIC in the United States to harness consumer devices so as to rapidly push forward the communication between vehicles and the roadside. It is uncertain whether the advent of SAFE TRIP-21 will supersede the use of DSRC, or whether indeed the safety-of-life applications of DSRC are in the near- or mid-term future. Safety, unless it is redefined to encompass the ancillary benefit of "congestion ahead" warning, may be put on the back shelf to mobility in this way of thinking; however, it need not be, as a multiband communication capability, to include use of the cellular network, emergent 802.16 (WiMax), WiFi, *and* DSRC may be appropriate.

The inclusion of DSRC somewhere in the gamut between near and long term will be contingent on the societal need for safety but also perhaps market forces. Can an approach and strategy that essentially leads to a demand for DSRC be determined? Take for example the SAE J2735 standard and the concept of the Basic Safety Message (BSM).

Consider in Table 3.2, the BSM, Part 1. The defined fields include standard data elements, most of them relating to position and heading; those data elements may be obtained through an aftermarket device with little or no connectivity to the car, especially if that device has a GPS receiver and perhaps, better yet, an accelerometer and nominal processing capability. Suffice it to say that the car may send via DSRC crucial "I am here" BSM data, not necessarily using an embedded system.

TABLE 3.2   Draft J2735 Basic Safety Message Definition

	Basic Safety Message						
Bit	All Multibyte Integer Fields are in Network Byte Order (i.e., Big-Endian)						
Byte \ 8 (MSB)	7	6	5	4	3	2	1 (LSB)
1	DSRC Message Id (*Set to the fixed value of 0 × 02*)						
2	Millisecond in Minute (MSB)						
3	Millisecond in Minute (LSBit = 1 ms)						
4	Temporary Id—Represents dynamic MAC or IP address of the radio						
5	-do-						
6	-do-						
7	-do-						
8	-do-						
9	-do-						
10	Latitude of the Host Vehicle Center Position (MSB) 32-bit signed						
11	-do-						
12	-do-						
13	LSBit = 1.25-e7						
14	Longitude of the Host Vehicle Center Position (MSB) 32-bit signed						
15	-do-						
16	-do-						
17	LSBit = 1.25-e7						
18	Elevation (MSB) 24-bit unsigned						
19	Range 0–1677721.6 m; Offset = 1 km (value 0 = 1 km below the WGS84 reference ellipsoid)						
20	LSBit = 0.1 m						
21	Vehicle speed (MSB); Range (−327.65 … 327.65) m/sec; Forward positive						
22	(LSBit = 0.01 m/sec)						
23	Heading (MSB); Range (0 … 65535); North is 0°, and Clockwise is positive angle						
24	(LSBit = 0.00549°)						
25	Longitudinal acceleration (MSB); Range (−327.65 … 327.65) m/sec^2; Forward positive						
26	(LSBit = 0.01 m/sec^2)						
27	Lateral acceleration (MSB); Range (−327.65 … 327.65) m/sec^2; Right positive						
28	(LSBit = 0.01 m/sec^2)						
29	Vertical acceleration; Range (−10.16 … 10.16) m/sec^2; LSB = 0.08 m/sec^2						
30	Yaw rate (MSB); Range (−327.68 … 327.67)°/sec; Clockwise positive						

Continued

TABLE 3.2 (continued)

Bit	Basic Safety Message							
Byte	All Multibyte Integer Fields are in Network Byte Order (i.e., Big-Endian)							
	8 (MSB)	7	6	5	4	3	2	1 (LSB)
31	(LSBit = 0.01°/sec)							
32	Brake active status			Traction control			ABS	
	LF = 0 × 01			00 = Not equipped			00 = Not equipped	
	RF = 0 × 02			01 = OFF			01 = OFF	
	LR = 0 × 04			10 = ON			10 = ON	
	RR = 0 × 08			11 = ENGAGED			11 = ENGAGED	
33	Steering wheel angle (MSB); Range (−655.36 ... 655.36)°; Clockwise positive							
34	(LSBit = 0.02°)							
35	Throttle position; Range (0–100)%; LSBit = 0.5%							
36	Vehicle exterior lights; 8-bit enumerated values is defined:							
	AllLightsOff = 0 × 00							
	LowBeamHeadlightsOn = 0 × 01							
	HighBeamHeadlightsOn = 0 × 02							
	LeftTurnSignalOn = 0 × 04							
	RightTurnSignalOn = 0 × 08							
	HazardSignalOn = 0 × 0C							
	AutomaticLightControlOn = 0 × 10							
	DaytimeRunningLightsOn = 0 × 20							
	FogLightOn = 0 × 40							
	ParkingLightsOn = 0 × 80							
37	Vehicle width (MSB); 10-bit; Range (0–1023) cm; LSBit = 1 cm							
38	Vehicle length (MSB)							
39	14-bit; Range (0–16383) cm; LSBit = 1 cm							

The aftermarket proliferation of devices that provide this may be sufficient to encourage vehicle manufacturers to develop embedded systems, allowing DSRC in VIC to become a reality. Granted, the consumer must have some motivation to purchase a device with DSRC, and the roadside operator must have the motivation to build the DSRC. Perhaps the argument that market forces such as some other safety services or the availability of high bandwidth access for telematics applications will suffice. How this may or may not happen is conjecture; however, the point is that there are conceivable pathways toward a DSRC and VIC future.

3.4.2 The European Future

The European future appears to be settled on the use of nomadic devices to deliver VIC, as discussed in Section 3.2.1.1. It is interesting to note that the future of VIC in the United States and VIC in the world may very well converge to the standards-based CALM architecture.

3.4.3 The Japanese Future

Can SmartWay lead to the widespread adoption of VIC in Japan (and elsewhere)? The Tokyo Metropolitan Expressway, with its myriad of geometrical problems, is relatively unique. It is an archaic road, expediently designed and constructed in advance of the 1964 Olympics. Moreover, the expressway is owned by a private company. Institutional issues, therefore, abound and include the following:

1. Are there a sufficient number of other roadways to justify a large-scale rollout of Smartway?
2. Can a public–private partnership be sustained, where the public invests and a private company receives SmartWay on a noncompetitive basis?
3. At what point is the chicken-and-egg conundrum overcome? That is, how many additional DSRC-equipped stations would be necessary for SmartWay vehicles and the road owner to realize overall benefit?

This being said, the Japanese are forging ahead. Following SmartWay 2007 are a series of "trial observations," or field tests, to verify system functions, assess SmartWay effectiveness, and to understand and assess usability. These tests will point the way to refinement and deployment, seeded by SmartWay's offspring being deployed in three urban areas within Japan. The horizon of telematics applications may broaden, as remote diagnostics, payment services, or infotainment and entertainment services may be viable. From subsequent lessons learned, a "nationwide deployment" in a short number of years is in the offing.

To consider the future of SmartWay outside Japan engenders a host of additional questions. Certainly, the Japanese surface transport infrastructure and governance are unique. The Japanese have not exported VICS, nor could such an export be possible unless another country provides the same high-quality and nearly ubiquitous mobility performance measurement system.

3.4.4 The U.S. Future

As with many things in the government, there must be a political imperative. Consider that a fully deployed VII in the United States would include OBE installed on every new vehicle manufactured after a specified date and RSE installed at all signalized intersections in major urban areas, at primary intersections in other areas, at all highway interchanges, and along major intercity and many rural highways. The coverage would be extensive and expensive: the proposed VII system has the potential to cover all urban roads within 2 min travel times, 70% of all signalized intersections in 454 urban areas and up to 15 million new vehicles per year would be DSRC- and therefore VII-equipped.[6] On the national level, the idea of a network and deployment strategy envisaged in Refs. 6 and 16, while alluring, is recognized as quite expensive, and in light of other transportation needs in our nation, it is increasingly difficult to imagine the resolve and funding (up to perhaps $8 billion, not including operations and maintenance costs) needed to install the 240,000 RSE "edge" or access points and extensive communications network. On the benefits side, this network would make it possible to collect transportation operations data of great breadth and depth, to implement sophisticated traffic safety systems, and to manage traffic flows with previously unthinkable precision.

Is there political will to incur the "Big Bang" cost? Really, how, in the context of scarce economic resources could such a communications cosmology evolve? How about seemingly mundane but very real issues of diversified and institutionalized state, regional, and local roadway operators, with different equipment, management philosophies, and above all, degrees of independence? A "Big Bang" may be resisted. In the end, it comes down to money. The next Transportation Reauthorization Act will likely apportion scarce resources to pressing capital investment needs, such as structurally deficient bridges, the collapse of which are widely published threats to public safety, and well-read amongst the electorate.

3.4.5 The World Future

High market penetration will encourage vehicle manufacturers and roadside operators to introduce a wide range of in-vehicle safety-of-life applications, such as intersection warnings, stopped vehicle ahead, and curve speed warnings. The VII vision will then have become a reality.

To maximize the VIC and some of the functionality, the following must happen:

1. Consideration of the existing regional public and private investment and partnerships, with their associated experience and expertise.
2. Use of the components of VIC and standards already in place or under way, for example, CALM.
3. Development of regional and national and international public–private alliances— or "outreach"—of suppliers, developers, and users of VIC. The objective is to develop a core of applications that will attract users.
4. Provision of the technological basis for VIC, such as developing small form factor gateways to link the vehicle to after-market devices (cellphones/PDA/PND), companion RSE, and an open platform to enable rapid distributed safety and congestion relief application development.

The future of VIC therefore envisions interoperation of a whole range of personal interface devices (cell phones, PDAs, PNDs), telecommunication services (WiFi, 2.5G, 3G, DSRC, Bluetooth), and data sources (on-road cameras, loop data, probe data). Through a multidevice, multi-industry, multispectrum, public–private combination, VIC should ride the convergence of GPS and Internet services towards a storm of intelligent transportation applications.

In the end, the VIC story could be this story:

> For Immediate Release (January 1, 2012): Amidst fanfare involving the rollout of thousands of "connected travelers" communicating to roadside hotspots across the globe, public authorities, X Inc. and Y Inc. have invited transportation safety, information, and management developers to participate and contribute more and more applications through an open standard. The unveiled plan is to marry the public sector's equipment in the roadside

with the private sector's consumer devices and services to create an open
environment for applications to be developed to connect these vehicles. The
ultimate vision for the connected traveler is to achieve unprecedented levels
of safety with large concentrations of communication-enabled vehicles in...

Lisa read the headline on her iPhone as she hurriedly ate breakfast, noting with some
pride that she and her company were among the initial participants. "Timmy, let's go,"
she said to her 12-year-old son. "We can't be late for school." As Timmy bounded down
the stairs, Lisa programmed the school destination from a list of potential routes into
her phone, and the quickest route to school and estimated arrival time was shown.

Upon entering her car, Lisa connected her phone to a gateway, automatically starting the
suite of applications. Jan, who sits in the cubicle next to Lisa at XYZ, starts her VIC a bit
differently. The gateway discovered and associated with Jan's Bluetooth-enabled personal
navigation device. However, the results are the same, as the potential *mobility applications*
are manifold: both Lisa and Jan receive *dynamic route advisories* on the way to their destinations by receiving *fused traffic and traveler information* from other "probe" vehicles, as
well as information from the myriad of existing traffic management resources. Lisa especially enjoys the *"congestion ahead" warning in advance of stopped traffic*, which is one of
several *immediate safety features*. Jan, who is always in a hurry, tells anyone who will listen
that, as she pulls into the XYZ parking lot, *smart parking information*, information on space
availability (with provision even to reserve slots ahead of time), is provided.

Importantly, with Lisa and Jan's consent given using the VIC service module on their
devices, the public or private service provider receives *arterial probe information* from cars
like Lisa's and Jan's. In addition, the roadway authority is able to implement *novel adaptive
signal control algorithms* in the vicinity of XYZ without expending capital or operational
costs; the road authority instead has access to the probe data from enough of a penetration
of equipped vehicles to *maximize the available capacity of arterial intersections*.

And what about Timmy? Indeed, he arrived at school in time. And in the future, after
the groundbreaking VIC initial applications and *open architecture* take root, VIC transceivers will carry the day, safely. By the time he is a driver, Timmy will enjoy a roadway
environment where nearly all drivers will be users of VIC.

References

1. Shladover, S.E., DSRC—It's Not Just for Toll Collection Any More, *Telematics
   Update*, 10, January/February, 16–17, 2002.
2. International Organization for Standardization, Intelligent Transport System—
   Continuous Air Interface Long and Medium—Medium Service Access Point, Draft
   International Standard ISO/DIS 21218, 2007.
3. Japan Ministry of Land, Infrastructure and Transport, Road Bureau, Smartway 2007
   Public Road Test, available at http://www.its.go.jp/ITS/topindex/topindex_sw2007.
   html, 2007.
4. IEEE Standards Association, IEEE P1609.1—Standard for Wireless Access in
   Vehicular Environments (WAVE)—Resource Manager, IEEE P1609.2—Standard

for Wireless Access in Vehicular Environments (WAVE)—Security Services for Applications and Management Messages, IEEE P1609.3—Standard for Wireless Access in Vehicular Environments (WAVE)—Networking Services, IEEE P1609.4— Standard for Wireless Access in Vehicular Environments (WAVE)—Multi-Channel Operations, adopted for trial-use in 2007, IEEE Operations Center, 445 Hoes Lane, Piscataway, NJ, 2007.

5. SAE International Surface Vehicle Standard, Draft SAE J2735 Dedicated Short Range Communications (DSRC) Message Set Dictionary, Unapproved Draft Rev 24,400 Commonwealth Driver, Warrendale, PA, May 16, 2008.

6. Federal Highway Administration, Vehicle Infrastructure Integration: All Architecture and Functional Requirements, Version 1.1, U.S. Department of Transportation, ITS Joint Program Office, July 20, 2005.

7. Florida Department of Transportation, District 5, Surface Transportation Security and Reliability Information System Model Deployment: iFlorida—Final Development Plan, prepared for U.S. DOT–FHWA, Cooperative Agreement Number DTFH61-03-H-00105, January 29, 2004.

8. Mixon/Hill of Michigan, Inc., VII Concept of Operations: VII Data Use Analysis and Processing, a report prepared for Michigan Department of Transportation, June 2007.

9. Shladover, S.E., Preparing the Way for Vehicle-Infrastructure Integration, California PATH Research Report UCB-ITS-PRR-2005-31, November 2005.

10. Misener, J.A. and Shladover, S.E., PATH Investigations in Vehicle-Roadside Cooperation and Safety: A Foundation for Safety and Vehicle-Infrastructure Integration Research, Intelligent Transportation Systems Conference, 2006, ITSC '06. IEEE, 2006, pp. 9–16.

11. Shladover, S.E., Probe Vehicle Data Needs and Opportunities, Proceedings of the Thirteenth World Congress on Intelligent Transportation Systems and Services, Paper 75, London, October 2005.

12. Research and Innovative Technology Administration, Proceedings from the December 2004 Workshop, Intelligent Transportation Systems, available at http://www.its.dot.gov/cicas/cicas_current_act.htm, U.S. Department of Transportation, accessed June 19, 2008.

13. Gumstix Inc., Product Information and Homepage, available at http://www.gumstix.com/, accessed June 19, 2008.

14. Dickey, S., Huang, C.L., and Guan, X., Field Measurement of Vehicle to Roadside Communication Performance, Proceedings of the Vehicular Technology Conference, 66th IEEE, October 2007.

15. U.S. Department of Transportation Research and Innovative Technology Administration, SAFE TRIP-21, Safe and Efficient Travel through Innovation and Partnerships in the 21st Century, available at http://volpedb.volpe.dot.gov/outside/owa/vntsc_outside.display.notespec?p_post_id=175, Volpe National Transportation Systems Center, 2007.

16. Henry D., VII Strategy for Safety and Mobility, ITS Michigan Wireless Communications Seminar, Brighton, Michigan, March 2007.

4

CAR-2-X Communication in Europe

Long Le, Andreas Festag, Roberto Baldessari, and Wenhui Zhang
Network Research Division NEC Laboratories Europe

Car-to-car and car-to-infrastructure communication (also known as CAR-2-X communication) has gained significant interest in Europe in the last few years as a promising technology to increase road safety. Its origins can be traced to the 1980s, when the PROMETHEUS (PROgramme for European Traffic with Highest Efficiency and Unprecedented Safety) project (1987–1995)[1] was initiated by the European car industry and performed in the framework of the EUREKA (European Research Coordination Agency) program. The original objective of the project PROMETHEUS was automated driving for private cars. Later, it shifted its focus to driver information using in-vehicle systems, such as the *intelligent copilot*. The follow-up project DRIVE (Dedicated Road Infrastructure for Vehicle Safety in Europe) I (1988–1991)[2] promoted advanced transport telematics (ATT), mainly considering road-side infrastructure. DRIVE II (1992–1994) concentrated on strategies for management and the introduction of telematics systems for communication and traffic control, including test projects and field tests. Overall, these projects have led to significant advances in European road transport, but the deployment of CAR-2-X communication was not possible due to lack of adequate and affordable communication technology.

Recent development and widespread deployment of wireless communication technologies, such as devices based on the IEEE 802.11 wireless technology, have finally made it possible to introduce intervehicular communication on a large scale. Technological development was supported by major European initiatives, notably the Transport Sector of the Telematic Applications Program (TAP, 1994–1998),[3] Information Society Technology (IST) actions in the 5th and 6th European Framework Programs (FWP),[4,5] and recently Information and Communication Technology (ICT) of the 7th European Framework Program (2007–2013).[6] Key projects were the EU projects DIATS, CHAUFFEUR-1 and -2, CarTalk2000, PReVENT, GST, and more recently CVIS, SAFESPOT, and COOPERS. The programs were complemented by national projects, such as in Germany (INVENT, FleetNet—Internet on the Road, NoW—Network on Wheels), Sweden (ISVV), and UK (CVHS). A historical overview of road transport and vehicular communication in Europe is illustrated in Figure 4.1. The various efforts are currently moving from pure research activities towards experimental evaluation and field trials.

This chapter provides an overview of the fast-changing field of vehicular communication from a European perspective. It addresses car-to-car and car-to-infrastructure communication, which we subsume as CAR-2-X communication. The chapter first gives an introduction to the Intelligent Car Initiative, the overarching framework for many R&D projects. It then provides a summary of recent and current R&D projects, and technologies developed in these projects. The chapter concludes with an overview of standardization activities and an outlook of CAR-2-X communication in Europe.

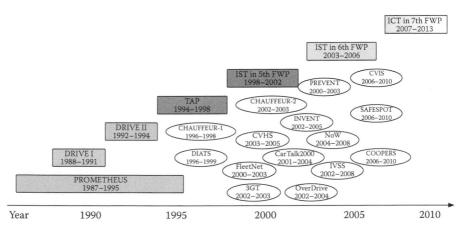

FIGURE 4.1   Historical overview of road transport and vehicular communication in Europe.

4.1 The Intelligent Car Initiative

The Intelligent Car Initiative[7] is one of the flagship initiatives of the European Union that explores the potential of *information and communication technologies* (ICT) in improving quality of life. The main purpose of the initiative is to facilitate the incorporation of advanced information and communication technologies into on-board intelligent vehicle systems to provide smarter, safer, and cleaner cars and roads. The initiative investigates innovative ICT-based solutions to many of today's problems in the transportation sector.

Generally, the Intelligent Car Initiative aims at drastic reductions of road accidents, traffic congestion, harmful effects on the environment and public health, and waste of energy. Intelligent cars can prevent, avoid, or mitigate road accidents by assisting drivers in their driving functions. They can eliminate or reduce traffic congestion by providing drivers with real-time information about the road conditions. Further, intelligent cars can improve overall energy efficiency by helping drivers select an optimal route and increasing engine performance.

The Intelligent Car Initiative targets three main goals. First, it coordinates and promotes the work of relevant stakeholders, such as car manufacturers, road operators, telecommunication companies, transport service providers, and regulators in Europe. Second, it facilitates research and development in information and communication technologies to provide smarter, safer, and cleaner vehicles and roads. Third, it creates awareness among consumers and decision makers of ICT-based solutions to remove political and societal hindrances that prevent intelligent cars from penetrating the European market.

The Intelligent Car Initiative has identified several core technologies for future intelligent transport systems[7] to be addressed in research activities. Of these challenges, the following are related to vehicular networking:*

1. *eCall.* eCall is a pan-European in-vehicle emergency call system. It aims at a drastic reduction in response time of emergency services in the event of a crash or an emergency situation. Motivation of eCall is that rapid assistance of emergency services is of utmost importance for saving lives or reducing the effects of injuries when an accident occurs. eCall can either be generated manually by a person or automatically by sensors located in a vehicle when an accident occurs. When activated, eCall sends important information such as time and location of the accident to emergency services. The European Commission is currently negotiating with European car markers for the deployment of eCall by 2010.

2. *Intersection assistant.* Many accidents occur due to drivers' inappropriate maneuvers, missing traffic signs or signals, and failure to anticipate and communicate with other drivers. At intersections, the risks of accidents are increased because the aforementioned factors can be combined. An intersection assistant could reduce these risks in two ways. First, an assistant can provide a right of way

---

* There are also challenges that are not directly related to vechicular communication technology but to other ICT research areas, such as sensor and human machine interface technologies.

warning, where a car detects other cars in its neighborhood by means of sensors and CAR-2-X communication based on IEEE 802.11, and warns the driver to stop for traffic from the right. The warning can be visual or an audio signal representing different risk levels. Second, a traffic light assistant can use a communication protocol based on IEEE 802.11 wireless technology in order to exchange information between vehicles and a traffic light. Safety applications inside vehicles provide drivers with the status of the traffic light and recommend a speed that the driver should adopt in order to achieve optimal energy efficiency.

3. *Wireless local danger warning.* This technology uses CAR-2-X communication based on IEEE 802.11 to detect local danger or hazards, and informs drivers about dangerous situations ahead. Local danger includes obstacles on the road, presence of emergency vehicles, or inordinately slow vehicles, electronic beacons warning about dangerous spots such as construction zones, and reduced friction or reduced visibility due to bad weather. In general, local danger warnings can be used to improve road safety and efficient traffic flow in the presence of local hazards due to accidents, congestion, and adverse weather.

4. *Lane change assistant and blind spot detection.* This technology provides a driver with a warning signal when he or she is about to switch to a lane in which another car is detected. Such critical situations frequently arise on the road and usually lead to an accident if the vehicle in the blind spot is overlooked. Using CAR-2-X communication based on IEEE 802.11, a vehicle can detect other vehicles in its neighborhood and provide a driver with a warning signal when he or she is about to change lanes.

5. *Dynamic traffic management.* This technology gathers real-time traffic information from various sources such as road-side sensors and GPS-equipped vehicle probes, and then combines the collected information with simulation models. The goal of this technology is to predict the effects of various management strategies on real-time traffic conditions. On the basis of these predictions, optimal traffic management strategies are selected to reduce congestion and delays. Further, this technology can also provide drivers with more accurate real-time traffic information and influence their speeds and route choices.

6. *Adaptive cruise control (ACC).* ACC enhances the function of standard cruise control by automatically adapting a vehicle's speed to the distance and speed of the vehicle ahead. Distance and speed can be estimated by means of sensors and CAR-2-X communication based on IEEE 802.11. If the vehicle detects a deceleration of the vehicle ahead, the ACC adjusts the vehicle's speed accordingly without any action on the part of the driver. If needed (e.g., in an emergency situation), ACC can slow the vehicle to a standstill by activating the brake system. When the road is clear again, ACC will increase the vehicle's speed back to the chosen value. As with standard cruise control, the driver can cancel ACC at any time.

4.2 Relevant R&D Projects

In order to reduce negative effects of transport, the European Commission developed a policy for sustainable mobility.[8,9] The policy defines a political framework to ensure a high level of mobility, protection of humans and the environment, technological innovation, and international cooperation. The policy also defines the ambitious goal to reduce road fatalities by 50% by 2010.

Based on this policy, the EU is funding major activities for technology research and development, coordination, and support actions. Many of the activities are organized as cooperative projects in the European FWP, which defines project objectives over time periods of typically 4–7 years. The projects are jointly initiated and carried out by industry and research organizations in Europe. A framework program comprises different types of projects, most importantly (1) flagship projects with a wide scope and large number of partners (typically 50 partners from industry and academia), referred to as integrated project (IP), (2) smaller projects dedicated to a certain topic and complementing an IP, called specific targeted research project (STREP), and (3) coordination actions (CA) and support actions (SA).

CAR-2-X communication is a key concept for future intelligent transport systems.* Major results have been achieved by projects in the 6th FWP, such as GST and PReVENT and national projects, such as FleetNet—Internet on the Road and NoW—Network on Wheels. Currently, the transitions from the 6th to 7th FWP are taking place, where large integrated projects of the 6th FWP (such as CVIS, SAFESPOT, COOPERS) are still running (until 2010), and projects for the 7th FWP have already started. Figure 4.2 gives an overview of projects related to CAR-2-X communication in Europe. These projects consolidate the current state-of-the-art for research and development, and build the basis for future deployment of such systems. It is expected that new projects will have a strong focus on field operational tests (FOTs) in order to validate the effectiveness of cooperative systems (an outlook is provided in Section 4.5).

In this section, we highlight some of the recent projects that are having a strong and sustainable impact on the European approach for vehicular communication. These projects are FleetNet—Internet on the Road, NoW—Network on Wheels, PReVENT, CVIS, SAFESPOT, COOPERS, SeVeCom, and COMeSafety.

4.2.1 FleetNet—Internet on the Road

FleetNet[10–12] was a pioneering research project that ran from 2000 to 2003. The project was initiated by Daimler, automotive suppliers, and research institutes, and was partially funded by the German Ministry of Education and Research (BMBF). The project can be regarded as an initial feasibility study for direct communication among cars. FleetNet considered three different classes of applications: cooperative driver-assistance for road safety, local floating car data applications, and user communication and information services. The project investigated different radio technologies: IEEE 802.11, UTRA-TDD,[†] and data transmission by radar communications. Later, it focused on IEEE 802.11. As one of the main results, the project developed a prototype of a car-to-car communication platform as a proof-of-concept.

---

* In Europe, these systems are often referred to *cooperative systems*, in analogy of a car, which cooperates with its environment.
† UTRA-TDD—UMTS terrestrial radio access—time division duplexing, a radio technology based on the European 3G cellular communication system.

AKTIV	**AIDE**	**COM2REACT**
2006–2010, Germany	2004–2008, EU	2006–2007, EU
Funding: 12.6 M€, 30 partners	Funding: 12.6 M€, 30 partners	Funding: 5.6 M€, 13 partners
Traffic management, active safety, cooperative cars using cellular nets	Integration of driver assistant system and nomadic devices	Road traffic control center
http://www.aktiv-online.org	*http://www.aide-eu.org*	*http://www.com2react-project.org*

COOPERS	**COMeSafety**	**CVIS**
2006–2010, EU	2006–2010, EU	2006–2010, EU
Funding: 16.8 M€, 37 partners	Funding: 1.5 M€, 6 partners	Funding: 41.2 M€, 59 partners
Vehicle-to-infrastructure, traffic management	Harmonization and promotion of efforts for vehicular communication	Cooperative vehicle-infrastructure systems
http://www.coopers-ip.eu	*http://www.comesafety.org*	*http://www.cvisproject.org*

CyberCars-2	**DAIDALOS-II**	**eImpact**
2006–2010, EU	2006–2008, EU	2006–2010, EU
Funding: 4.0 M€, 10 partners	Funding: 21.4 M€, 36 partners	Funding: 2.5 M€, 12 partners
Automated urban driving	Heterogeneous access to network infrastructure	Assessment of socioeconomic effects of safety systems
http://www-c.inria.fr/cybercars2	*http://www.com2react-project.org*	*http://www.eimpact.info*

FRAME	**GST**	**INTERSAFE-I (PReVENT)**
2001–2004, EU	2004–2007, EU	2004–2007, EU
Funding: 0 M€, 19 partners	Funding: 22.1 M€, 53 partners	Funding: 5.0 M€, 12 partners
Framework for national European ITS architectures	Open architecture for telematics, framework for telematics services	Intersection safety based on sensors and communication
http://www.frame-online.net	*http://www.gstforum.org*	*http://www.prevent-ip.org*

i-Way	**MORYNE**	**NoW—Network on Wheels**
2006–2009, EU	2006–2008, EU	2006–2010, Germany
Funding: 4.6 M€, 13 partners	Funding: 3.8 M€, 10 partners	Funding: 4 M€, 11 partners
Integration of in-vehicle sensors, road infrastructure and vehicles comm	Public transport, vehicle–traffic management center communication	Reliable and secure communication for safety and infotainment
http://www.iway-project.eu	*http://www.fp6-moryne.org*	*http://www.network-on-wheels.de*

PReVENT	**REPOSIT**	**SAFESPOT**
2004–2008, EU	2006–2007, EU	2006–2010, EU
Funding: 55 M€, 60 partners	Funding: 0.9 M€, 4 partners	Funding: 37,6 M€, 45 partners
Frame program for preventive and active safety	Improving position accuracy with vehicular communication	Vehicle-to-vehicle communication, road safety, local dynamic map
http://www.prevent-ip.org	*http://www.ist-reposit.org*	*http://www.safespot-eu.org*

SeVeCom	**Watchover**	**WILLWARN (PReVENT)**
2006–2008, EU	2006–2008, EU	2004–2007, EU
Funding: 4.5 M€, 7 partners	Funding: 5.9 M€, 12 partners	Funding: 3.3 M€, 11 partners
Security and privacy for vehicular communication	Communication among vehicles and vulnerable road users (pedestrians)	Emergency warnings based on wireless communication
http://www.sevecom.org	*http://www.watchover-eu.org*	*http://www.prevent-ip.org*

FIGURE 4.2 Overview of European R&D projects. [Bold frames indicate projects with a wide scope and a large number of partners, such as European Integrated Projects (IP) or national projects.]

Because IEEE 802.11 wireless technology is limited in its communication range, the project investigated multihop communication. Owing to high dynamics and frequent topology changes of vehicular ad hoc networks, routing is a challenging task. An additional requirement in the FleetNet project was scalability to ensure functioning of the system with a potentially high number of cars in dense deployment scenarios, such as highway or city environments. In order to meet these requirements, position-based routing was chosen as the basic approach for vehicular ad hoc networks.

In position-based forwarding, each data packet carries its destination's location. A forwarding node selects the next hop for a data packet so that the geographical distance to the destination is reduced (a special algorithm is *greedy forwarding*, where a forwarding node always makes a greedy, locally optimal forwarding decision in choosing a data packet's next hop). Position-based routing was attractive in vehicular ad hoc networks because it did not require route setup and route maintenance between nodes. Simulation studies conducted in the project indicated that for highway scenarios, position-based routing significantly outperformed topology-based routing protocols, such as dynamic source routing (DSR), in terms of packet delivery ratio, in particular when the communication distance between the source and the destination was larger than the one-hop communication range; in other words, multihop communication was required.

Based on the position-based routing approach, the project FleetNet implemented a Linux-based router, a prototype for a car-to-car communication platform. The router used a IEEE 802.11a network interface with external planar antenna. The router was connected to application systems via a standard Ethernet interface. The router also used the Ethernet interface to obtain GPS information from the on-board navigation system. Application systems in different cars could exchange messages via the FleetNet routers and provided drivers with an audiovisual interface. Real-world field trials for the prototype were successfully conducted and showed promising results.

4.2.2 NoW—Network on Wheels

NoW (Network on Wheels)[13,14] was the successor of the FleetNet project and ran from 2004 to 2008. NoW was initiated by the major German car manufacturers and partially funded by the German Ministry of Education and Research (BMBF). The project objectives were to leverage results of the FleetNet project and to develop an open communication platform for road safety, traffic efficiency, and infotainment applications. Besides technical aspects, NoW also analyzed strategies for market introduction and supported the Car-2-Car Communication (C2C-C) Consortium[15] (see Section 4.4.1), an industry consortium promoting an open European standard for CAR-2-X communication.

A particular characteristic of the NoW system is the integration of applications for road safety and infotainment in a single system. Clearly, safety and infotainment applications have very different requirements and characteristics. Infotainment applications typically establish sessions and exchange unicast data packets over a long period of time. On the other hand, safety applications typically broadcast messages to neighboring nodes. The project distinguished between two different types of safety messages. Periodic status information, such as *beacons* or *heartbeats*, achieve cooperative awareness among vehicles. Event-driven safety messages are nonperiodic messages, triggered by the detection of an abnormal condition or imminent danger. Periodic messages are broadcasted to the neighbor nodes and not distributed via multihop forwarding by the network protocol. In contrast, *event-driven messages* are distributed via multihop forwarding to a certain geographical area specified by the source node.

Because critical safety messages need to be delivered with low latency, overload should be avoided on the wireless medium. Two important issues regarding load control in multihop ad hoc communication were investigated in the NoW project: congestion control and prevention of the broadcast storm problem. A distributed algorithm called Distributed Fair Power Assignment in Vehicular environments (D-FPAV) addressed the congestion problem by adjusting the transmit power for the periodic messages and retained the total bandwidth consumed by these messages below a predefined threshold.

Another distributed algorithm called Emergency Message Dissemination in Vehicular environments (EMDV) prevented the broadcast storm problem by applying a contention-based forwarding approach. In EMDV, all nodes receiving a broadcast event-driven message acted as potential forwarders of this message and entered a contention period. The length of a contention period at each potential forwarder was inversely proportional to the progressed distance from the sender toward the destination. This strategy gave a potential forwarder closer to the destination a shorter contention period. When its contention period expired, a potential forwarder rebroadcast the event-driven message. Other potential forwarders could receive the rebroadcast message and cancel their own contention period.

The network architecture developed in the NoW project consisted of a dual network protocol stack that provided support for road safety and infotainment applications. For road safety applications, novel network and transport protocols provided ad hoc car-to-car and car-to-infrastructure communication. Infotainment applications could use the traditional IP protocol stack. In this case, the NoW network protocol acted as a sub-IP layer and provided ad hoc multihop communication. An information connector in the NoW network architecture followed a cross-layer approach to support efficient and structured information exchange across the protocol layers (more technical details are provided in Section 4.3). The network architecture used an IEEE 802.11p radio interface for road safety and an IEEE 802.11a/b/g radio interface for infotainment applications. Based on this network architecture, the project implemented a Linux-based communication system as prototype of a CAR-2-X communication platform. The implementation has been used in a number R&D projects such as the European projects PReVENT WILLWARN,[16,17] SAFESPOT,[18] and the German project AKTIV.[19]

4.2.3 PReVENT

PReVENT was an R&D integrated project co-funded by the European Commission to increase road safety by investigating preventive safety technologies and applications.[20] Preventive safety systems make use of information, communication, sensor, and positioning technologies to help drivers avoid or mitigate an accident. PReVENT ran from 2004 to 2008 and consisted of 13 subprojects. Two subprojects of PReVENT investigated research issues related to vehicular networking: Wireless Local Danger Warning (WILLWARN) and Intersection Safety (INTERSAFE). (The other subprojects of PReVENT conducted research in other areas such as sensor technologies, radar technologies, laser technologies, positioning technologies, digital map technologies, data fusion techniques, control and decision systems, human machine interface, and strategies for market introduction.) The subprojects WILLWARN[16,17] and INTERSAFE[21,22] are briefly summarized below.

4.2.3.1 WILLWARN

This subproject ran from 2004 to 2007 and developed a distributed system for a decentralized hazard warning application. The developed hazard warning system provided intelligent warning of dangerous situations and extended a driver's horizon. WILLWARN relied on sensor technologies and data fusion techniques to implement hazard detection. Upon detecting a hazard, a vehicle generated a message that consisted of three parts: distribution information, event information, and position information. Distribution information controlled the forwarding process of the message and contained a number of parameters such as unique message identification, priority, generation time, expiration time, and address area. Event information described the hazard—it contained an event type and a reliability level of the message. Position information described the location where the event was detected. The generated message was stored and forwarded using the communication system developed in the NoW project (i.e., the NoW communication system).

4.2.3.2 INTERSAFE

This subproject ran from 2004 to 2007 and combined sensor and communication technologies to increase safety at intersections.[21,22] The project INTERSAFE's goal was to reduce or even eliminate fatal accidents at intersections. Equipped with advanced sensor technologies, a vehicle could detect other vehicles and objects when approaching an intersection. Further, a vehicle could communicate with infrastructure, such as traffic lights, to exchange additional information about the status of the traffic light and weather, traffic, and road conditions. Combining sensor technology and car-to-infrastructure communication, INTERSAFE developed an intersection driver warning system.

4.2.4 CVIS

The project Co-operative Vehicle-Infrastructure Systems (CVIS) is an R&D integrated project cofunded by the European Commission and is to run from 2006 to 2010.[23] CVIS's high-level goal is to enable continuous communication and cooperative services between vehicles and infrastructure to increase road safety and traffic efficiency. CVIS complements the projects SAFESPOT (car makers' view) and COOPERS (road operators' view), and focuses on core technologies. To this end, CVIS provides the following four enabling services:

1. COMM provides a network architecture that allows transparent and continuous CAR-2-X communication.*
2. POMA provides positioning, map services, and location referencing services.

---

* CVIS, SAFESPOT, SeVeCom, and ComeSafety actually use the terminology *vehicle-to-vehicle* and *vehicle-to-infrastructure communication* to include motor vehicles other than cars such as trucks and motorbikes as well. However, we use the terminology CAR-2-X (car-to-car and car-to-infrastructure) communication for the sake of consistency throughout in this book chapter.

3. COMO provides cooperative monitoring services. COMO allows both local and central data fusion.
4. FOAM provides an open end-to-end framework for in-vehicle systems, roadside units, and backend systems. This framework enables a link between vehicles and infrastructure.

CVIS seeks to develop a unified network architecture for various link technologies that are needed for car-to-car and car-to-infrastructure communication. Unlike NoW and PReVENT WILLWARN, which focused on the IEEE 802.11 wireless technology, CVIS considers a wide range of communication technologies such as infrared technologies, cellular technologies (GPRS or UMTS), short-range microwave beacons (dedicated short-range communications), and local wireless technologies (IEEE 802.11 and IEEE 1609). Access to these interfaces is based on the standard ISO TC 204 (Section 4.4.4) for Continuous Air interface for Long and Medium Range (CALM)[24] via standardized CALM service access points. In CVIS, IPv6 plays a central role in providing continuous and transparent communication on top of different link technologies, but non-IP communication is also possible for fast safety-related applications. CVIS also supports network mobility (NEMO).[25]

4.2.5 SAFESPOT

SAFESPOT is an R&D integrated project cofunded by the European Commission, running from 2006 to 2010.[18] SAFESPOT is developing a "Safety Margin Assistant" by combining car-to-car and car-to-infrastructure communication. The Safety Margin Assistant not only detects potentially dangerous situations in advance but also extends a driver's awareness of the surrounding environment in space and time. Core technologies developed in SAFESPOT are ad hoc dynamic networking, accurate and real-time relative positioning, dynamic local traffic maps, and infrastructure-based sensing techniques.

A key technological challenge in SAFESPOT is to develop a communication platform that enables reliable, fast, secure, and efficient CAR-2-X communication. In SAFESPOT, the communication is mostly based on the IEEE 802.11 wireless technology (both car-to-car and car-to-infrastructure). SAFESPOT cooperates with the C2C-C Consortium (Section 4.4.1) and partially relies on results of the NoW project. For types of communication other than IEEE 802.11, SAFESPOT cooperates with CVIS to define a common communication architecture that is in line with the standard ISO CALM. Another technological challenge in SAFESPOT is to develop a highly accurate and reliable relative positioning system in order to enable the advanced cooperative behaviors. A number of techniques are considered including GNSS-based positioning, such as GPS and Galileo, communication-based positioning such as UWB, WLAN, and GSM, image-based positioning such as Laserscanner and Landmarks, and the cooperative use of two or more technologies.

To handle various static and dynamic information, SAFESPOT has developed the concept of Local Dynamic Map (LDM).[26] The LDM is a multilayered dynamic representation of the environment surrounding a vehicle or a road side unit, where all the information acquired via sensors and communication are collected after fusion processes. An LDM consists of four different layers: a static map, similar to the ones currently used by navigation systems; a landmarks layer, used for referencing; a temporary objects layer, where accidents or other situations are marked; and finally a layer including the dynamic objects, in other words, the vehicles. Applications access the LDM as a source of information to be distributed as well as information to be provided to the driver.

Based on the developed communication platform, SAFESPOT provides vehicles with improved range, quality, and reliability of safety-related information. Further, SAFESPOT provides vehicles with extended cooperative awareness. This gives drivers and on-board systems more time and space to react safely and appropriately to potential hazards. Examples of safety-related applications to be developed by SAFESPOT are safe lane-changing maneuvers, road departure prevention, cooperative maneuvering (e.g., highway merging), cooperative tunnel safety, hazard and incident warning, safe urban and extraurban intersection.

4.2.6 COOPERS

COOPERS is an R&D integrated project cofunded by the European Commission, running from 2006 to 2010.[27] The project investigates new safety-related services using continuous two-way wireless communication between vehicles and infrastructure from a traffic management's perspective. In COOPERS, bidirectional car-to-infrastructure communication will be used to provide real-time location-based safety-related information as well as infrastructure's status information, such as traffic jams. The project also seeks to improve the reliability and accuracy of information provided by road operators to drivers. COOPERS plans to provide different information services: accident warning, weather condition warning, roadworks information, lane utilization information, speed limit information, traffic congestion warning, and route navigation. The project follows a three-step approach:

1. The project seeks to improve road sensor infrastructure and traffic control applications to achieve accurate, up-to-date, and situation-based traffic information.
2. The project develops a communication architecture that satisfies the requirements of car-to-infrastructure communication in terms of reliability, robustness, and real-time capability. A combination of different wireless communication technologies will be considered: digital audio broadcast (DAB), GPRS, and UMTS, wireless LAN (IEEE 802.11 and IEEE 1609), infrared and microwave technologies (DSRC).
3. The project deploys and tests the results on important parts of busy European highways in Belgium, Netherlands, Germany, Austria, Italy, and France. The test results will serve as input for a future large-scale deployment.

4.2.7 SeVeCom

SeVeCom is an R&D project in the 6th Framework Program, cofunded by the European Commission, running from 2006 to 2008.[28] The project addresses security and privacy aspects of CAR-2-X communication. It defines a security architecture for CAR-2-X communication systems and provides a road map for the progressive introduction and deployment of security functions. Based on the definition of various attacker models, SeVeCom identifies potential vulnerabilities, in particular attacks against the radio channel and communication protocols, but also attacks against a vehicle's on-board unit and car internal devices, such as a car's data bus. Further, the project also specifies the security architecture and security mechanisms that provides the right level of protection. This is achieved by striking a balance between liability and privacy.

In a first step, the project addresses key and identity management, secure communication protocols, tamper-proof devices, and privacy. In a later phase, the project will develop solutions for in-vehicle intrusion detection, malfunction detection and data consistency, and secure positioning.

4.2.8 COMeSafety

Communications for eSafety (COMeSafety) is a specific support action of the European Commission.[29] The project, running from 2006 to 2009, supports the eSafety Forum[30] as a joint platform for all road-safety stakeholders; including the European Commission, Member States of the European Union, road and safety authorities, the automotive industry, the telecommunications industry, service providers, user organizations, the insurance industry, technology providers, research organizations, and road operators. COMeSafety supports the eSafety Forum issues related to CAR-2-X communication as the basis for cooperative intelligent transport systems.

To achieve its goal, COMeSafety provides coordination and consolidates results from major R&D projects: CVIS, SAFESPOT, and COOPERS. COMeSafety has also been exchanging information with other R&D projects: GeoNet, PReVENT, and SeVeCom. COMeSafety supports coordinated development and the interoperability of technical components between these projects. Further, COMeSafety provides the C2C-C Consortium and the eSafety Forum with appropriate recommendations. COMeSafety also seeks worldwide harmonization with standards for CAR-2-X as developed in the U.S. and Japan.

4.3 Vehicular Communication Systems and Geocast

This section presents the CAR-2-X system architecture and details of Geocast protocols that serve as a basic building block for CAR-2-X communication in many European R&D projects. As shown in Figure 4.3, the CAR-2-X communication system consists of three domains: the in-vehicle domain, the ad hoc domain, and the infrastructure domain.

The in-vehicle domain refers to a network logically composed of an on-board unit (OBU) and (potentially multiple) application units (AUs). The OBU is responsible for CAR-2-X communication. It also provides communication services to AUs and forwards data on behalf of other OBUs in the ad hoc domain. An OBU is equipped with at least a single network device for short-range wireless communications based on IEEE 802.11p radio technology, and may also be equipped with more network devices, for example, for nonsafety communications, based on other radio technologies such as IEEE 802.11a/b/g/n. An AU is typically a dedicated device that executes a single or a set of applications and uses the OBU's communication capabilities. An AU can be an integrated part of a vehicle and be permanently connected to an OBU. It can also be a portable device such as laptop, PDA, or game pad that can dynamically attach to (and detach from) an OBU. An AU and an OBU are usually connected with a wired

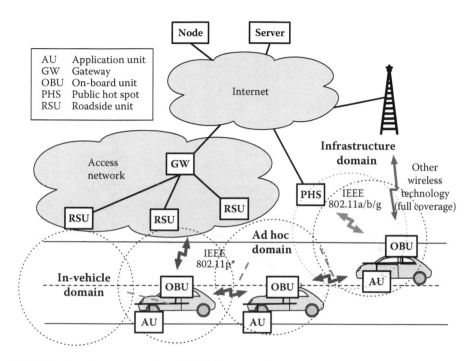

FIGURE 4.3 Systems overview of CAR-2-X communication.

connection, but the connection can also be wireless, using Bluetooth, WUSB, or UWB, for example. The distinction between AU and OBU is logical; they can also reside in a single physical unit.

The ad hoc domain, or vehicular ad hoc network (VANET), is composed of vehicles equipped with OBUs and stationary units along the road, termed road-side units (RSUs). OBUs form a mobile ad hoc network (MANET), which allows communications among nodes in a fully distributed manner without the need for centralized coordination. OBUs directly communicate if wireless connectivity exists among them. In the case of no direct connectivity, dedicated routing protocols allow multihop communications, where data are forwarded from one OBU to another, until they reach the destination.

An RSU can be attached to an infrastructure network, which in turn can be connected to the Internet. As a result, RSUs may allow OBUs to access the infrastructure. In this way it is possible for AUs registered with an OBU to communicate with any host on the Internet, when at least one infrastructure-connected RSU is available.

4.3.1 Geocast Protocols

Geocast is essentially an ad hoc routing protocol using geographical positions for data transfer. Its basic principles were originally proposed as an alternative to pure topology-based internetworking[31] and in mobile ad hoc networks.[32] Geocast assumes that vehicles acquire information about their position (i.e., geodetic coordinates) via GPS or any other positioning system. Every vehicle periodically advertises this information to its neighbor vehicles and a vehicle is thus informed about all other vehicles located within its direct communication range. If a vehicle intends to send data to a known target geographic location, it chooses another vehicle as a message relay, which is located in the direction towards the target position. The same procedure is executed by every vehicle on the multihop path until the destination is reached. This approach does not require establishment and maintenance of routes. Instead, packets are forwarded "on the fly" based on the most recent geographic positions.

In detail, Geocast assumes that every node knows its geographical position and maintains a location table containing other nodes and their geographical positions as soft state. Geocast supports point-to-point and point-to-multipoint communication. The latter case can be regarded as group communication, where the endpoints are inside a geographical region.

Core protocol components of Geocast are beaconing, a location service, and forwarding. With beaconing, nodes periodically broadcast short packets with their ID, current geographical position, speed, and heading. On reception of a beacon, a node stores the information in its location table. The location service resolves a node's ID to its current position. When a node needs to know another node's position that is currently not available in its location table, it issues a location query message with the sought node ID, sequence number, and hop limit. Neighboring nodes rebroadcast this message until it reaches the sought node (or the hop limit). If the request is not a duplicate, the sought node answers with a location reply message carrying its current position and a time stamp. On reception of the location reply, the originating node updates its location table. Forwarding basically means relaying a packet towards the destination.

The most innovative method for distribution of information enabled by geographical routing is to target data packets to certain geographical areas. In practice a vehicle can select and specify a well delimited geographic area to which the messages should be delivered. Once again, intermediate vehicles serve as message relays and only the vehicles located within the target area process the message and further send it to corresponding applications. In this way, only vehicles that are actually affected by a dangerous situation or a traffic notification are notified, whereas vehicles unaffected by the event are not targeted.

In summary, *Geocast* comprises the following forwarding schemes:

1. *GeoUnicast.* Figure 4.4 provides packet delivery between two nodes via multiple wireless hops. When a node wishes to send a unicast packet, it first determines the destination's position (by location table look-up or the location service) and then forwards the data packet to the node towards the destination, which in turn reforwards the packet along the path until the packet reaches the destination.

2. *GeoBroadcast.* Figure 4.5 distributes data packets by flooding, where nodes rebroadcast the packets if they are located in the geographical region determined by the packets. This simple flooding scheme is enhanced with techniques based on packet numbering to alleviate the effects of so-called broadcast storms. Broadcast storms (a typical problem in wireless ad hoc networks) occur when multiple nodes simultaneously rebroadcast a data packet that they have just received. GeoAnycast is similar to GeoBroadcast but addresses a single node (i.e., any node) in a geographical area.

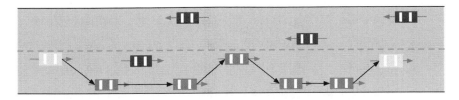

FIGURE 4.4   Geographical unicast.

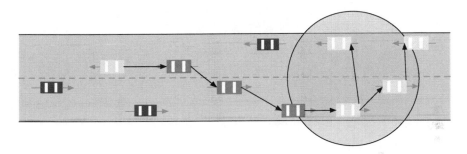

FIGURE 4.5   Geographical broadcast.

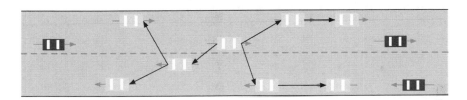

FIGURE 4.6   Topologically scoped broadcast.

3. *Topologically scoped broadcast (TSB).*   Figure 4.6 provides rebroadcasting of a data packet from a source to all nodes in the *n*-hop neighborhood. Single-hop broadcast are a specific case of TSB and are used to send periodic messages (*beacons* or *heartbeats*).

The concepts of geographical networking serve as a basis for CAR-2-X communication. The CAR-2-X protocol architecture is depicted in Figure 4.7. Based on *Geocast*, a number of advanced solutions have been developed to address numerous challenges in this field. Advanced solutions for robustness, reliability, security and privacy are highlighted below.

4.3.1.1 Robustness and Reliability

Several enhanced mechanisms have been developed for improving robustness and reliability of packet delivery in vehicular communication. The main enhanced mechanisms for robustness and reliability are detection and suppression of duplicated

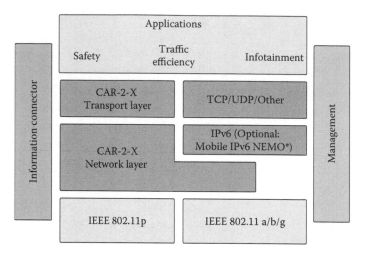

FIGURE 4.7   CAR-2-X protocol architecture.

packets, temporary caching of data packets, and transmit power control for beacon and data packets.

In order to alleviate the broadcast storms, each node caches the sequence numbers of data packets that it receives. This caching mechanism allows nodes to detect and drop duplicated packets.

Temporary caching of data packets can potentially improve the reliability of packet delivery when the density of vehicles is low. For unicast communication, a node temporarily caches a packet when no vehicles closer to the destination are available. When a new neighbor arrives, the node forwards the cached packet to the neighbor if this new neighbor is closer to the destination. A similar technique is also applied for GeoBroadcast where a node retransmits data packets upon detection of new neighbors. Reliability is further improved by combining geographical routing with contention-based routing and controlling dynamically the level of redundancy.[33]

To increase the system's robustness, the IEEE 802.11 radio's transmit power can be controlled on a per-packet basis. This mechanism allows control of the transmit power for every individual data packet, which in turn enables the implementation of distributed transmit power control algorithms[34] and efficient coupling with the routing scheme.[35] These schemes reduce the physical radio interference among the nodes, which results in better overall network performance.

4.3.1.2 Security and Privacy

A solution for data security based on asymmetric cryptography and digital certificates has been developed.[36] This solution guarantees authentication and nonrepudiation of the information exchanged between vehicles and protects against malicious users. Further, plausibility checks based on time stamp and communication ranges are performed for every received packet. Packets that fail the plausibility checks are simply dropped. Each node also uses results of the plausibility checks as input for confidence assessment of its neighbors. A node originating too many packets that fail the plausibility checks is classified as untrusted. Packets from untrusted nodes are simply dropped. Privacy protection has also been studied and a method based on variable network identifiers (pseudonyms) has been developed.[37]

4.3.2 Internet Integration

Solutions for the integration of the Internet protocol suite and geographical routing have been developed. This enables vehicles to use traditional applications based on the Internet Protocol (IP), as well as to gain Internet connectivity directly or even via multi-hop communication. The usage of IP solutions for mobility like Network Mobility (NEMO) support is also investigated.

In principle, the Internet protocol suite does not target specific application fields like the automotive sector. Instead, it is a set of protocols and mechanisms used by every Internet user for very different applications like web browsing, email, audio and video streaming, and so on. Nevertheless, the suite was designed with an infrastructure-based, topological view of the network (reflecting, in fact, the Internet model of interconnection of different networks) and with static nodes as user terminals. On the other hand, vehicular networks are intrinsically extremely mobile and often lacking an infrastructure. Furthermore, as pointed out in Section 4.3.1, the geographic position of vehicles is a key criterion for automotive applications that determines how information should be distributed. For these and other reasons, existing solutions developed by the IETF cannot be used as-is in vehicular applications. Thus, the car industry has been following an approach that consists of defining specific protocols whose scope is restricted to automotive applications with particular requirements, making sure at the same time that the Internet protocol suite is integrated in the protocol stack providing interoperability with the Internet. This allows for certain separation between mechanisms that are purely automotive specific and others that serve multiple purposes.

In order to allow for the integration of the Internet protocol suite in vehicular networks based on geographical routing, several protocol design issues are being addressed. First of all, in order to become part of the Internet, vehicles need to configure an Internet address. Several approaches have been proposed for general purpose MANET.[38] Nevertheless, none of them has been identified as the basis for a standardized solution.* Moreover, only a few existing proposals, like Refs. 39 and 40, explicitly target VANET and automotive requirements like high scalability, low complexity, security, and so on.

Once vehicles are able to obtain Internet addresses, their high mobility plays a big role in how Internet applications can be used by drivers and passengers. In fact, without support for IP mobility, standard applications would constantly experience interruptions, and new applications that can handle mobility might cause unnecessary traffic

---

* To date, the IETF AUTOCONF work group has not finalized the problem statement and is not expected to provide a standard solution in the short term.

overhead, as well as potential privacy issues. The IETF Network Mobility Basic Support protocol (NEMO BS[25]) can handle changes of attachment points to the Internet for a whole set of devices attached to a vehicle. As analyzed in Ref. 41, NEMO BS can be applied to vehicular networks in different ways, obtaining different impacts in terms of economic, functional, and performance requirements. An approach that is in line with automotive requirements consists in applying IPv6 and NEMO BS on top of the geocast mechanisms described in Section 4.3.1. This approach is illustrated and evaluated in Ref. 42. For issues related to NEMO Route Optimization and the current standardization activities within IETF refer to Section 4.4.5.

4.3.3 Frequency and Channel Allocation

In Europe, 30 MHz of spectrum in the range from 5875 to 5905 MHz will be designated to safety-related ITS applications (for details, refer to Section 4.4.6). A recent paper provided a comprehensive overview of existing approaches for the usage of the 30 MHz frequency band dedicated for safety-related car-to-car and car-to-infrastructure communication.[43] Advantages and disadvantages of these approaches were analyzed based on an extensive set of evaluation criteria. These criteria are summarized as follows:

1. *Usability.* This criterion represents the main requirements for safety-related car-to-car and car-to-infrastructure communication: low latency and high reliability for critical safety messages. The analysis focuses on latency, because network and/or applications should be responsible for reliability.
2. *Robustness.* This criterion evaluates the wireless link's robustness in two aspects: it has to be robust in terms of bit errors (e.g., the bit error rate should be as low as possible) and it has to be robust in terms of interference.
3. *Cost.* This criterion considers the material costs for mass production and deployment. Obviously, an inexpensive solution is preferred in order to reduce the market barrier.
4. *Efficiency.* This criterion evaluates the effectiveness of channel allocation in terms of bandwidth usage. Given the scarcity of available bandwidth allocated for car-to-car and car-to-infrastructure communication, this precious resource must be used effectively.
5. *Scalability.* This criterion evaluates the impact of channel allocation on the flexibility of the overall car-to-car and car-to-infrastructure communication system in different scenarios such as highways, cities, and rural areas.
6. *Development effort.* This criterion considers development costs apart from material costs. A solution for channel allocation that allows a simple design and implementation of the car-to-car and car-to-infrastructure communication system is clearly preferred.

Taking into account the basic channel usage scenarios from C2C-C Consortium, some scenarios are eliminated based on qualitative observations. First, currently available WLAN chipsets only support 10 and 20 MHz channels. Thus, a 30 MHz channel would require new hardware and incur significant development cost. For this reason, the usage of a 30 MHz channel is ruled out. Second, a measurement study for channel

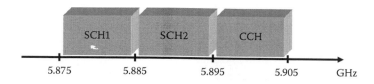

FIGURE 4.8 Proposed channel allocation for safety-related frequency band in Europe.

interference[44] indicates that the simultaneous usage of two adjacent channels caused significant packet loss and hence is unacceptable for ITS safety applications. Although measurements were conducted only for 10 MHz channels, similar effects are expected for simultaneous usage of 10 and 20 MHz channels. For this reason, it is not possible to divide the allocated spectrum into a 10 and 20 MHz channel. Further, because measurement results[45] show that 20 MHz channels were more susceptible to bit error rates than 10 MHz channels, the usage of 20 MHz channels is not further considered.

Although the usage of adjacent channels is possible, certain mechanisms such as WAVE channel switching have to be in place to prevent simultaneous transmissions on these channels. Although interference also occurs between non-adjacent channels, it is much lower than in the case of adjacent channels. However, it is expected that packet losses caused by interference between nonadjacent channels can be recovered by reliability mechanisms at the network layer and/or applications.

From recent measurement results[44,45] and the reasoning above, two channel usage schemes are analyzed and compared: SCH1 + SCH2 + CCH (Scheme A) and CCH + 2 * SCH with WAVE channel switching (Scheme B), where SCH1 is the first service channel, SCH2 is the secondary service channel, and CCH is the control channel.

Based on a detailed analysis, it is shown that the recommended channel allocation scheme shown in Figure 4.8 outperforms a channel allocation scheme that requires one transceiver performing WAVE channel switching. This analysis compares these two schemes based on an extensive set of evaluation criteria.[43] This recommendation requires two transceivers and uses low transmit power on SCH2.

4.4 Standardization Activities

The previous sections have given an overview of R&D projects for CAR-2-X communication in Europe. This section summarizes regulation and standardization activities for CAR-2-X communication in Europe.

For CAR-2-X communication, the European Telecommunications Standards Institute (ETSI), with its newly created technical committee ITS, plays a particular important role. ETSI, has strong liaisons with ISO and CEN. The relationships between ETSI and other standardization bodies in Europe are depicted in Figure 4.9.

4.4.1 Car-2-Car Communication Consortium

The Car-to-Car Communication (C2C-C) Consortium is an industry consortium initiated by major European vehicle manufacturers in 2002. By 2008, the C2C-C Consortium

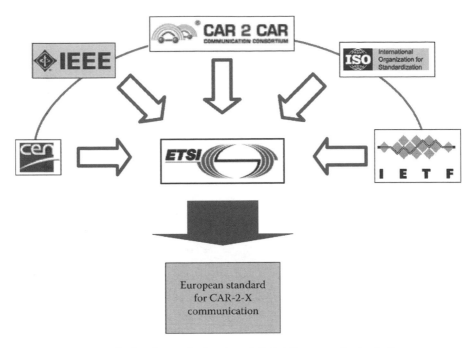

FIGURE 4.9　Major bodies for standardization of CAR-2-X communication in Europe.

had about 50 partners from industry and research. The main target is the creation of an open European industry standard for CAR-2-X communication based on wireless LAN technology, to support efforts to achieve interoperability, to push for harmonization of worldwide standards, and to develop deployment strategies. The major focus of the consortium is road safety and traffic efficiency applications. The C2C-C Consortium is open for car manufacturers, suppliers, research organizations, and other partners. Currently, the work in the C2C-C Consortium is being carried out in the following working groups:

1. *WG PHY/MAC.* This working group investigates technical issues of CAR-2-X communication at the protocol layers PHY and MAC. A particular focus is on the adoption of the IEEE 802.11p standard for road safety, whereas other topics are also being investigated such as frequency allocation and channel usage.

2. *WG NET.* This working group investigates algorithms and protocols for data networking and transport. The considered protocols cover high-frequency beaconing (heartbeats) and multihop communication for safety applications. They provide scalable and reliable communication and feature security and privacy-enhancing functions. Another focus is on integration of IP protocols.

3. *WG ARCH.* This working group defines the system and protocol architecture for CAR-2-X communication.

4. *WG APP.* The working group identifies the application requirements and specifies application protocols, such as Cooperative Awareness Message or Decentralized Environmental Message exchange.

5. *WG SEC.* The working group analyzes security threats and specifies security concepts and mechanisms, including key and identity management, secure routing protocols, tamper resistance of devices, data consistency, and others.
6. *WG STA.* The working group focuses on standardization issues and interaction with other standardization bodies. A particular focus has been on the frequency allocation for CAR-2-X communication in Europe.

The C2C-C Consortium has close relation to various R&D projects in Europe and attempts to harmonize the different project approaches and outcome. Also, the C2C-C Consortium is linked to ongoing standardization efforts, including ETSI, ISO TC 204, and CEN TC 278. A major step in the C2C-C Consortium's work was the publication of the Car-2-Car Communication Consortium Manifesto.[46]

4.4.2 ETSI

The European Telecommunications Standards Institute (ETSI) is a European standardization organization of the telecommunication industry in Europe, well known for its standards for GSM, TISPAN, and others. By the end of 2007, it created a new technical committee for Intelligent Transport Systems, TC ITS. The TC will develop standards and specifications for ITS service provision in Europe. The TC ITS is organized in the following five working groups:

1. WG 1 User and application requirements
2. WG 2 Architecture and cross-layer issues
3. WG 3 Transport and network
4. WG 4 Media and related issues
5. WG 5 Security

The work of the committee has just started, and it is commonly expected that ETSI will play a major role in system and protocol specification in Europe. The scope of the committee is not limited to CAR-2-X communication, but encompasses the various transport modes (car and motorbike, pedestrians, railway, aeronautics, and so on) and wireless communication technologies (5 GHz, 60 GHz, infrared, and GSM).

Starting the work, the groups have already agreed on a number of work items for various aspects of vehicular communication including media, networking, security, and safety applications. In WG3, the current focus is on specification of ad hoc networking based on geographical addressing and routing. In order to allow for the use of different media, the specification distinguishes between media-independent and media-dependent network functions. The specifications are backed by other work groups, which specifically address media and security issues, such as a European profile standard of IEEE 802.11 for ITS. The technical committee is developing a road map for standardization developments for the coming years in order to achieve a complete set of standards ranging from communication architecture to protocol specifications together with formal test procedures.

4.4.3 CEN TC 278

CEN is the European Committee for standardization, a private, nonprofit organization with the mission to foster European economy and welfare of European population and environment. In 1991, CEN established the technical committee TC 278[47] for the field of road transport and traffic telematics (RTTT). CEN TC 278 focuses on services and system aspects and mainly works on electronic fee collection, dedicated short-range communication (DSRC),* automatic identification of vehicles and equipment. The CEN/TC 278 is organized into the following working groups:

1. WG 1 Electronic fee collection
2. WG 2 Freight and fleet management systems
3. WG 3 Public transport
4. WG 4 Traffic and traveler information
5. WG 5 Traffic control systems
6. WG 7 Geographic data files (dormant working group)
7. WG 8 Road databases
8. WG 9 Dedicated short-range communications
9. WG 10 Human–machine interfaces
10. WG 12 Automatic vehicle and equipment identification
11. WG 13 Architecture
12. WG 14 Recovery of stolen vehicles
13. WG 15 eSafety

4.4.4 ISO TC 204

The work of ISO TC 204 (Intelligent Transport Systems),[24] started in 1993, covers standardization of systems for various types of transport and includes system and infrastructure aspects. Currently it has the following working groups:

1. WG 1 Architecture
2. WG 3 TICS database technology
3. WG 4 Automatic vehicle and equipment identification
4. WG 5 Fee and toll collection
5. WG 7 General fleet management and commercial freight
6. WG 8 Public transport/emergency
7. WG 9 Integrated transport information, management and control
8. WG 10 Traveler information systems
9. WG 11 Route guidance and navigation systems
10. WG 14 Vehicle/roadway warning and control systems
11. WG 15 Dedicated short-range communications for TICS applications
12. WG 16 Wide area communications, protocols and interfaces
13. WG 17 Nomadic devices in ITS systems

WG 16 is particularly important for CAR-2-X communication. It defines a framework for a communication system dedicated for ITS, referred to as Communication Access for Land Mobiles (CALM). Key characteristics of the system are: (1) Support of multiple wireless technologies (5.8 GHz DSRC, 5.9 GHz ITS, 60 GHz millimeter wave, infrared,

---

* This is the so called 'CEN DSRC' for electronic toll service.

GSM, 3G, DAB and so on), (2) interface abstraction for plug-and-play-like use of wireless interfaces via an adaptation layer, (3) dynamic selection of the optimal interface for a communication session, and (4) support of IP version 6 and IP mobility extensions (Mobile IPv6, NEMO, MONAMI, and others). The focus of WG 16 is on nonreal-time communication for information and entertainment applications based on communication sessions and on communication between vehicles and infrastructure.

Recently, WG 16 has started to work on safety applications and appropriate protocols in order to include them into the overall CALM system architecture. In this context, Fast Applications and Communication Enabler (FAST) is being developed. FAST is a CALM subsystem dedicated to nonsession-based, non-IPv6, real-time communication and provides direct communication among vehicles for beaconing and a newly introduced Groupcast service.

ISO TC 204 WG 16 works on a large set of standards for specification of the CALM system. Some of these standards have already been published. Also, ISO TC 204 WG16 has a close cooperation with the European project CVIS. In fact, the outcome of the project is regarded as a reference implementation of ISO CALM.

4.4.5 IETF

The Internet Engineering Task Force (IETF) is an open standards organization that develops and promotes Internet standards. A large number of protocols, which create the basis for the today's Internet, were developed by this organization as part of the IP suite. The IETF's open and pragmatic approach, based on publication of draft specifications, review, and independent testing by participant and republication, has strongly contributed to the achievement of interoperability among different vendors and consequently to the wide deployment of Internet protocols.

The current most remarkable activity within IETF related to automotive applications involves the Network Mobility Basic Support protocol (NEMO BS[25]). This protocol has been successfully applied to automotive applications in many research projects, such as the Japanese project InternetCar,[48] the EU FP 6 project Overdrive,[49] and CVIS[23] (Section 4.2.4) and is part of automotive protocol architectures of ISO CALM (Section 4.4.3) and the C2C-C Consortium (Section 4.4.1). Considering a car as a collection of interconnected communication devices, it is natural to look at it as a communication network and, as such, it requires an efficient mechanism to handle changes in the attachment point to the Internet. In NEMO BS, an entity running in the vehicle, called the "mobile router," handles the mobility for all the car's devices by updating a binding at a fixed node in the infrastructure network, called the "home agent." The home agent assures global reachability and session continuity for the mobile network by forwarding data packets coming from Internet hosts to the current vehicle's location. It can also be observed that NEMO BS is a natural extension of Mobile IPv6[50] for entire moving networks.

The IETF working group MEXT is currently working on extensions for NEMO BS protocol with the goal of designing mechanisms to overcome the suboptimal packet routing and overhead due to the tunneling through the home agent. Owing to the increased complexity as compared to host mobility, for the design of route optimization techniques, the working group decided to gather requirements from industries involved in the deployment of NEMO and to address these requirements possibly with multiple solutions. Requirements are currently being provided by aviation, automotive,[51] and industries involved in deployment of personal mobile routers. The automotive requirements cover aspects like security, privacy protection, and multihoming, among others. Furthermore, ref. 51 is the first document submitted to the IETF that represents automotive consortia, which testifies that cars are going to be the next new elements of the global network.

4.4.6 Frequency and Channel Allocation Scheme

In Europe, ETSI has presented requirements for European-wide harmonization of spectrum to CEPT (European Conference of Postal and Telecommunications Administrations) and European telecommunication administrations for deployment of ITS within the 5.9 GHz band. The frequency band 5875 to 5925 MHz has been requested for deployment of safety-related ITS applications, which require protection against interference from other services and the frequency band. The frequency band 5855 to 5875 MHz has been requested for nonsafety-related ITS, which can be operated on a nonprotected/noninterference basis.

CEPT has performed extensive compatibility studies[52] and concludes that within the frequency band 5875 to 5905 MHz, ITS applications will not suffer from excessive interference resulting from other services/systems and that ITS in this band is compatible with all other services, providing that the unwanted emission levels have the following properties:

1. Less than −55 dBm/MHz below 5850 MHz in order to protect the Radiolocation Services.
2. Less than −65 dBm/MHz below 5815 MHz in order to protect the RTTT applications.
3. Less than −65 dBm/MHz above 5925 MHz in order to protect the Fixed Service.

In early 2008, CEPT finalized the frequency allocation in the 5.9 GHz band for ITS. The Electronic Communications Committee (ECC) within CEPT has published documents concerning the frequency allocation on the server of the European Radio Office (ERO),[53] among which, two documents provide a good overview of the current status:

1. ECC Recommendation ECC/REC/(08)01[54] considers 20 MHz of spectrum in the range 5855 to 5875 MHz for nonsafety ITS application.
2. ECC Decision ECC/DEC/(08)01[55] designates 30 MHz of spectrum in the range 5875 to 5905 MHz to safety related ITS applications with the band 5905 to 5925 MHz to be considered for future ITS extension.

The development of the EC Decision is still in progress and adoption is expected by the middle of 2008.

Within the C2C-C Consortium, there have been technical in-depth discussions on how to partition and utilize the available 30 MHz spectrum in the band 5875 to 5905 MHz for road safety applications. Currently it is proposed to have a dual-transceiver solution without the need for WAVE channel switching. The preliminary proposal of channel allocation is shown in Figure 4.8.

The frequency band is divided into 10 MHz channels; the upper channel is used as a control channel (CCH) and the lower channel is used as the first service channel (SCH1). Considering the compatibility studies from CEPT,[52] in the future, it will also be possible to swap the center frequency of the CCH and SCH1 in order to allow higher transmit power on the CCH. This can be easily done by reconfiguring the channel center frequency. Dual transceivers may run simultaneously on the two channels with limited interference with each other. The inner channel is used as a secondary service channel (SCH2), which could be used for low-power, short-distance communication. This channel usage scheme has been adopted by the European projects COMeSafety, SAFESPOT, COOPERS, and CVIS in order to allow interoperability. It is planned that this proposal will be further fed into the standardization at ETSI TC ITS.

4.5 Outlook

Recent years have witnessed many R&D activities in vehicular communication in Europe. Preliminary results suggest that vehicular communication has great potential for increasing road safety. Nevertheless, a significant amount of work still has to be done. It is expected that new and upcoming European R&D projects in the area of vehicular communication will target large-scale development and field operation tests. This section provides a summary of relevant new and upcoming European R&D projects. It concludes with a long-term vision for CAR-2-X communication in Europe.

4.5.1 GeoNet

GeoNet is an R&D project cofunded by the European Commission, running from 2008 to 2010.[56] GeoNet aims to define a reference specification and develop two prototype implementations for a geographic-based vehicular communication protocol with IPv6 support. The protocol will be used to support vehicular safety applications with the aim of increasing road safety.

Although the C2C-C Consortium has invested significant effort into the specification of CAR-2-X communication for safety applications, the solution is not standardized. At the same time, ongoing projects, such as SAFESPOT, CVIS, and COOPERS, are also involved in CAR-2-X communication. It has also been observed that there is a high risk of technological divergence in their respective implementations, arising from their different objectives.

The goal of GeoNet is thus to implement and formally test a networking mechanism as a standalone software module that supports safety applications and enables transparent IP connectivity between vehicles and the infrastructure with single-hop and multihop communication. GeoNet will produce two prototype implementations and will disseminate them to existing consortia, particularly SAFESPOT, CVIS, COOPERS, and C2C-C Consortium.

The key challenges of GeoNet consist firstly in precisely defining and specifying the new geographical routing protocol that fulfills the new requirements of VANET applications, especially with respect to reliability, latency, and dissemination area for safety applications. Another challenge is how to integrate the geographical routing with IPv6 in a seamless way. GeoNet will also investigate the security schemes that are raised by the integration of geographical routing and IPv6.

GeoNet will take the basic results of CAR-2-X communication from the work of the C2C-C Consortium and further improve them and create a baseline software implementation interfacing with IPv6. From an IPv6 networking perspective, geographic networking provides assistance with message delivery in a vehicular environment of quickly changing topology. GeoNet plans to integrate geographic routing with IPv6 NEMO architecture by running IPv6 on top of the geographic networking layer.

4.5.2 INTERSAFE-2

INTERSAFE-2 is an R&D project cofunded by the European Commission, running from 2008 to 2011.[57] It is the successor of the project INTERSAFE discussed in Section 4.2.3. INTERSAFE-2's goal is to develop and demonstrate a Cooperative Intersection Safety System that can significantly reduce fatal accidents at intersections. INTERSAFE-2 goes beyond the scope of INTERSAFE in that it integrates warning and intervening systems in the vehicles (INTERSAFE only provided warning systems). Further enhancements of INTERSAFE-2 are right-turning assistance and protection for vulnerable road users such as pedestrians and bicyclists. INTERSAFE-2 plans to equip RSUs at selected important and dangerous intersections to provide vehicles with warning signals via CAR-2-X communication. This strategy will boost the penetration rate of vehicular communication and offer immediate benefits for early adopters of CAR-2-X communication.

In INTERSAFE-2, objects (vehicles, pedestrians, and bicyclists) at intersections are tracked and classified based on both on-board and infrastructure sensing, and relative localization techniques based on digital maps. Different kinds of sensing technologies are considered in INTERSAFE-2: radar, lasers, and video cameras. CAR-2-X communication will provide scalable, robust, and secure mechanisms for information exchange among vehicles and also between vehicles and infrastructure. A number of research topics related to CAR-2-X communication will be investigated in INTERSAFE-2:

1. *Efficient and responsive congestion control mechanisms.* Because intersections can have potentially dense network topologies, controlling the channel load is an important issue. INTERSAFE-2 plans to investigate congestion control mechanisms for both single-hop and multihop CAR-2-X communication.
2. *Efficient data aggregation.* Unlike in traditional communication systems where applications are usually independent of each other, vehicular safety applications have to coordinate and share information to achieve a common goal: safety for road users. INTERSAFE-2 plans to design and implement an information dispatcher that coordinates applications and aggregates their data elements into common messages. On the receiving end, applications can register with the information dispatcher and subscribe to the data in which they are interested. When a message arrives, the information dispatcher decomposes it into multiple data elements and delivers them to the subscribing applications.
3. *Hybrid and multipath routing.* In previous R&D projects, safety messages were forwarded only in the ad hoc domain, that is, among cars. INTERSAFE-2 plans to

investigate multipath routing that makes use of both the infrastructure domain and the ad hoc domain for multihop CAR-2-X communication.

4.5.3 Demonstrations of the C2C-C Consortium

The C2C-C Consortium performed its first demonstration during the yearly C2C-C Consortium Forum in October 2008. The demonstration was a big success because it showed not only a working system but also interoperability between different car manufacturers and hardware suppliers. This demonstration will be followed by a series of demonstrations in the C2C-C Consortium. The goals of these demonstrations are provided below:

1. The demonstrations will increase the visibility of the C2C-C Consortium and create awareness among consumers and decision-makers.
2. The demonstrations will show that car manufacturers in Europe share a common vision, and they are working together to achieve road safety and a significant reduction of road accidents.
3. The demonstrations will show that different hardware suppliers are working together to achieve interoperability in CAR-2-X communication technology in Europe.
4. The demonstrations will serve as a proof of concept and demonstrate the maturity of CAR-2-X communication technology. They will also be used as an intermediate step in preparing other field-operational tests, as discussed in Sections 4.5.4 and 4.5.5.

4.5.4 SIM-TD

SIM-TD[*,58] is a field trial in Germany, involving major German car manufacturers, automotive and telecommunication suppliers, road operators, fleet operators, test region municipalities, and scientific organizations. It is a four-year project starting in 2008. SIM-TD is regarded as a large-scale feasibility study of CAR-2-X communication. Although the project lays the technological basis in a first phase, it plans to equip about 500 cars and 500 road-side units with prototypes for CAR-2-X communication and conduct field tests in the second phase. SIM-TD can be regarded as a predecessor of European Field Operational Tests (FOTs) on a European scale.

4.5.5 PRE-DRIVE C2X

PRE-DRIVE C2X is a two-year R&D project started in 2008, initiated by European car manufacturers and suppliers and cofunded by the European Commission in the 7th FWP. PRE-DRIVE C2X stands for PREparation for DRIVing implementation and Evaluation of C2X communication technology. Its main objective is the development of an overall

---

[*] German: Sichere Intelligente Mobilität—Testfeld Deutschland or Safe Intelligent Mobility Test area Germany.

description of a common European architecture for CAR-2-X communication, in close cooperation with the COMeSAFETY project (Section 4.2.8). Based on the architecture, the project prepares FOTs for cooperative systems in a later phase as a follow-up of PRE-DRIVE C2X. The FOT on cooperative system is seen as a complementary activity to a FOT for driver assistant systems (noncooperative systems). EUROFOT started in 2008.

In the context of preparation of FOTs, the project enhances existing prototypes for CAR-2-X communication by functions needed for FOTs (such as remote monitoring and configuration functions), verifies their functionality, prepares test centers, and develops testing tools. Eventually, the prototypes will be used for initial tests in laboratory environments, on test tracks and on real roads for "friendly user tests." The project also develops advanced simulation tools and conducts extensive performance evaluations by means of simulations. The simulation results support the planning of field tests and the selection of proper scenarios. Eventually, the simulations shall predict the estimated benefits of CAR-2-X communication and validate initial tests.

4.5.6 European Vision for CAR-2-X Communication

CAR-2-X communication based on short-range radio with ad hoc networking capability will become a mature technology in the next few years. There are two main reasons for this optimism. First, IEEE 802.11 technology, the technological basis of CAR-2-X communication, has become mature, inexpensive, and widely available. Second, CAR-2-X communication benefits from products of advanced research such as security and geographical routing. Beyond achieving safety purposes, CAR-2-X communication in combination with infrastructure access will be an important technological basis for future intelligent vehicles and road-side infrastructure. In this area, CAR-2-X communication also benefits from other advanced technologies such as security and Internet integration, for example, NEMO. Overall, CAR-2-X communication is now considered a cornerstone of Intelligent Transport Systems (ITS). In the long run, CAR-2-X communication will change the landscape of ITS. It will enable a wide range of services for smarter, safer, and cleaner roads in Europe:

1. Future cars will be equipped with communication interfaces based on positioning devices (GPS, Galileo) and IEEE 802.11 (such as IEEE 802.11p for safety and 802.11a for non-safety applications). The usage of safety services is free of charge.
2. Ad hoc networking allows for fast and direct communication among cars. Cars exchange periodic status messages with their neighbors, disseminate data (such as hazard warnings) in geographical regions, and establish communication sessions with other cars in their vicinity.
3. Similarly, cars can also communicate with RSUs in their neighborhood using the IEEE 802.11 technology. Although communication between cars and RSUs already exists today, this communication is mostly based on specialized communication technologies to realize specific purposes, for example, electronic toll collection. In the future, IEEE 802.11 will be the underlying technology for communication between cars and RSUs. A wide range of services can be provided on top of IEEE 802.11-based communication technology such as location-based

search and advertisements. Further, if RSUs are connected to the Internet, RSUs can also provide cars with Internet access for free or for a reasonable service fee.

4. In-vehicle OBUs will have networking protocols that integrate the IP protocol stack with full support of IP version 6. Mobile IPv6 enables global reachability and session continuity.

5. A Geocast protocol provides efficient, scalable, and reliable multihop communication in sparse and dense network scenarios. Geocast will be the underlying mechanism to achieve collision warning and distribute emergency messages among cars.

6. CAR-2-X communication is protected by means of asymmetric cryptography, providing authentication, integrity and nonrepudiation, and other security measures. Anonymity is ensured by the use of revocable pseudonyms provided by a trusted third party.

7. Cars can be equipped with additional wireless interfaces such as 3G/4G technologies. Unlike IEEE 802.11-based CAR-2-X communication that is free of charge but only provides intermittent Internet access, 3G/4G technologies will provide communication with almost full coverage anywhere and anytime and at reasonable costs. Multihop communication and Internet integration may allow cars within a neighborhood to share Internet access via 3G/4G technologies and further reduce access fees.

8. A network interface can be dynamically chosen based on availability of the connectivity, service type, costs, and other criteria. For example, drivers and passengers will prefer to use free WLAN-based Internet access provided by RSUs (either through direct or multihop communication) if this service is available. Alternatively, a car may switch the network interface to 3G/4G technologies until a new RSU providing WLAN-based Internet access is found. The new standard IEEE 802.21 for media independent handover will allow seamless connectivity across different communication technologies.[59]

References

1. Gillan, W.J., Prometheus and Drive: Their Implications for Traffic Managers, in In Vehicle Navigation and Information Systems Conference, September 1989.
2. DRIVE I, available at http://cordis.europa.eu/telematics/tap_transport/research/16. html.
3. Transport Sector of the Telematics Applications Programme, available at http:// cordis.europa.eu/telematics/tap_transport/home.html.
4. IST in EU 5th Framework Program, available at http://cordis.europa.eu/ist/ist-fp5. html.
5. IST in EU 6th Framework Program, available at http://cordis.europa.eu/ist.
6. ICT in EU 7th Framework Program, available at http://cordis.europa.eu/fp7/ict/ home_en. html.
7. The Intelligent Car Initiative, available at http://www.ec.europa.eu/intelligentcar.
8. European Commission, White Paper: European Transport Policy for 2010: Time to Decide, 2001.

9. European Commission, White Paper: Keep Europe Moving. Sustainable Mobility for our Continent—Mid Term Review of the European Commission's 2001 Transport White Paper, 2006.

10. FleetNet—Internet on the Road, available at http://www.et2.tu-harburg.de/fleetnet.

11. Festag, A., Füssler, H., Hartenstein, H., Sarma, A., and Schmitz, R., FleetNet: Bringing Car-to-Car Communication into the Real World, in 11th ITS World Congress and Exhibition, October 2004.

12. Franz, W., Hartenstein, H., and Mauve, M., editors, Inter-Vehicle-Communications Based on Ad Hoc Networking Principles—The Fleetnet Project, Universitätsverlag Karlsruhe, available at http://www.uvka.de/univerlag/volltexte/2005/89, November 2005, ISBN 3-937300-88-0.

13. NoW—Network on Wheels, available at http://www.network-on-wheels.de.

14. Festag, A. et al., NoW—Network on Wheels: Project Objectives, Technology, and Achievements, in 5th International Workshop on Intelligent Transportation (WIT), March 2008.

15. Car-2-Car Communication Consortium, available at http://www.car-to-car.org.

16. WILLWARN, available at http://www.prevent-ip.org/willwarn.

17. Hiller, A., Hinsberger, A., Strassberger, M., and Verburg, D., Results from the WILLWARN Project, in European ITS Congress, June 2007.

18. SAFESPOT, available at http://www.safespot-eu.org.

19. Adaptive and Cooperative Technologies for the Intelligent Traffic, available at http://www. aktiv-online.org/englisch/projects.html.

20. PReVENT, available at http://www.prevent-ip.org.

21. INTERSAFE, available at http://www.prevent-ip.org/intersafe.

22. Fuerstenberg, K., Intersection Safety, in 3rd International Workshop on Intelligent Transportation (WIT), March 2006.

23. CVIS, Cooperative Vehicle-Infrastructure Systems, available at http://www. cvisproject.org.

24. ISO TC204 WG16, available at http://www.isotc204.com.

25. Devarapalli, V., Wakikawa, R., Petrescu, A., and Thubert, P., Network Mobility (NEMO) Basic Support, RFC 3963, January 2005.

26. Brignolo, R., SAFESPOT Integrated Project—Co-operative Systems for Road Safety, in Proceedings of Transport Research Arena Europe Conference (TRA), Goteborg, Sweden, June 2006.

27. COOPERS, Cooperative Systems for Intelligent Road Safety, available at http:// www.coopers-ip.eu.

28. SeVeCom, Secure Vehicular Communication, available at http://www.sevecom.org.

29. COMeSafety, Communications for eSafety, available at http://www.comesafety.org.

30. eSafetySupport, available at http://www.esafetysupport.org.

31. Finn, G., Routing and Addressing Problems in Large Metropolitan-Scale Internetworks, ISI Research Report ISI/EE-87-180, University of Southern California, March 1987.

32. Karp, B.N., and Kung, H.T., GPSR: Greedy Perimeter Stateless Routing for Wireless Networks, in Proceedings of the 6th Annual ACM/IEEE International Conference on Mobile Computing and Networking (MobiCom'00), Boston, MA, USA, August 2000, pp. 243–254.

33. Torrent-Moreno, M., Inter-Vehicle Communication: Achieving Safety in a Distributed Wireless Environment, Challenges, Systems and Protocols, PhD thesis, University Karlsruhe, Karlsruhe, July 2007, available at http://www.uvka.de/univerlag/volltexte/2007/263/.

34. Torrent-Moreno, M., Santi, P., and Hartenstein, H., Distributed Fair Transmit Power Adjustment for Vehicular Ad Hoc Networks, in Proceedings of 3rd Annual IEEE Communications Society Conference on Sensor, Mesh., and Ad Hoc Communications and Networks (SECON 2006), Reston, VA, USA, September 2006.

35. Festag, A., Baldessari, R., and Wang, H., On Power-Aware Greedy Forwarding in Highway Scenarios, in Proceedings of 4th International Workshop on Intelligent Transportation (WIT), 31–36, Hamburg, Germany, March 2007.

36. Harsch, C., Festag, A., and Papadimitratos, P., Secure Position-Based Routing for VANETs, in VTC Fall, Baltimore, MD, USA, October 2007.

37. Fonseca, E., Festag, A., Baldessari, R., and Aguiar, R., Support of Anonymity in VANETs—Putting Pseudonymity into Practice, in Proceedings of IEEE Wireless Communications and Networking Conference (WCNC), Hong Kong, March 2007.

38. Bernardos, C.J., Calderon, M., and Moustafa, H., Survey of IP Address Auto-Configuration Mechanisms for MANETs, Internet Draft, draft-bernardos-manetautoconf-survey-03.txt, work-in-progress, April 2008.

39. Baldessari, R., Bernardos, C.J., and Calderon, M., Scalable Address Autoconfiguration for VANET Using Geographic Networking Concepts, in Proceedings of IEEE International Symposium on Personal, Indoor, and Mobile Radio Communications (PIMRC), Cannes, France, September 2008.

40. Fazio, M., Palazzi, C.E., Das, S., and Gerla, M., Vehicular Address Configuration, in Proceedings of the 1st IEEE Workshop on Automotive Networking and Applications (AutoNet), GLOBECOM, 2006.

41. Baldessari, R., Festag, A., and Abeille, J., NEMO meets VANET: A Deployability Analysis of Network Mobility in Vehicular Communication, in Proceedings of 7th International Conference on ITS Telecommunications (ITST 2007), Sophia Antipolis, France, June 2007, pp. 375–380.

42. Baldessari, R., Festag, A., Zhang, W., and Le, L., A MANET-Centric Solution for the Application of NEMO in VANET Using Geographic Routing, in Proceedings of Trident Com, Innsbruck, Austria, March 2008.

43. Le, L., Zhang, W., Festag, A., and Baldessari, R., Analysis of Approaches for Channel Allocation in Car-to-Car Communication, in 1st International Workshop on Interoperable Vehicles, March 2008.

44. Rai, V., Bai, F., Kenney, J., and Laberteaux, K., Cross-Channel Interference Test Results: A Report from the VSC-A project, Technical Report, IEEE 802.11 technical report, July 2007.

45. Stancil, D., Cheng, L., Henty, B., and Bai, F., Performance of 802.11p Waveforms over the Vehicle-to-Vehicle Channel at 5.9 GHz, Technical Report, IEEE 802.11, September 2007.

46. Car-to-Car Communication Consortium, C2C-CC Manifesto, Version 1.1, August 2007, available at http://www.car-to-car.org.

47. CEN Technical Committee 278, available at http://www.nen.nl/cen278.

48. InternetCar, available at http://www.sfc.wide.ad.jp/InternetCAR/.

49. OverDRiVE, available at http://www.ist-OverDRiVE.org.
50. Johnson, D., Perkins, C., and Arkko, J., Mobility Support in IPv6, RFC 3775 (Proposed Standard), June 2004.
51. Baldessari, R., Ernst, T., Festag, A., and Lenardi, M., Automotive Industry Requirements for NEMO Route Optimization, Internet Draft, draft-ietf-mext-nemo-ro-automotive-req-00, work in progress, February 2008.
52. ECC Report 101, Compatibility Studies in the Band 5855–5925 MHz Between Intelligent Transport Systems (ITS) and Other Systems, 2007, available at http://www.erodocdb.dk.
53. ERO—European Radio Office, available at http://www.erodocdb.dk.
54. ECC/REC/(08)01, Use of the Band 5855–5875 MHz for Intelligent Transport, 2008, available at http: //www.erodocdb.dk.
55. ECC/DEC(08)01, ITS in 5 GHz Band ECC, Decision of 14 March 2008 on the Harmonized Use of the 5875–5925 MHz Frequency Band for Intelligent Transport Systems (ITS), 2008, available at http://www.erodocdb.dk.
56. GeoNet, available at http://www.geonet-project.eu.
57. INTERSAFE-2, available at http://www.intersafe-2.eu.
58. SIM—TD—Sichere Intelligente Mobilität Testfeld Deutschland, available at http://www.aktiv-online.org.
59. IEEE 802.21—Media Independent Handover Services, available at http://ieee802.org/21.

III

Applications

5

Safety-Related Vehicular Applications

Mohammad
S. Almalag
*Department of
Computer Science
Old Dominion University*

5.1 Introduction

Mobile computing and wireless communication have experienced large improvements that have lead to the development of Intelligent Transportation Systems (ITS). In such systems the focus is on improving safety on the roads and providing comfort applications. This chapter introduces safety applications in vehicular ad hoc networks (VANET). It classifies safety applications in different categories and describes each application individually.

According to the U.S. Department of Transportation (U.S. DOT),[1] in 2005, more than 43,000 people were killed and more than 2.6 million were injured in car accidents in the United States. This high number of fatalities and injuries cost billions of dollars in healthcare, more than any other type of injury or disease. Such issues make traffic safety a major concern to government agencies and automotive manufacturers, as well as researchers in related fields.

The emergence of vehicular networking has encouraged researchers to study how such communications could be used to enhance driver safety. In the past several years, government agencies have partnered with car manufacturers to design and prototype different types of safety-related vehicular applications.

The remainder of this chapter is organized as follows. In Section 5.2, the communications used in VANET safety applications are presented. Section 5.3 illustrates the message dispatcher, which is a new concept that will improve channel utilization. Section 5.4 outlines the different types of safety applications, with a short description for each individual application. Section 5.5 will explain in more detail two of the most important safety applications: Extended Emergency Brake Light (EEBL), which uses vehicle-to-vehicle (V2V) communications and Cooperative Intersection Collision Avoidance System (CICAS), which uses infrastructure-to-vehicle (I2V) and vehicle-to-infrastructure (V2I) communications.

5.2 Communications Overview

Communication in safety applications is based on dedicated short-range communications (DSRC).[2] DSRC is designed for vehicular wireless communications and works in the 5.9 GHz band in the United States. The range of communication using DSRC is up to 1000 m. Using this protocol, vehicles can communicate with each other as well as with roadside infrastructure.

DSRC is divided into two types of communication: V2V and V2I. V2V communication is used when vehicles need to exchange data among themselves in order for safety applications to work properly, whereas V2I communication is used when roadside units are part of the safety application.

In safety applications, some applications are required to send messages periodically (e.g., every 100 msec), whereas other safety applications send messages when an event occurs.[3] All safety applications have a range of communications between 100 and 1000 m, and a minimum frequency between 1 and 50 Hz.

5.2.1 Types of Messages

All safety applications are DSRC-based applications, which require the exchanging of messages with other vehicles. These applications obtain data from sensors, other vehicles, or both, depending on the application's functionality. Each application processes the data and then sends messages to nearby vehicles or to infrastructure. Sending messages in safety applications is needed for one of two reasons: awareness of the environment or detection of an unsafe situation. Messages that are sent because of awareness of the environment are called periodic messages, whereas those triggered by an unsafe situation are called event-driven messages.

5.2.1.1 Periodic Messages

Periodic messages are generated to inform nearby vehicles about the vehicle's current status, for example, speed, position, and direction. They may also include other non-safety application data. By processing data received in periodic messages, other vehicles will be able to avoid unsafe situations even before they occur. Because information in periodic messages is important to all vehicles surrounding the sender and needs to be broadcasted frequently, periodic messages may cause the broadcast storm problem,[4] leading to contention, packet collisions, and inefficient use of the wireless channel. Torrent-Moreno and colleagues[5] proposed a method to limit the load sent to a channel using a strict fairness principle among the nodes, which will solve the problem of a broadcast storm caused by periodic messages.

5.2.1.2 Event-Driven Messages

Event-driven messages are emergency messages sent to other vehicles based on unsafe situations that have been detected. If there is no emergency situation, event-driven messages will not be disseminated. This type of message has a very high priority. Event-driven messages contain the location of the vehicle, the time, and the event type. The challenge with this type of message is that the sender needs to make sure that all vehicles intended to benefit from these messages receive them correctly and quickly.

5.3 Message Broadcast

Most of the safety applications running on vehicles exchange similar data elements among themselves. These data elements (e.g., speed, acceleration, position) are needed by different safety applications to complete their jobs. Table 5.1 shows several data elements and some applications that use them. For the efficient exchange of data elements, vehicles running multiple safety applications need a special method to avoid redundancy. Also, some data elements that change their values quickly (e.g., position) must be sent frequently. These data elements need to be gathered and compressed in one packet to be sent to nearby vehicles and infrastructure. By using data element compression and single-hop broadcast communication, the number of messages to be sent will be reduced significantly, improving channel utilization.

TABLE 5.1   Message Dispatcher

Data Element	Signal Violation	Curve Warning	Emergency Brake	Precrash Warning	Collision Warning	Turn Assistant	Lane Warning	Stop Sign Assist	No. of Uses
Acceleration	✓	✓	✓	✓	✓	✓	✓	✓	8
Airbag count					✓				1
Antilock brake state	✓		✓		✓				3
DSRC message ID	✓	✓	✓	✓	✓	✓	✓	✓	8
Elevation		✓		✓					2
Heading	✓	✓	✓	✓	✓	✓	✓	✓	8
Speed	✓	✓		✓	✓	✓	✓	✓	7
Vehicle length				✓	✓	✓	✓	✓	5
Vehicle mass		✓	✓	✓	✓				4

5.3.1 Message Dispatcher

The message dispatcher (MD)[6] is an interface that is added between different safety applications running on a vehicle and lower-layer protocols. So, the communication layers will be divided into three parts: safety applications, MD, and the rest of the communication layers (e.g., link layer and physical layer), as shown in Figure 5.1.

When the safety applications need to send data elements to lower layers to be broadcast to nearby vehicles or infrastructure, the MD will receive them, eliminate the redundant ones, and then put them all together in one packet. This packet, which includes all the nonredundant data elements, will be delivered to the other communication layers to be broadcast (DSRC radio). In Figure 5.2, two applications are running on the vehicle. The first application is a Curve Warning system, which has acceleration, speed, position, and traction control state as outputs. The second application is an Emergency Brake system, which has acceleration, speed, position, BrakeAppliedStatus, and AntiLockBrakeStatus as outputs. These applications need to send their data to the nearby vehicles and/or infrastructure. Because both applications need to send similar data elements (e.g., acceleration, speed, and position), the MD will send all data elements in a single packet without redundancies.

When another vehicle receives the packet, the MD in that vehicle will do the reverse job. It will get all the data elements from the packet and deliver them to the appropriate safety applications (Figure 5.3). The receiving vehicle or infrastructure may receive one data element and deliver it to many safety applications at the same time, if needed.

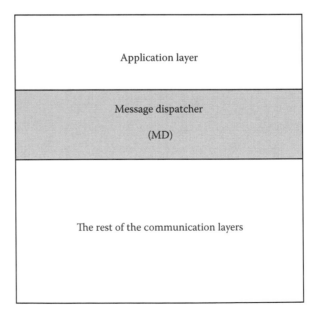

FIGURE 5.1 The MD resides between the application layer and the rest of the communication layers.

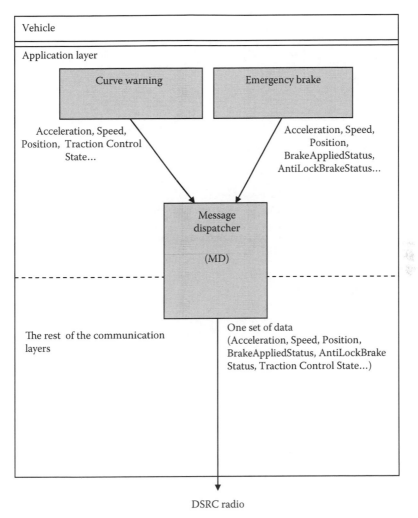

FIGURE 5.2 The MD receives all data elements from all applications on board and then puts them all in one packet. This packet will be delivered to other network layers to be transmitted to other vehicles/infrastructure.

5.3.1.1 Data Element Dictionary

The way that data elements are identified and formatted is based on the Society of Automotive Engineers (SAE) standard.[7] In the SAE standard dictionary, there are about 70 elements that have been identified. More data elements can be added to the SAE standard data element dictionary, if needed in the future.[6] For the elements described in the SAE data element dictionary, several fields are used to define each element:

1. Standard name
2. Unique identifier

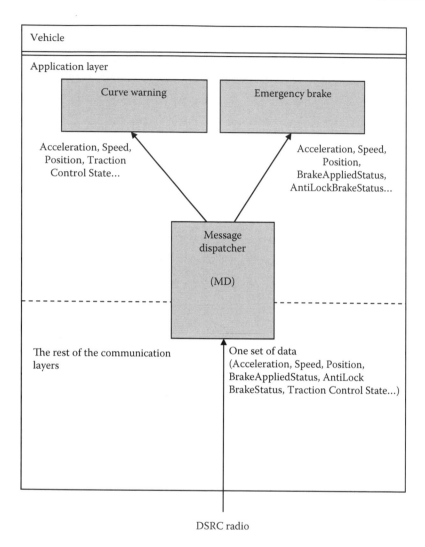

FIGURE 5.3 The MD receives the packet from the standard communication layers, and then delivers each data element to the appropriate application.

3. Unit of measure
4. Accuracy of measure
5. Range of measure
6. Description

An example of a data element is latitude. In the data element dictionary, latitude has a standard name, a unique identifier, a unit of measure, accuracy of measure, range of measure, and a description. At the same time, latitude is used with another data element, longitude, to calculate the position of the vehicle to avoid sending both elements in separate data elements. Both latitude and longitude can be put into a single data frame called "position." In this case, we reduce the overhead of having two elements to be sent by combining them in one data frame that will be sent as one element. This concept of data frame can be applied to other data elements that are used to calculate a single value.

The SAE standard has defined many data frames, in which each one of them has a unique ID. When sending a data frame, its ID will be sent, followed by the data elements of that frame. Data elements in each data frame are organized in a specific order so that the receiver will be able to recognize each data element based on the ID of the data frame.

5.3.1.2 Message Construction

The MD divides each message into three parts: data frames, data elements, and newly defined terms. The first part, data frames, includes the entire data frame needed to be sent in the message. Each data frame in the message starts with its own ID, followed by its data elements. The MD may not send any data frames in the message, if it does not need to. The second part, data elements, includes all the data elements that are not part of any data frame in this message. The last part is designed to include the newly defined terms.[6]

5.4 Safety Applications

Current safety technologies in vehicles are single-vehicle-based technologies (e.g., parking sensors). Vehicles with such technologies are not able to share data with nearby vehicles, which limits their capabilities. Also, they do not work unless other vehicles are in direct line of sight. These issues have led to the development of other techniques to improve safety on the road.

The Vehicle Infrastructure Integration (VII) project[8] engages communication between vehicles and with infrastructures. VII is a coalition between U.S. DOT, ten state DOTs and automotive manufactures. The main goal of this coalition is to have communication technologies in new vehicles and in roads. Having such communication technologies will lead to the development of more advanced safety applications.

Vehicles will be able to communicate with each other, which will allow them to share data that will be helpful to some in-vehicle applications in order to increase safety. All safety applications described in this chapter are based on V2V communication, V2I communication, or both.

The Vehicle Safety Communications Project was tasked with identifying intelligent safety applications enabled by DSRC.[3] These applications were classified in five different categories:

1. Intersection collision avoidance
2. Public safety
3. Sign extension
4. Vehicle diagnostics and maintenance
5. Information from other vehicles

Each of these categories will be explained with some examples in the following sections.

5.4.1 Intersection Collision Avoidance

Intersection collision avoidance systems are the work of several government agencies and automobile manufacturers. According to the U.S. DOT,[9] in 2003 more than 9000 Americans died and about 1.5 million Americans were injured in accidents that are related to intersections. Improving such systems will help drivers avoid many fatal accidents.

Generally, intersection collision avoidance systems are based on I2V and/or V2I communication.[10] The infrastructure has sensors around intersections to collect data about the movement of nearby vehicles. All data collected from sensors are processed and analyzed to determine if there is any unsafe situation that may lead to an accident. If there is a risk of an accident, a warning message will be sent to vehicles in the concerned area.

There are many different intersection collision avoidance applications[3]:

1. Traffic signal violation warning
2. Stop sign violation warning
3. Left turn assistant
4. Stop sign movement assistant
5. Intersection collision warning
6. Blind merge warning
7. Pedestrian crossing information designated intersection

All of these applications use I2V communications with periodic messages. The minimum frequency for all of these applications is 10 Hz, and the required range of communication is between 200 and 300 m (Table 5.2). A brief description of each individual intersection collision avoidance application follows.

5.4.1.1 Traffic Signal Violation Warning

The traffic signal violation warning system is designed to warn a vehicle when the system detects that the vehicle is in danger of running the traffic signal. The system will make a decision (sending a warning message to the vehicle) based on the traffic signal status and timing and the vehicle's speed and position. Some other factors are considered in such situations, such as the road surface and weather conditions.

5.4.1.2 Stop Sign Violation Warning

The stop sign violation warning system is designed to warn the vehicle when the system detects that the vehicle is in danger of running the stop sign. The system will make a decision (sending a warning message to the vehicle) based on the distance to the stop sign and whether the driver will be able to reach a complete stop at the stop sign. Some other factors are considered in such situations, such as road surface and weather conditions.

5.4.1.3 Left Turn Assistant

The left turn assistant system helps a driver to make a safe left turn at signalized intersections with no left turn arrow phase. The system will collect data about vehicles coming from the opposite direction that are approaching the intersection, using sensors located at the intersection or DSRC communications, and then inform vehicles that are going to make the left turn. One way to implement such system is by having data collected

TABLE 5.2 Intersection Collision Avoidance

Application	Types of Communication	Transmission Mode	Minimum Frequency (Hz)	Allowable Latency (msec)	Data to be Transmitted and/or Received	Maximum Required Range (m)
Traffic signal violation warning	I2V One-way One-to-many	Periodic	~10	~100	Traffic signal status, timing, directionality, position of the traffic signal stopping location, weather condition, road surface type near traffic signal	~250
Stop sign violation warning	I2V One-way One-to-many	Periodic	~10	~100	Directionality, position of the stopping location, weather condition, road surface type near the stop sign	~250
Left turn assistant	V2I and I2V One-way One-to-many	Periodic	~10	~100	Traffic signal status, timing, and directionality; road shape and intersection information; vehicle position, velocity, and heading	~300
Stop sign movement assistance	V2I and I2V One-way One-to-many	Periodic	~10	~100	Vehicle position, velocity, and heading; warning	~300
Intersection collision warning	I2V One-way One-to-many	Periodic	~10	~100	Traffic signal status, timing, and directionality; road shape and intersection information; vehicle position, velocity, and heading	~300
Blind merge warning	I2V One-way One-to-many	Periodic	~10	~100	Velocity, position, heading, and acceleration	~200
Pedestrian crossing information at designated intersections	I2V One-way One-to-many	Periodic	~10	~100	Presence of a pedestrian	~200

continuously, and when there is a vehicle with its left turn signal on, the system will send a message to that vehicle about the traffic traveling in the opposite direction of the vehicle. The other way is to have an in-vehicle system that sends a request to be notified about the traffic in the opposite direction when the left turn signal is activated. The system will then collect data about the traffic in the opposite direction and send a message to the vehicle that requested the information. In both ways, the driver will be informed and warned about the traffic coming in the opposite direction. This system is different to most of the other intersection collision avoidance systems because it can use both V2I and I2V communications.

5.4.1.4 Stop Sign Movement Assistant

The stop sign movement assistant system is designed to avoid accidents at stop sign intersections. The system uses data collected by sensors or DSRC communications to inform the driver if it is unsafe to go through an intersection, so it warns drivers if there is traffic coming through the intersection at the same time. This system uses both V2I and I2V communications.

5.4.1.5 Intersection Collision Warning

The intersection collision warning system is designed to warn drivers that are approaching the same intersection if there is a possibility of an accident occurrence. Data about all vehicles approaching an intersection are gathered using sensors and/or DSRC communications and then processed to see if there is a chance of a collision at the intersection. If there is such a situation, messages will be sent to vehicles that might be involved in this unsafe situation. The data required by the system includes information regarding the vehicles that are approaching the intersection and other data about the road. The data needed about vehicles are position, velocity, acceleration, and turning status. The data needed about the road are shape and surface. Data about the weather status is helpful in calculating the chance of a collision occurrence.

5.4.1.6 Blind Merge Warning

The blind merge warning system is designed to help drivers avoid accidents when a vehicle is approaching a merge point. The problem of merging is that visibility is often very poor. When the driver is attempting to merge, this system will warn if it is unsafe. The system will also warn the other vehicles in the main road if there is any danger in such situations. The system collects data about all vehicles at the intersection and then processes the data to determine the risk of the merge. If there is a possibility of collision occurrence, all vehicles that might be involved will receive messages to warn the drivers.

5.4.1.7 Pedestrian Crossing Information Designated Intersection

This system is designed to alert drivers when a pedestrian is attempting to cross the road. The system works by collecting data using sensors that are located on the sidewalk near the road. These sensors will be able to detect the movement of a pedestrian, if there are any nearby, and then send the data to the system. The system is also able to obtain data when the "walk" button on the pedestrian crossing signal is pressed. If there is an unsafe situation, the system will send warning messages to vehicles in the area.

5.4.2 Public Safety

Public safety applications are designed to assist emergency teams in providing their services as well as help drivers when in need of aid. According to *USA Today*,[11] the national reported average response time for emergency vehicles is between 6 and 7 min. In their investigation, they saw cases where the response time was up to 25 min. Most of the response time is wasted in traveling from the emergency center to the scene, while the turnout time in most cases is less than 60 sec.

Minimizing the travel time for emergency teams is the focus of most public safety applications. Other applications in this category focus on requesting help when drivers get into accidents and avoiding potential second accidents.

The popular public safety applications are as follows[3]:

1. Approaching emergency vehicle warning
2. Emergency vehicle signal preemption
3. SOS services
4. Postcrash warning

These applications use V2V communications, V2I communications, or both. The messages that are used in this type of application are event-driven messages with a minimum frequency of 1 Hz. The required range of communication is between 300 and 1000 m (Table 5.3). A brief description of each individual public safety application follows.

TABLE 5.3   Public Safety

Application	Types of Communication	Transmission Mode	Minimum Frequency (Hz)	Allowable Latency (sec)	Data to be Transmitted and/or Received	Maximum Required Range (m)
Approaching emergency vehicle warning	V2V One-way One-to-many	Event-driven	~1	~1	Emergency vehicle position, lane information, speed, and intended path/route	~1000
Emergency vehicle signal preemption	V2I Two-way One-to-one	Event-driven	N/A	~1	Emergency vehicle position, speed, direction of travel, and intended path/route	~1000
SOS services	V2I or V2V One-way One-to-many	Event-driven	~1	~1	Position, vehicle status, vehicle description, and time	~400
Postcrash warning	V2I or V2V One-way One-to-many	Event-driven	~1	~0.5	Position, heading, and vehicle status	~300

5.4.2.1 Approaching Emergency Vehicle Warning

Every second in an emergency situation counts. So, having an emergency vehicle arrive to the scene faster may actually save lives. The approaching emergency vehicle warning system is designed to help emergency vehicles make their travel time as short as possible by having the roads clear. When an emergency vehicle is traveling in the road, it will send messages to other vehicles in its route to tell them to clear the road. The message being sent to other vehicles has some information about the emergency vehicle's position, lane information, speed, and intended path. When other vehicles receive the message, if the message is related to them, they will warn the drivers. This system uses one-way V2V communications.

5.4.2.2 Emergency Vehicle Signal Preemption

To make the trip of an emergency vehicle even shorter and safer, emergency vehicle signal preemption sends messages to all traffic lights in its path to be green phased just before the emergency vehicle gets to the intersection. This means that the emergency vehicle does not need to stop or slow down when the traffic light is red phased at the intersection. This system works by having the infrastructure at each intersection collect data from messages that are sent by emergency vehicles. This system can be incorporated with the approaching emergency vehicle warning system, because they have the same goal of helping emergency vehicles reduce their travel times.

5.4.2.3 SOS Services

This system is designed to send SOS messages when a driver is in a life-threatening situation. The system will work automatically and send messages in the case of collisions or by initiation by the driver in other situations. For instance, when the vehicle is involved in an accident the system will detect that the airbag has been deployed. If the emergency situation is not related to a car collision, the driver can activate the system to send SOS messages.

This system uses both V2I and V2V communications. When the vehicle is in the range of infrastructure, an SOS message will be sent and help will be directed to the location. If the vehicle is far away from infrastructure, the vehicle will send an SOS message to another nearby vehicle. The receiving vehicle will relay the message when it gets close to infrastructure.

5.4.2.4 Postcrash Warning

The postcrash warning system is designed to help drivers avoid secondary collisions. When a vehicle is not moving because of an accident or any other reason, it will warn other vehicles coming in the same direction or other directions about its situation. This can be very helpful in certain situations, especially when visibility is poor.

This system uses both types of communication, V2I and V2V. In V2I communications, the disabled vehicle sends a warning message that contains the vehicle's location, heading, and status to nearby infrastructure. This message will then be broadcast to all vehicles that are approaching. In V2V communications, the disabled vehicle sends a message to all vehicles moving toward it. When other vehicles receive a postcrash warning message, their location and direction will determine if this message is important and warn the driver, or ignore the message.

5.4.3 Sign Extension

In the past few years, new technologies, such as cell phones, have become commonly used while driving. These technologies may distract drivers, which can lead to reckless driving and accidents. Keeping drivers alert while driving is the focus of sign extension applications. Drivers will be alerted about all types of signs on the side of the roads as well as structures, such as bridges in the surrounding area.

There are many different sign extension applications[3]:

1. In-vehicle signage
2. Curve speed warning
3. Low parking structure warning
4. Wrong way driver warning
5. Low bridge warning
6. Work zone warning
7. In-vehicle AMBER alert

Sign extension applications use I2V communications with mostly periodic messages. The minimum required frequency is 1 Hz, with the range of communication between 100 and 500 m (Table 5.4). A brief description of each individual sign extension application follows.

TABLE 5.4 Sign Extension

Application	Types of Communication	Transmission Mode	Minimum Frequency (Hz)	Allowable Latency (sec)	Data to be Transmitted and/or Received	Maximum Required Range (m)
In-vehicle signage	I2V One-way One-to-many	Periodic	~1	~1	Condition, position, and direction of travel	~200
Curve speed warning	I2V One-way One-to-many	Periodic	~1	~1	Curve location, curve speed limits, curvature, bank, and road surface condition	~200
Low parking structure warning	I2V One-way One-to-many	Periodic	~1	~1	Clearance height and location of parking structure	~100
Wrong way driver warning	V2V One-way One-to-many	Periodic	~10	~100	Position, direction, and warning	~500
Low bridge warning	I2V One-way One-to-many	Periodic	~1	~1	Height of bridge and distance to bridge	~300
Work zone warning	I2V One-way One-to-many	Periodic	~1	~1	Distance to work zone and reduced speed limits	~300
In-vehicle AMBER alert	I2V One-way One-to-many	Event-driven	~1	~1	AMBER alert information	~250

5.4.3.1 In-Vehicle Signage

The in-vehicle signage system is designed to inform drivers about different types of signs along the road. Signs that might use such a system include school zones, hospital zones, animal crossing zones, and so on. This system is based on having roadside units located at the areas where the drivers want to be warned. Roadside units will send warning messages to all vehicles near their zones to inform drivers about the situation at that time.

5.4.3.2 Curve Speed Warning

The curve speed warning is designed to alert drivers approaching a curve if they are going too fast. The way this system works is by having some roadside units before the curve. By using I2V communications, units will send periodic messages to vehicles approaching the curve. The messages that are sent by the system contain the curve location, curve speed limit, curvature, and road surface condition. Vehicles approaching the curve will receive these messages and then process the data if the driver is driving too fast, and the driver will be alerted.

5.4.3.3 Low Parking Structure Warning

The low parking structure warning is designed to inform drivers approaching parking garages about the clearance height of the parking garage. The way this system works is by having units that are placed in or near the parking structures. These units use I2V communication to send periodic messages to inform drivers that are going into the parking garage about the clearance height of the parking garage. The in-vehicle system will receive these messages, and if the clearance height is too low, the driver will be alerted.

5.4.3.4 Wrong Way Driver Warning

The wrong way driver warning system is designed to warn drivers if they are driving in the wrong direction. The data elements this system needs are the position of the vehicle and map database data. If the driver is driving in the wrong direction, the driver will be warned. The wrong-way vehicle will then send information messages to other vehicles around it to inform them about the situation. The communication type in this system is V2V communication.

5.4.3.5 Low Bridge Warning

The low bridge warning is similar to the low parking structure warning. This system works by having units placed on or near bridges to inform vehicles going underneath the bridge about how low the bridge is. It uses I2V communication to send periodic messages about the height of the bridge to all vehicles that are approaching the bridge in both directions. The in-vehicle system will receive the message and then warn the driver if the bridge is too low.

5.4.3.6 Work Zone Warning

The work zone warning system is similar to the in-vehicle signage system in a certain way. The system is designed to warn drivers when they approach a work zone area.

Roadside units will be located near work zones. By using I2V communications, these roadside units will send periodic messages to inform vehicles approaching such areas.

5.4.3.7 In-Vehicle AMBER Alert

This system is designed to send AMBER (America's Missing: Broadcast Emergency Response) alert messages to vehicles. An AMBER alert is a notification to the general public that is issued when police confirm that a child has been abducted. The vehicle that is involved in such crime can be excluded from receiving the alert. The communication type in this system is I2V communication.

5.4.4 Vehicle Diagnostics and Maintenance

Vehicle diagnostic and maintenance applications provide alerts and reminders to vehicle owners about safety defects and maintenance schedules for their vehicles. The National Highway Traffic Safety Administration (NHTSA)[12] is the agency that is in charge of issuing recall information about vehicles. This information is based on safety-related defects or a violation of the federal motor vehicle safety standards. Such defects could be related to the vehicle equipment, such as tires, and so on.

The two main vehicle diagnostic and maintenance applications[3] are as follows:

1. Safety recall notice
2. Just-in-time repair notification

I2V is the type of communication that is used in this kind of application. Also, all messages are event-driven messages, and the required range of communication is about 400 m (Table 5.5). A brief description of each of the above applications follows.

5.4.4.1 Safety Recall Notice

The safety recall notice system is designed to notify vehicle owners when a recall is issued for their vehicles. It uses I2V communications to send messages to recalled vehicles. When the vehicle receives the message, the driver will be alerted by a warning lamp or some other method.

TABLE 5.5 Vehicle Diagnostics and Maintenance

Application	Types of Communication	Transmission Mode	Minimum Frequency	Allowable Latency (sec)	Data to be Transmitted and/or Received	Maximum Required Range (m)
Safety recall notice	I2V One-way One-to-one	Event-driven	N/A	~5	Safety recall message	~400
Just-in-time repair notification	V2I and I2V Two-way One-to-one	Event-driven	N/A	N/A	Position, heading, fault code information; location of nearest services	~400

5.4.4.2 Just-in-Time Repair Notification

This system is designed to detect if there is any flaw in the vehicle. It is an in-vehicle system that sends a message to the infrastructure about the defect. The message will then be received by OEM technical support centers, which will advise the driver about the best way to handle such a problem. The system uses V2I communication to send the defect report and I2V to send the advice to the driver.

5.4.5 Information from Other Vehicles

Information from other vehicles applications uses short-range communication between a host vehicle and other nearby vehicles. This information, for example, position heading, are used by in-vehicle applications to complete their functionalities.

There is a lot of different information from other vehicles applications[3]:

1. Cooperative forward collision warning
2. Vehicle-based road condition warning
3. Emergency electronic brake lights
4. Lane change warning
5. Blind spot warning
6. Highway merge assistant
7. Visibility enhancer
8. Cooperative collision warning
9. Cooperative vehicle–highway automation system
10. Cooperative adaptive cruise control
11. Road condition warning
12. Precrash sensing
13. Highway/rail collision warning
14. V2V road feature notification

All of these applications use V2V communication, I2V communication, or both. They also use either periodic or event-driven messages with a minimum frequency of 2 to 50 Hz. The required range of communication for such applications is between 50 and 400 m (Table 5.6). A brief description of each of the above applications follows.

5.4.5.1 Cooperative Forward Collision Warning

The cooperative forward collision warning is designed to help a driver avoid collisions with the rear-end of vehicles ahead of him/her. The system will achieve its goal by warning the driver about such situations. This system uses V2V communication to pass the message. It requires multihop operation to cover the zone of relevance going beyond the communication range of one hop. Having an efficient broadcast algorithm will guarantee the delivery of the message. The message that will be sent contains data such as the position, velocity, heading, yaw-rate, and acceleration of all vehicles in the surrounding area. When the vehicle receives the message, the vehicle processes the content of the message with the information it has about itself to determine if the situation is dangerous. After that, the vehicle will resend the message along with its data to other vehicles around it.

TABLE 5.6 Information from Other Vehicles

Application	Types of Communication	Transmission Mode	Minimum Frequency (Hz)	Allowable Latency (sec)	Data to be Transmitted and/or Received	Maximum Required Range (m)
Cooperative forward collision warning	V2V One-way One-to-many	Periodic	~10	~100	Position, velocity, acceleration, heading, and yaw-rate	~150
Vehicle-based road condition warning	V2V One-way One-to-many	Event-driven	~2	~0.5	Position, heading, and road condition parameters	~400
Emergency electronic brake lights	V2V One-way One-to-many	Event-driven	~10	~100	Position, heading, velocity, and deceleration	~300
Lane change warning	V2V One-way One-to-many	Periodic	~10	~100	Position, heading, velocity, acceleration, and turn signal status	~150
Blind spot warning	V2V One-way One-to-many	Periodic	~10	~100	Velocity, position, heading, acceleration, and turn signal status	~150
Highway merge assistant	V2V One-way One-to-many	Periodic	~10	~100	Position, speed and heading	~250
Visibility enhancer	V2V One-way One-to-many	Periodic	~2	~100	Velocity, position, and heading	~300

Continued

TABLE 5.6 (continued)

Application	Types of Communication	Transmission Mode	Minimum Frequency (Hz)	Allowable Latency (sec)	Data to be Transmitted and/or Received	Maximum Required Range (m)
Cooperative collision warning	V2V One-way One-to-many	Periodic	~10	~100	Position, velocity, acceleration, heading, and yaw-rate	~150
Cooperative vehicle–highway automation system (Platoon)	V2V and I2V One-way and two-way One-to-one and one-to-many	Periodic	~50	~20	Position, velocity, acceleration, heading, and yaw-rate	~100
Cooperative adaptive cruise control	V2V and I2V One-way One-to-many	Periodic	~10	~100	Position, velocity, acceleration, heading, and yaw-rate	~150
Road condition warning	I2V One-way One-to-many	Event-driven	~1	~1	Road condition warning message	~200
Precrash sensing	V2V Two-way One-to-one	Event-driven	~50	~20	Vehicle type, position, velocity, acceleration, heading, and yaw-rate	~50
Highway/rail collision warning	I2V or V2V One-way One-to-many	Event-driven or periodic	~1	~1	Position, heading, and velocity	~300
V2V road feature notification	V2V One-way One-to-one	Event-driven	~2	~0.5	Position, heading, and road feature parameters	~400

5.4.5.2 Vehicle-Based Road Condition Warning

The vehicle-based road condition warning system is designed to inform the driver and other vehicles on the road about the road condition. This system uses V2V communication. It works by having sensors on the vehicle that will collect data about the road condition, that is, stability control. The collected data will be processed by the in-vehicle system which will provide the driver with results. The result could be warning the driver about the maximum safe speed on the road, for example. The result will also be sent to other vehicles around to warn them about the road condition.

5.4.5.3 Emergency Electronic Brake Lights (EEBL)

This system is designed to alert drivers when a vehicle in front of them is braking hard. This system uses V2V communications. The main goal of this system is to notify drivers about such an action even when visibility is poor. EEBL will be explained in more detail at the end of this chapter (Section 5.5.1).

5.4.5.4 Lane Change Warning

The lane change warning system is designed to avoid a crash occurrence when the driver is trying to make an unsafe lane change. It uses V2V communications to send a warning message to the driver. The system will hold information such as speed, direction, and position of surrounding vehicles. When the driver signals for a lane change, the system will process the data it has, and then determine if the gap between vehicles is safe enough to make the lane change. If the gap is not safe, the system will warn the driver about the risk of the lane change.

5.4.5.5 Blind Spot Warning

The blind spot warning system is designed to avoid collisions when the driver is trying to make lane changes and there is another vehicle in the blind spot. This system uses V2V communications to send periodic messages. It works in a very similar way to the lane change warning system.

5.4.5.6 Highway Merge Assistant

The highway merge assistant system is designed to warn drivers in the highway if there is a vehicle in their way trying to merge. The merging vehicle could be on a highway ramp and the other vehicles could be the blind spot of the merging vehicle. When the merging vehicle is on an on-ramp, it starts sending periodic messages to other vehicles in the highway containing its position, speed, and heading. Other vehicles in the highway will be alerted about the merging vehicle.

5.4.5.7 Visibility Enhancer

The visibility enhancer system is designed to enhance poor visibility situations, such as caused by heavy rain, fog, snow storms, and so on. This system senses such situations automatically or through inputs from the driver.

5.4.5.8 Cooperative Collision Warning

The cooperative collision warning system is designed to warn drivers if an accident is about to occur. The warning will be issued based on information collected from nearby vehicles. This system uses V2V communications to exchange messages. Messages contain information about each vehicle individually. Information such as the position, heading, acceleration, and yaw-rate of other nearby vehicles are compared with the same information about the vehicle. If an accident is about to occur, the driver will be warned.

5.4.5.9 Cooperative Adaptive Cruise Control

The cooperative adaptive cruise control system is designed as an improvement of adaptive cruise control (ACC). This system makes driving safer by achieving the appropriate speed of the vehicle based on the speed of the other vehicles ahead. To do that, V2V communication is used in this system to exchange messages between the lead vehicle and the following ones. Messages that have to be exchanged include the position, velocity, acceleration, heading, and yaw-rate. I2V communication can be used in this system in order to maintain a speed limit in low-speed-limit zones.

5.4.5.10 Road Condition Warning

The road condition warning system is designed to warn drivers when it is unsafe to drive at full speed on that specific road. The way this system works is by having sensors on the side of the roads. These sensors will detect the road condition, for example, if it is icy or slippery. After that, messages will be sent to alert vehicles in that area and to cause them to reduce their speed. The communication type that is used in this system is I2V communication.

5.4.5.11 Precrash Sensing

The precrash sensing system is designed to prepare vehicles when an accident is about to happen. The way this system works is by having sensors in the vehicle that can measure the harshness of the accident that is about to happen. Along with information from sensors, the system can use V2V communications to obtain more information from other vehicles that might get involved in such an accident. The precrash sensing system may perform actions to increase the safety of people in the vehicle, such as, pretightening of seatbelts.

5.4.5.12 Highway/Rail Collision Warning

The highway/rail collision warning system is designed to keep vehicles away from accidents with trains at highway/rail intersections. The way this system works is by having roadside units that are placed near such intersections send warning messages to nearby vehicles. If the driver is approaching the intersection and there is a message that has been sent, he/she will be alerted about such a risk. Another option to implement this system is by having the train send messages to other vehicles approaching the same intersection.

5.4.5.13 V2V Road Feature Notification

The V2V road feature notification system is designed to pass information about road features from V2V in order to be used in some other VANET applications. Using V2V communications, vehicles can sense road features, such as, a curve, and then send messages with such information to other vehicles nearby. This system can be the basis for many other safety applications because of the information it sends.

5.5 Safety Applications Examples

This section illustrates two of the most important safety applications. It further describes how these applications work and how they successfully increase safety in vehicles. The first application is extended emergency brake light (EEBL), which is based on V2V communications. The second application is the Cooperative Intersection Collision Avoidance System (CICAS), which is based on V2I and I2V communications.

5.5.1 The EEBL System

The EEBL application was an OEM-funded effort between BMW, Daimler Chrysler, Ford, GM, Nissan, and Toyota.[13] It is considered to be the first safety application that uses pure V2V communication. EEBL as a project started in June 2005 and was completed in March 2006. The main idea of EEBL is to alert the driver when a preceding vehicle performs severe braking (Figure 5.4).

In real life, rear tail lights alert other drivers driving behind, but there are some situations in which this system is not sufficient. For example, in bad weather where

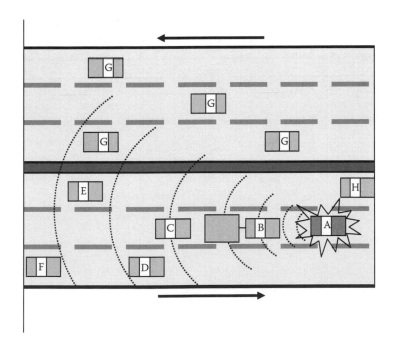

FIGURE 5.4 A scenario of the EEBL application where vehicle A is braking hard.

visibility is poor (fog, heavy rain, or snow), drivers behind the braking vehicle will not be able to see the rear tail lights until they come too close. So, the requirement for having another way to alert drivers led to the development of EEBL.

Referring to Figure 5.4, suppose vehicle A is breaking hard for some reason. The EEBL system will send an alert message to all vehicles in vehicle A's range. This alert message will include some information, including vehicle ID, position, speed, deceleration, and direction. All information in the alert message will be used by the vehicles that received it to decide whether to take action or to ignore it.

If the receiving vehicles are behind vehicle A, their drivers will be notified about the brake even if they cannot see vehicle A. For example, the driver of vehicle C cannot see vehicle A because vehicle B is blocking his/her vision. The driver of vehicle C will receive the alert because it is in the range of vehicle A. Other vehicles like D and E will receive the alert message from vehicle A as well, and the drivers will be notified. Because vehicle F is too far from the scene, it will not be alerted by vehicle A. If the receiving vehicle is in the other direction, the alert message will be ignored, and the driver will not be notified (e.g., vehicle G). The same scenario will exist for vehicles ahead of the braking vehicle, such as vehicle H.

5.5.2 CICAS

CICAS was a four-year project sponsored by U.S. DOT.[14] The main goal of CICAS is to prevent accidents between vehicles approaching or crossing intersections. The way to do that is by determining if a vehicle is about to break the traffic rights of other vehicles or traffic control. When such an unsafe situation is detected, other vehicles that might get involved will be alerted.

Intersections that CICAS focuses on are traffic light intersections and stop sign intersections. This system relies on V2I and I2V communications using roadside units that are located at each intersection to send/receive messages to/from vehicles near the intersections. The messages sent by CICAS are used by some in-vehicle applications to warn drivers about unsafe situations at intersections they are approaching.

According to Holfelder,[13] CICAS is divided into two phases:

1. *Phase I.* This started on May 1, 2006, and ended on April 30, 2008, and phase focused on developing the field operational test (FOT).
2. *Phase II.* This started on May 1, 2008, and is to end on April 30, 2010. This phase is focused on conducting the FOT and analyzing the data.

According to O'Connor,[10] there are five different scenarios of vehicle accidents at intersections:

1. Left turn across path—opposite direction (Figure 5.5)
2. Left turn across path—lateral direction (Figure 5.6)
3. Left turn into path (Figure 5.7)
4. Right turn into path (Figure 5.8)
5. Straight crossing path (Figure 5.9)

Any of these scenarios could happen in either traffic light or stop sign intersections.

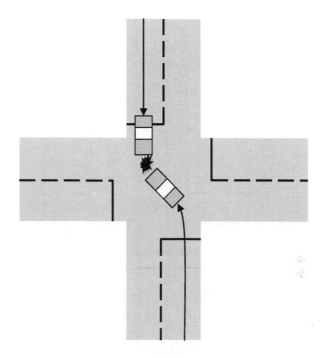

FIGURE 5.5 Left turn across path—opposite direction.

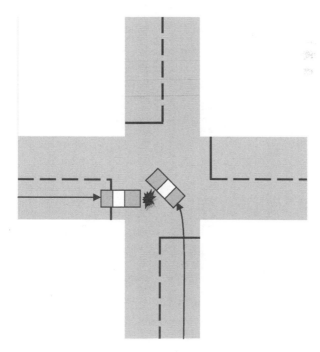

FIGURE 5.6 Left turn across path—lateral direction.

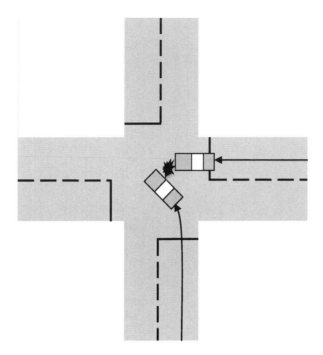

FIGURE 5.7   Left turn into path.

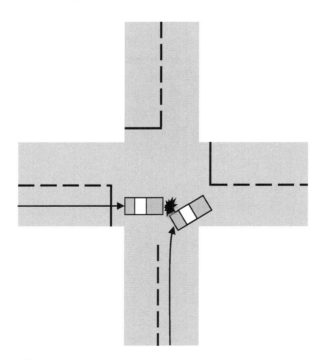

FIGURE 5.8   Right turn into path.

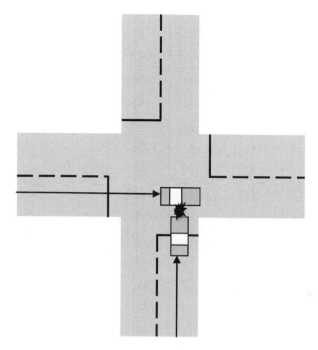

FIGURE 5.9 Straight crossing path.

CICAS works using two systems, one is an in-vehicle system and the other is a roadside system. Both systems exchange messages in order achieve the purpose of CICAS. The in-vehicle system will send messages about the speed, acceleration, position, and heading of the vehicle. The roadside system will receive the messages and process the data. In the case of traffic light intersections, the system will find out if the vehicle will be able to pass through the intersection safely or not. If it is going to be unsafe, other vehicles in the intersection will be warned. The roadside system determines the unsafe situations by calculating all of the information received from the vehicle within a certain distance of the intersection. Also, the system will send a warning message to the vehicle that is about to violate the traffic control. In the case of stop sign intersections, the roadside system will act in the same way as in traffic light intersections.

5.6 Summary

This chapter summarizes the current state-of-the-art safety applications in VANETs. It shows some official statistics about deaths and injuries that are related to car accidents as well as the motivation of developing safety applications. Communications in safety applications were explained regarding the protocol and messages that have been used in such applications. The MD was clarified as a method of sending data elements between vehicles. It eliminates the redundant data elements that are caused by running different safety applications that use similar data elements.

Safety applications are classified into five different categories: intersection collision avoidance, public safety, sign extension, vehicle diagnostics and maintenance, and information from other vehicles. Each category has several applications, which were explained. Two of the most important safety applications, EEBL and CICAS, were explained in more detail.

References

1. U.S. Department of Transportation, Traffic Safety Facts 2005, A Compilation of Motor Vehicle Crash Data from the Fatality Analysis Reporting System and the General Estimates System, DOT HS 810 631, National Highway Traffic Safety Administration, Washington, DC, 2005.
2. ASTM E2213-03, Standard Specification for Telecommunications and Information Exchange Between Road-side and Vehicle Systems—5 GHz Band Dedicated Short-Range Communications (DSRC) Medium Access Control (MAC) and Physical Layer (PHY) Specifications, ASTM International, July 2003.
3. National Highway Traffic Safety Administration, CAMP Vehicle Safety Communications, Vehicle Safety Communications Project, Task 3 Final Report, Identify Intelligent Vehicle Safety Application Enabled by DSRC, DOT HS 809 859, National Highway Traffic Safety Administration, Washington, DC, March 2005.
4. Tonguz, O.K. et al., On the Broadcast Storm Problem in Ad hoc Wireless Networks, in Proceedings of the IEEE International Conference on Broadband Communications, Networks and Systems (BROADNETS), 2006, pp. 1–11.
5. Torrent-Moreno, M., Santi, P. and Hartenstein, H., Fair Sharing of Bandwidth in VANETs, in Proceedings of the ACM International Workshop on Vehicular Ad Hoc Networks (VANET), Cologne, Germany, 2005, pp. 49–58.
6. Robinson, C.L., Caminiti, L., Caveney, D., and Laberteaux, K., Efficient Coordination and Transmission of Data for Cooperative Vehicular Safety Application, in Proceedings of the ACM International Workshop on Vehicular Ad Hoc Networks (VANET), Los Angeles, California, 2006, pp. 10–19.
7. SAE International, Dedicated Short Range Message Set (DSRC) Dictionary, Technical Report Standard J2735, 2006.
8. Resendes, R., Vehicle Infrastructure Integration Program Status, National Highway Traffic Safety Administration, 2005.
9. U.S. Department of Transportation, Cooperative Intersection Collision Avoidance Systems, available at http://www.its.dot.gov/cicas/index.htm, 2008.
10. O'Connor, T., Intersection Avoidance, available at http://www.calccit.org/its decision/serv_and_tech/Collision_avoidance/intersection.html, California Center for Innovative Transportation (CCIT), April 2004.
11. Davis, R., Cities Rise to EMS Challenge, *USA Today*, December 23, 2003.
12. National Highway Traffic Safety Administration, Recall Information, available at http://www.recalls.gov/nhtsa.html, 2008.
13. Holfelder, W., Vehicle Safety Communications in the US, presented at 2007 ITS World Congress, London, UK, October 2006.
14. Laberteaux, K., CAMP and the Vehicle Safety Communication 2 Consortium, presented at the C2C CC Security Workshop, Berlin, Germany, 2006.

6

Emerging Vehicular Applications

Uichin Lee,
Ryan Cheung, and
Mario Gerla
*Department of
Computer Science
University of California*

6.1 Introduction

Safe navigation support through wireless car-to-car and car-to-curb communications has become an important priority for car manufacturers as well as transportation authorities and communications standards organizations. New standards are emerging for car-to-car communications (DSRC and more recently IEEE 802.11p).[1] There have been several well-publicized testbeds aimed at demonstrating the feasibility and effectiveness of car-to-car communication safety; for instance, the ability to rapidly propagate accident reports to oncoming cars, the awareness of unsafe drivers in the proximity, and the prevention of intersection crashes.

Although safe navigation has always been the prime motivation behind vehicle-to-vehicle (V2V) and vehicle-to-infrastructure (V2I) communications, vehicular networks provide a promising platform for a much broader range of large-scale, highly mobile applications. Given the automobile's role as a critical component in peoples' lives, embedding software-based intelligence into cars has the potential to drastically improve the user's quality of life. This, along with significant market demand for more reliability, safety, and entertainment value in automobiles, has resulted in significant commercial development and support of vehicular networks and applications.

These emerging applications span many fields, from office-on-wheels to entertainment, mobile Internet games, mobile shopping, crime investigation, civic defense, and so on. Some of these applications are conventional "mobile Internet access" applications, such as downloading files, reading email while on the move. Others involve the discovery of local services in the neighborhood (e.g., restaurants, movie theaters), using the vehicle grid as an ad hoc network. Yet others demand close interaction among vehicles, such as interactive car games.

To support more advanced services, new brands of functions must be deployed such as the creation/maintenance of distributed indices, "temporary" storage of sharable content, "epidemic" distribution of content, and indices. Examples include the collection of "sensor data" in mobile vehicular sensor platforms, the sharing and streaming of files in a BitTorrent fashion, and the creation/maintenance of massively distributed databases with locally relevant commercial, entertainment, and cultural information (e.g., movies, hotels, and museums). Typically, these applications are distributed and follow a P2P collaboration pattern.

In this chapter, we review such emerging applications. We address potential vehicular networking architectures by reviewing various wireless access methods such as DSRC, 3G, and WiMAX. We discuss the unique characteristics of vehicular communications and briefly analyze various VANET routing protocols that are essential to supporting applications. VANET applications are classified by the vehicle's role in managing data: as a data source, data consumer, source and consumer, or intermediary. Based on this, we review various emerging applications proposed in the research community.

6.2 Background

We overview various wireless communication methods available in vehicular networks, namely DSRC/WAVE, Cellular, WiFi, WiMAX, and so on. We then outline key differences that distinguish the vehicular platform from the traditional mobile wireless ad hoc networks (MANETs). Finally, we review various routing protocols to gain better insight into the application protocol design.

6.2.1 Wireless Access Methods in Vehicular Networks

6.2.1.1 DSRC/WAVE

Dedicated short-range communication (DSRC) is a short- to medium-range communication technology operating in the 5.9 GHz range.[1] The Standards Committee E17.51 endorses a variation of the IEEE 802.11a MAC for the DSRC link. DSRC supports vehicle speeds up to 120 mi/h, nominal transmission range of 300 m (up to 1000 m), and default data rate of 6 Mb/sec (up to 27 Mb/sec). This will enable operations related to the improvement of traffic flow, highway safety, and other intelligent transport system (ITS) applications in a variety of application environments called DSRC/WAVE (wireless access in a vehicular environment). DSRC has two modes of operations: (1) ad hoc mode characterized by distributed multihop networking (vehicle-vehicle), (2) infrastructure mode characterized by a centralized mobile single-hop network (vehicle-gateway). Note that, depending on the deployment scenarios, gateways can be connected to one another or to the Internet, and they can be equipped with computing and storage devices, for example, Infostations.[2,3] Readers can find a detailed overview of the DSRC standards in Ref. 4.

6.2.1.2 Cellular Networks

Cellular systems have been evolving rapidly to support the ever increasing demands of mobile networking. 2G systems such as IS-95 and GSM support data communications at the maximum rate of 9.6 kb/sec. To provide higher rate data communications, GSM-based systems use GPRS (<171 kb/sec) and EDGE (<384 kb/sec), and IS-95-based CDMA systems use 1xRTT (<141 kb/sec). Now 3G systems support much higher data rates.[*] UMTS/HSDPA provides maximum rates of 144 kb/sec, 384 kb/sec, and 2 Mb/sec under high-mobility, low-mobility, and stationary environments, respectively. CDMA2000 1xEvDO (Rev. A) provides 3 Mb/sec and 1.8 Mb/sec for down and up links, respectively. The average data rate perceived by users is much lower in practice: <128 kb/sec for GSM/EDGE and <512 kb/sec for 3G technologies. In the U.S. Verizon and Sprint provide 1xEvDO, and AT&T and T-Mobile provide GSM/EDGE.

The behavior of 3G services (i.e., 1xEvDO) in a vehicular environment has been evaluated by Qureshi et al.[5†] They reported that (1) the average RTT was consistently high (~600 ms) with high variance ($\rho = 350$ ms); (2) there were a small number of short-lived (<30 sec) disconnections during their experiments; (3) the download throughput varied, ranging from 100 to 420 kb/sec, and the peak upload throughput was less than 140 kb/sec; and (4) they found no correlation between the vehicle's speed and the achieved throughput, but geographic location is the dominant factor leading to variations.

6.2.1.3 WiMAX/802.16e

802.16e or WiMAX (Worldwide Interoperability for Microwave Access) aims at enabling the delivery of last mile wireless broadband access (<40 Mb/sec) as an alternative to cable and xDSL, thus providing wireless data over long distances. This will fill the gap between 3G and WLAN standards, providing the data rate (tens of Mb/sec), mobility (<60 km/h), and coverage (<10 km) required to deliver the Internet access to mobile clients. In its part, WiBro, developed in Korea based on 802.16e draft version 3, provides 1 km range communications at the maximum rate per user of 6 and 1 Mb/sec for down and up links.[‡] It also supports several service levels including guaranteed Quality of Service (QoS) for delay sensitive applications, and an intermediate QoS level for delay tolerant application that requires a minimum guaranteed data rate. Han et al.[6] measured the performance of

---

[*] http://en.wikipedia.org/wiki/3G.
[†] Readers can find the evaluation of UMTS/HSDPA systems in a static environment.[7]
[‡] The peak sector (or cell) throughput is 18 Mbps and 6 Mbps for downlink and uplink, respectively.

WiBro networks in a subway whose maximum speed is 90 km/h, and showed that (1) the average uplink and downlink speeds were 2 and 5.3 Mb/sec, respectively, and (2) the average packet delay (half RTT) was less than 100 msec, and almost all packets experienced delay below 200 msec, except the case when handoffs happened (>400 msec).

6.2.1.4 WLAN

WiFi or WLAN can also support broadband wireless services. 802.11 a/g provides 54 Mb/sec and has a nominal transmission range of 38 m (indoor) and 140 m (outdoor). Despite its short radio range, its ubiquitous deployment makes WLAN an attractive method to support broadband wireless services. It has long been used as a means of Internet access in vehicles, known as Wardriving.* Also, open WiFi mesh networking has received a lot of attention; for example, Meraki sells $50 WiFi access points and provides Internet access for free by forming a mesh network over those access points.† Readers can find a thorough evaluation of WiFi performance in a vehicular environment in Ref. 8.

6.2.1.5 Possible Vehicular Networking Scenarios

Given the above wireless access methods, we now summarize possible vehicular networking scenarios. If vehicles are only equipped with DSRC, we can have an infrastructure-free mode (V2V only), infrastructure mode (V2I), and mixed mode (V2V and V2I), as shown in Figure 6.1a. Note that this can also be done with commercial WiFi devices. The mixed mode has been extensively studied in the research communities in terms of routing and network capacity, and readers can find the details in Ref. 9. If vehicles are only equipped with other broadband wireless access (i.e., cellular and WiMAX), we can have a scenario where vehicles can talk to each other via the Internet as in Figure 6.1b. For instance, people with iPhone or other Smart Phones with internet access can form a P2P overlay network via the internet. Finally, when vehicles have both DSRC and other broadband wireless access methods, we can have a mixed access scenario as in Figure 6.1c. Researchers mostly focused on the first scenario, yet the second scenario has recently received a lot of attention due to the widespread usage of Smart Phones, or WiBro.[10] Thus far, the third scenario has not yet received attention, but it has the potential to enable novel applications in the future.

6.2.2 Characteristics of VANET Environments

In designing protocols for the next generation vehicular network, we recognize that nodes in these networks have significantly different characteristics and demands from those in traditional wireless ad hoc networks deployed in infrastructureless environments (e.g., sensor field and battlefield). These differences have a significant impact on application infrastructures:

1. Vehicles have much higher power reserves than a typical mobile computer. Power can be drawn from on-board batteries and recharged as needed from a gasoline or alternative fuel engine.

---

* http://en.wikipedia.org/wiki/Wardriving.
† http://meraki.com.

(a)

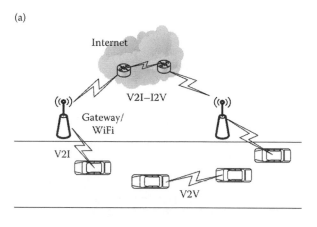

(b)

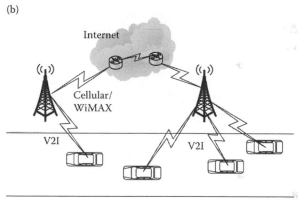

(c)

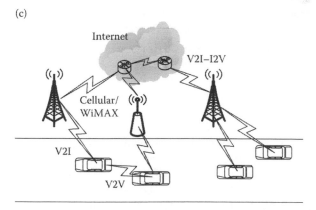

FIGURE 6.1 Possible wireless vehicular networking scenarios: (a) DSRC/WiFi; (b) Cellular/WiMax; (c) DSRC/WiFi and Cellular/WiMax.

2. Vehicles are orders of magnitude larger in size and weight compared to traditional wireless clients and can therefore support significantly heavier computing (and sensorial) components. This, combined with plentiful power, means vehicular computers can be larger, more powerful, and equipped with extremely large storage (up to terabytes of data), as well as powerful wireless transceivers capable of delivering wire-line data rates.

3. Vehicles travel at speeds up to 100 mi/h, making sustained, consistent V2V communication difficult to maintain. However, existing statistics of vehicular motion, such as tendencies to travel together or traffic patterns during commute hours, can help maintain connectivity across mobile vehicular groups.

4. Vehicles in a grid are always a few hops away from the infrastructure (WiFi, cellular, satellite, and so on). Thus, network protocol and application design must account for easy access to the Internet during "normal" operations.

6.2.3 VANET Routing Protocols

Several VANET applications critically rely on VANET routing protocols (unicast, broadcast, geocasting, and so on). These protocols originate from prior ad hoc network architectures but have been extensively redesigned by targeting the unique characteristics and needs of VANET scenarios and applications. We review the VANET routing protocols first as this offers an initial insight into VANET application characteristics.

6.2.3.1 Broadcasting

Safety-related applications (e.g., forward/backward collision warnings and lane change assistance) call for the delivery of messages to all nodes located close to the sender (reliable single-/multihop broadcasting) with high delivery rate and short delay. Recent research addressed this issue by proposing reliable broadcasting strategies.[11,12] Xu et al.[12] studied the impact of rapid repetition of broadcast messages on the packet reception failure in random access protocols. Torrent-Moreno et al.[11] showed channel access time and reception probability under deterministic and statistical channel models. Yin et al.[13] detailed the DSRC PHY layer model and incorporated the model into a VANET simulator to support generic safety application models. ElBatt et al.[14] modeled Cooperative Collision Warning (CCW) applications that broadcast a fixed-size packet at a certain rate. They measured the quality of reception using packet inter-reception time (IRT), which captures the effect of successive packet collisions on the perceived latency. Urban Multihop Broadcast (UMB)[15] supports directional broadcast in VANETs. UMB tries to improve reliability of broadcast by alleviating a hidden terminal problem through an RTS/CTS-style handshake, and broadcast storms through black-burst signals to select a forwarding node that is farthest from the sender using location information. Unlike UMB, Broadcast Medium Window (BMW)[16] and Batch Mode Multicast MAC (BMMM)[17] require all the receiving nodes to send back an ACK to the sender in order to achieve reliability. BMMM has also adapted to directional MAC in VANETs.[18]

6.2.3.2 Unicast Routing

There are many MANET routing protocols: proactive routing (e.g., DSDV and OLSR), reactive routing (e.g., AODV and DSR), geographic routing (e.g., GPSR), and hybrid geographic routing (e.g., Terminode), and yet they cannot directly be used due to high mobility and nonuniform distribution of vehicles, which causes intermittent connectivity. In VANETs, geographic or hybrid geographic routing protocols are often preferred. Also, the carry-and-forward strategy is used to overcome intermittent connectivity; when disruption happens, a node stores a packet in its buffer and waits until connectivity is available. Chen et al.[19] considered a "straight highway" scenario and evaluated two ideal strategies: pessimistic (i.e., synchronous), where sources send packets to destinations only as soon as a multihop path is available, and optimistic (i.e., carry-and-forward), where intermediate nodes hold packets until a neighbor closer to the destination is detected. In such a highway scenario, they showed that the latter scheme has demonstrated the achievement of a lower delivery delay. However, in more realistic situations (i.e., Manhattan-style urban mobility and buffer constraints), carry-and-forward protocols call for careful design and tuning. MaxProp,[20] part of the UMass DieselNet project,* has a ranking strategy to determine packet delivery order where precedence is given in the following order: (1) packets destined to the neighboring nodes, (2) packets containing routing information, (3) acknowledgement packets of delivered data, (4) packets with small hop-counts, and (5) packets with a high probability of being delivered through the other party. VADD[21] rests on the assumption that most node encounters happen in intersection areas. Effective decision strategies are proposed to reduce packet delivery failures and delay. Naumove et al.[22] proposed a hybrid geographic routing protocol, called connectivity-aware routing (CAR). Route discovery finds a set of anchor points (i.e., junctions) to the destination via flooding. Geographic greedy forwarding is used to deliver packets over the anchored path.

6.2.3.3 Geocast

Applications for distributed data collection in VANET call for geographic dissemination strategies that deliver packets to all nodes belonging to target remote areas (or geocasting), despite possibly interrupted paths.[23-25] MDDV[23] exploits geographic forwarding to the destination region, favoring paths where vehicle density is higher. In MDDV, messages are carried by head vehicles, that is, those that are best positioned toward the destination with respect to their neighbors. As an alternative, Sormani et al.[24] proposed several strategies based on virtual potential fields generated by propagation functions: a node estimates its position in the field and retransmits packets until nodes placed in locations with lower potential values are found; this procedure is repeated until minima target zones are detected. Maihöfer et al.[25] proposed abiding geocast, a time-stable geocast where messages are delivered to all nodes that are inside a destination region within a certain period of time and discussed design space, semantics, and strategies for abiding geocast.

6.3 Vehicular Application Classification

The major departure of vehicle networks from conventional ad hoc networks is the opportunity to deploy, in addition to traditional applications, a broad range of innovative content-sharing applications [typically referred to as peer-to-peer (P2P) applications].

---

* UMass' DieselNet. http://prisms.cs.umass.edu/dome.

Whereas the popularity of P2P applications has been well documented, these applications have been thus far confined to the wired Internet (e.g., BitTorrent). The much increased storage and processing capacity of VANETs with respect to personal or sensor-based ad hoc networks make such applications feasible. Moreover, the fact that car passengers are a captive audience provides incentive for content distribution and sharing applications at a scale that would be unsuitable for other ad hoc network contexts. We describe a representative set of VANET P2P applications and classify them by the vehicle's role in managing data: as a data source, data consumer, data source and consumer, or intermediary.

First, the vehicles provide an ideal platform for mobile data gathering, especially in the context of monitoring urban environments (i.e., vehicular sensor networks).[26-30] Each vehicle can sense events (e.g., images from streets or the presence of toxic chemicals), process sensed data (e.g., recognizing license plates), and route messages to other vehicles (e.g., forwarding notifications to other drivers or police officers). Because vehicular sensors have few constraints on processing power and storage capabilities, they can generate and handle data at a rate impossible for traditional sensor networks. These applications require persistent and reliable storage of data for later retrieval. In addition, they require networking protocols (including sophisticated query processing) to efficiently locate/retrieve data of interest (e.g., finding all the vehicles at a certain time and location).

Second, the vehicles can be significant consumers of content. Their local resources are capable of supporting high-fidelity data retrieval and playback. For the duration of each trip, drivers and passengers make up a captive audience for large quantities of data. Examples include locality-aware information (map-based directions) and content for entertainment (streaming movies, music, and ads).[31-34] These applications require high-throughput network connectivity and fast access to desired data.

In a third class of compelling applications, vehicles are both producers and consumers of content. Examples include services that report on road conditions and accidents, traffic congestion monitoring, and emergency neighbor alerts; for example, my brakes are malfunctioning.[35-39] Also, interactive applications (e.g., voice over V2V and online gaming) belong to this category. These applications require location-aware data gathering/dissemination and retrieval. In particular, interactive applications require real-time communication among vehicles.

Finally, all of the above applications will need to rely on vehicles in an intermediary role. Individual vehicles in a mobile group setting can cooperate to improve the quality of the applicant experience for the entire network. Specifically, vehicles will provide temporary storage (caching) for others, as well as forwarding of both data and queries for data. In this capacity, they require reliable storage as well as efficient location of and routing to data sources and consumers.

The demands of these applications give us a list of requirements and challenges for vehicular applications (note that we can leverage them to simplify the applications infrastructure):

1. *Time sensitivity.* Time-sensitive data must be retrieved or disseminated to the desired location within a given time window. Failure to do so renders the data useless. This mirrors the needs for multimedia streaming across traditional networks, and we can leverage relevant research results from related areas.

2. *Location awareness.* Both data gathered from vehicles and data consumed by vehicles are highly location-dependent. This property has direct implications for the design of data management and security components. Data caching and indexing should focus on location as a first-order property, while data dissemination must be location-aware in order to maintain privacy and prevent tampering.

Most applications require methods of storing/retrieving such location-/time-sensitive information. As in MANETs, we can use structured approaches such as geographic hashing[40] and DHT,[41] or structureless approaches such as epidemic dissemination.[42] However, it is nontrivial to maintain structure in VANETs due to the high-mobility, nonuniform distribution of vehicles, and intermittent connectivity. Thus, most application protocols rely on variants of epidemic data dissemination such that the produced information is disseminated to nodes in an area where the information is produced.[34-37,43]

6.4 Data Source (Vehicular Sensor) Applications

Vehicular networks are emerging as a new network paradigm of primary relevance, for example, for proactive urban monitoring using sensors and for sharing and disseminating data of common interest. In particular, we are interested in urban sensing for effective monitoring of environmental conditions and social activities in urban areas using vehicular sensor networks (VSNs). The major departure from traditional wireless sensor nodes is that vehicles are not strictly affected by the energy constraints and the size of sensor units. Thus, vehicles can easily be equipped with powerful processing units, wireless communication devices, GPS, and sensing devices such as chemical detectors, still/video cameras, and vibration/acoustic sensors. Figure 6.2 shows an application scenario.

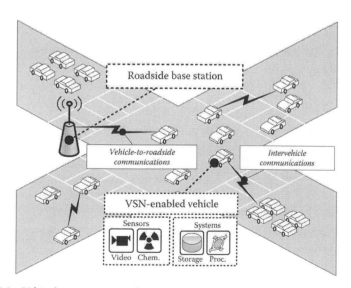

FIGURE 6.2 Vehicular sensor network.

6.4.1 MobEyes: Proactive Urban Monitoring Services

MobEyes aims at provisioning proactive urban monitoring services where vehicles continuously monitor events from urban streets, maintain sensed data in their local storage, process them (e.g., recognizing license plate numbers), and route messages to vehicles in their vicinity to achieve a common goal (e.g., to allow police agents to find the trajectories of specific cars). However, this requires the collection, storage, and retrieval of massive amounts of sensed data. In conventional sensor networks, sensed data is dispatched to "sinks" and is processed for further use (e.g., directed diffusion),[44] but that is not practical in VSNs due to the sheer size of generated data. Moreover, it is impossible to filter data *a priori* because it is usually unknown which data will be of use for future investigations. Thus, the challenge is to find a completely decentralized VSN solution, with low interference to other services, good scalability, and tolerance to disruption caused by mobility and attacks.

MobEyes is a novel middleware that supports VSN-based proactive urban monitoring applications.[26–28] Each sensor node performs event sensing, processing/classification of sensed data, and periodically generates metadata of extracted features and context information such as timestamps and positioning coordinates. Metadata are then disseminated to other regular vehicles, so that mobile agents, for example, police patrolling cars, move and opportunistically harvest metadata from neighboring vehicles. As a result, agents can create a low-cost opportunistic index that enables agents to query the completely distributed sensed data storage. This enables us to answer questions such as: (1) which vehicles were in a given place at a given time?; (2) which route did a certain vehicle take in a given time interval?; and (3) which vehicles collected and stored the data of interest?

6.4.1.1 Metadata Diffusion

Any regular node periodically advertises a packet with a set of newly generated metadata to its current neighbors. Each packet is uniquely identified (generator ID and locally unique sequence number). This advertisement to neighbors provides more opportunities for the agents to harvest the metadata packets. Note that the duration of periodic advertisement is configured to fulfil the desired latency requirements, because harvesting latency depends on it. Neighbors receiving a packet store it in their local metadata databases. Therefore, depending on node mobility and encounters, packets are opportunistically diffused into the network.

MobEyes is usually configured to perform "passive" diffusion; only the packet source advertises its packets. Two different types of passive diffusion are implemented in MobEyes: single-hop passive diffusion (packet advertisements only to single-hop neighbors) and k-hop passive diffusion (advertisements travel up to k-hop as they are forwarded by j-hop neighbors with $j < k$). MobEyes can also adopt other diffusion strategies, for instance single-hop "active" diffusion, where any node periodically advertises all packets (generated and received) in its local database at the expense of a higher traffic overhead. In a usual urban VANET, it is sufficient for MobEyes to exploit the lightweight k-hop passive diffusion strategy, with very small k values, to achieve needed diffusion.

Figure 6.3 depicts the case of a VSN node C1 encountering other VSN nodes while moving (for the sake of readability, only C2 is explicitly represented). Encounters occur

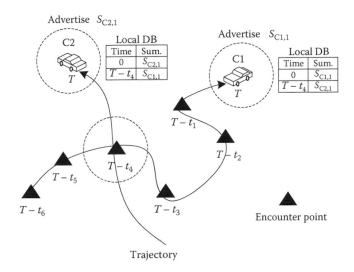

FIGURE 6.3 MobEyes single-hop passive diffusion. (From Lee, U. et al., *IEEE Wireless Communications*, 13(5); 51–57, 2006. With permission.)

when two nodes exchange metadata, that is, when they are within their radio ranges and have a new metadata packet to advertise. In the figure dotted circles and timestamped triangles represent radio ranges and C1 encounters, respectively. In particular, the figure shows that C1 (while advertising $S_{C1,1}$) encounters C2 (advertising $S_{C2,1}$) at time $T-t_4$. As a result, after $T-t_4$ C1 includes $S_{C2,1}$ in its storage, and C2 includes $S_{C1,1}$.

6.4.1.2 Metadata Harvesting

In parallel with diffusion, metadata harvesting can take place. A MobEyes police agent can request the collection of diffused metadata packets by proactively querying its neighbor regular nodes. The ultimate goal is to collect all the metadata packets generated in a given area. Obviously, a police agent is interested in harvesting metadata packets it has not collected so far. To focus only on missing packets, a MobEyes agent compares its already collected packets with the packet list at each neighbor (a set difference problem), by exploiting a space-efficient data structure for membership checking, that is, a Bloom filter. A Bloom filter for representing a set of n elements consists of m bits, initially set to 0. The filter applies k independent random hash functions $h_1, \ldots, h_k$ to MobEyes packet identifiers and records the presence of each element into the m bits by setting k corresponding bits. To check the membership of the element x, it is sufficient to verify whether all $h_i(x)$ are set.

Therefore, the MobEyes harvesting procedure consists of the following steps:

1. The police agent broadcasts a "harvest" request with its Bloom filter.
2. Each neighbor prepares a list of "missing" packets from the received Bloom filter.
3. One of the neighbors returns missing packets to the agent.

4. The agent sends back an acknowledgment with a piggybacked list of just-received packets. Upon listening or overhearing this, neighbors update their missing packet lists for the agent.

5. Steps 3 and 4 are repeated until there are no missing packets.

Note that Bloom filter membership checking is probabilistic. In particular, false positives may occur and induce MobEyes regular nodes not to send packets still missing to the agent. The probability of a false positive depends on m and n.[45] Nevertheless, in MobEyes, the agent can obtain a missing packet with high probability, because it is highly probable that other nodes have the packets as time passes, and the harvesting procedure is repeated as the agent moves. For example, in usual VSN deployment scenarios (e.g., with 10 neighbors on average), the probability of missing one packet due to false positives after repeating the procedure multiple times is very low.

6.4.2 Related Mobile Sensor Platform Projects

In CarTel,[29] users submit their queries about sensed data on a portal hosted on the wired Internet. Then, an intermittently connected database is in charge of dispatching queries to vehicles and of receiving replies when vehicles move in the proximity of open access points to the Internet. Eriksson et al.[30] proposed a system called Pothole Patrol (P^2) that uses the mobility of vehicles, opportunistically gathering data from vibration, and GPS sensors to access road surface conditions. Yoon et al.[46] proposed a method of identifying traffic conditions on surface streets using the GPS location traces collected from vehicles.

In general, a vehicular sensor network can be considered as a form of "opportunistic mobile sensing platform". The opportunistic mobile sensing area has been extremely productive recently, providing a wealth of related work. ZebraNet[47] addresses remote wildlife tracking, for example, zebras in Mpala Research Center in Kenya, by equipping animals with collars that embed wireless communication devices, GPS, and biometric sensors. As GPS-equipped animals drift within the park, their collars opportunistically exchange sensed data, which must make its way to the base station (the ranger's truck). SWIM[3] addresses sparse mobile sensor networks with fixed Infostations as collecting points. Sensed data is epidemically disseminated via single-hop flooding to encountered nodes and offloaded whenever they encounter an Infostation. Eisenman et al.[48] proposed a three-tier architecture called MetroSense, in which servers in the wired Internet are in charge of storing/processing sensed data, Internet-connected stationary sensor access points (SAPs) act as gateways between servers and mobile sensors (MSs), and MSs move in the field opportunistically delegating tasks to each other, and "muling"[49] data to SAPs. MetroSense requires infrastructure support, including Internet-connected servers and remotely deployed SAPs. Wang et al.[50] proposed data delivery schemes in a delay/fault-tolerant mobile sensor network (DFT-MSN) for human-oriented pervasive information gathering. The trade-off between data delivery ratio/delay and replication overhead is mainly investigated in terms of buffer and energy resource constraints. CENS' Urban Sensing project[51] addresses "participatory" sensing, where people of the same interest participate in an urban monitoring campaign. Intel IrisNet[52] and Microsoft SenseWeb[53] investigate the integration of heterogeneous sensing platforms in the Internet via a common data publishing architecture. Urbanet[54] proposes application programming models for opportunistic sensing.

6.5 Data Consumer Applications

6.5.1 Content Distribution

Content distribution to vehicles ranges from multimedia files to road condition data and to updates/patches of software installed in the vehicle. Nandan et al.[31] proposed SPAWN, a BitTorrent-like file swarming protocol in a VANET. In SPAWN, a file is divided into pieces and is uploaded into an Internet server. Each file has a unique ID (e.g., hash value of the file content), and each piece has a unique sequence number. Users passing by the access points (APs) download parts of the file. Once out of the range of APs, they cooperatively exchange missing pieces as in Figure 6.4.

SPAWN is composed of the following components: peer/content discovery, and peer/content selection. Owing to the intermittent presence of APs, SPAWN cannot use a centralized server as in BitTorrent that keeps track of all the peers. Instead, SPAWN uses a decentralized "gossiping" mechanism for peer/content discovery that leverages the broadcast medium of the wireless networks. A gossip message of a node contains a file ID, a list of pieces that the node has, a hop-count, and so on. For efficient gossiping, SPAWN uses gossiping methods, namely probabilistic spawn and rate-limited spawn. In the probabilistic spawn, nodes forward gossip messages with a certain probability, whereas in the rate-limited spawn, nodes forward gossip messages in their buffer with a certain rate, for example, forwarding a random gossip message in the buffer every 2 sec. The hop-count of a gossip message is incremented, whenever a gossip message is forwarded. For a given file, there are three types of users in the network: those who are

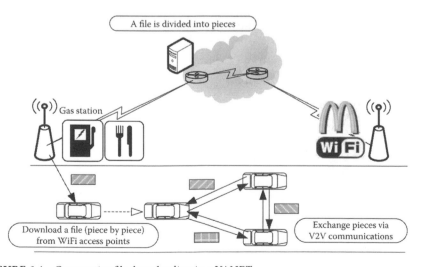

FIGURE 6.4 Cooperative file downloading in a VANET.

interested in downloading the files, those who are uninterested in downloading the files, and those who do not understand the SPAWN protocol. These roles are considered in the gossiping. For instance, interested users may have a higher probability of packet forwarding than uninterested users.

After the peer/content discovery, a node has to select a peer to download a piece. Given that TCP connections spanning fewer hops perform better in multihop wireless networks, SPAWN uses proximity-driven piece/peer selection strategies where the proximity is estimated by the hop-count in the gossip messages: (1) *rarest-closest first* chooses the rarest piece among all the peers in one's peer list, and breaks the tie based on proximity, and (2) *closest-rarest first* selects the rarest piece among all the closest peers. Recall that BitTorrent uses a rarest piece first selection strategy where the rarest piece among all the peers in its list is selected. After peer selection, the node finally downloads pieces by setting up a TCP connection. Any routing protocol such as AODV and DSR can be used for this purpose.

By simplifying SPAWN, Lee et al.[55] proposed CarTorrent. Given that proximity is the key factor of peer selection, CarTorrent uses k-hop limited probabilistic gossiping and closest-rarest first is used for piece/peer selection. CarTorrent uses a crosslayer approach in that route discovery of underlying on-demand protocols is utilized for gossiping. Lee et al.[32] proposed CodeTorrent, a network-coding-based content distribution protocol. Recall that BitTorrent-like protocols suffer from a coupon collection problem; that is, as a node collects more pieces, it will take a progressively longer time to collect a new piece. It is known that network coding can mitigate this problem.[56,57]

Eriksson et al.[58] proposed techniques to improve data delivery throughput. QuickWiFi, a streamlined WiFi client, reduces the end-to-end link establishment delay to a WiFi access point, and Cabernet Transport Protocol (CTP) improves the data throughput by differentiating congestion in wired links and packet loss in wireless links. Recently, Yoon et al.[59] proposed Mobile Opportunistic Video-on-demand (MOVi), a mobile peer-to-peer (P2P) video-on-demand application. As switching WiFi modes (between infrastructure and ad hoc modes) takes time, MOVi exploits the opportunistic mixed usage of roadside WiFi access points and direct P2P communications using Direct Link Service (DLS) in 802.11 standards that enables direction communications between nodes within a single BSS.

6.5.2 Location-Aware Advertisements

Advertisements are one of the most important sources of revenue for Internet-based companies. Similarly, cars with wireless communications will become lucrative targets for advertisements. As an extension of the physical billboards, Nandan et al.[33] proposed AdTorrent, which delivers location-sensitive commercial advertisements to the vehicles using digital billboards (or Ad Stations). With AdTorrent, business owners in the vicinity of the billboards can subscribe to digital billboards. Advertisements include simple text-based ads or multimedia ads, for example, trailers of movies playing at the nearby theater, virtual tours of hotels in a 5 mi radius, or conventional television advertisements relevant to local businesses.

AdTorrent aims at allowing drivers to download the advertisements of interest. So, it has a location-aware distributed mechanism to *search, rank*, and *deliver* relevant advertisements. Each advertisement has metadata information (e.g., keywords and advertisement ID). For a keyword search, a client builds a personal inverted index that links keywords to advertisements. As it is expensive to disseminate the raw inverted index, AdTorrent uses a special hash table that is based on a Bloom filter, a space-efficient membership checking data structure. Keywords are stored in a Bloom filter, and the resulting hash table is disseminated to k-hop neighbors. After receiving a set of Bloom filters, a node aggregates them by performing the logical AND operation on Bloom filters. To resolve a query, a node first searches its local index (i.e., its aggregated Bloom filter). When failing to retrieve ℓ advertisements (where ℓ is set by the query originator), the node tries to search more results via m-hop scoped flooding. After collecting ℓ matched advertisements, the node downloads all the metadata information of those advertisements. The node ranks each advertisement based on its relevance, location, stability, and so on, and it starts downloading the best advertisement using CarTorrent, a content distribution protocol in VANETs.

Note that some of the products advertised to cars may be very dependent on car navigation and management. Caliskan et al.[34] proposed a parking spot information dissemination protocol where infrastructure (e.g., parking meters) periodically broadcasts parking spot information, and vehicles disseminate this information via periodic single-hop broadcasting. For efficient dissemination, they used spatiotemporal characteristics of parking spot information. Related to this is the advertising of "lanes" (on a time shared basis) to vehicles interested in bidding to the service.

6.6 Data Producer/Consumer Applications

6.6.1 Emergency Video Streaming

6.6.1.1 Vehicle-to-Vehicle Live Video (V3) Streaming Architecture

V3 supports location-aware video streaming so that users can watch videos originating from remote regions of interest.[38] V3 assumes that vehicles are equipped with on-board computing, wireless communication devices, and GPS devices to keep track of their locations; and some of the vehicles have video cameras and have enough storage to buffer videos.

V3 is composed of a video triggering subsystem and a video transmission subsystem. The video triggering subsystem is responsible for forwarding video trigger message to the destination region. The video transmission subsystem sends video data back to the receiver. In Figure 6.5, a vehicle R that wants to receive streaming video from the region A, sends a triggering message to that region. A trigger message contains the requester ID, query time, destination region information, deadline, and so on. The trigger message signals vehicles in a region of interest to start capturing videos and to send back the captured video to the requestor. In Figure 6.5 two vehicles observing the accidents stream the captured videos to the receiver using multihop routing.

One of the key challenges of implementing V3 is to overcome intermittent connectivity due to the dynamic nature of a VANET (i.e., high-mobility, nonuniform vehicle

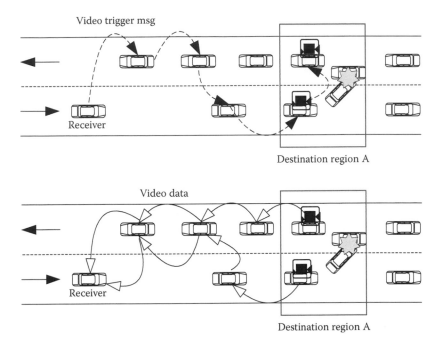

FIGURE 6.5   An illustration of a typical V3 service process.

distribution) and the low penetration ratio of V3-enabled vehicles. For triggering message delivery, V3 uses a *directional* flooding method based on a trigger message forwarding zone (TMFZone). For a given node v, TMFZone defines a set of potential forwarders that are closer to the destination region than node v or are moving towards the destination region. After a request is disseminated to nodes in the destination region, a video source vehicle must be selected. V3 selects a node that will possibly stay in the destination region the longest. Note that to handle the case that a source vehicle moves away from the region, V3 uses continuous trigger methods that forward the trigger message to the incoming vehicles to the region. For video transmission, V3 mainly uses a store-carry-and-forward approach to overcome intermittent connectivity. Like TMFZone, a node v selects a set of candidate forwarders that are closer to the requester or are moving towards the requester. As it may not be efficient to send a packet to all of the candidate forwarders, the node v selects a subset of candidate forwarders, based on the amount of knowledge it has.

6.6.1.2 Reliable Video Streaming Using Network Coding

Park et al.[39] proposed a *network coding* based emergency video streaming protocol. Unlike V3, it "pushes" urgent video streams regarding emergency situations such as natural disaster, traffic accidents, and terrorist attacks in order to help drivers effectively avoid the danger, and random linear network coding is used to provide reliable and robust video streaming.

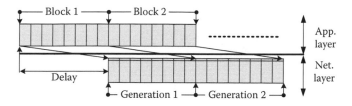

FIGURE 6.6 Multimedia streaming: each block has eight frames.

Suppose a multimedia data source generates a stream of frames $\mathbf{p}_1, \mathbf{p}_2, \mathbf{p}_3, \ldots$ where subscripts denote unique and consecutive sequence numbers (Figure 6.6). A tuple (*blockid, blocksize*) is used to indicate a block of frames with sequence numbers greater than or equal to *blockid* and smaller than (*blockid + blocksize*) (i.e., $\mathbf{p}_{blockid}, \ldots, \mathbf{p}_{blockid + blocksize - 1}$) belong to the block. A *coded packet* $\mathbf{c}_{(blockid, blocksize)}$ is a linear combination of frames in (*blockid, blocksize*). That is, $\mathbf{c}_{(blockid, blocksize)} = \sum_{k=1}^{blocksize} e_k \mathbf{p}_{(k-1+blockid)}$ where e_k is an element in a Galois field $\mathbb{F}$. Here, every arithmetic operation (i.e., addition and multiplication) is over the Galois field $\mathbb{F}$. Data frames $\mathbf{p}$s and coded packets $\mathbf{c}$s are also regarded as vectors over the field. In the header of a coded packet, $\mathbf{e} = [e_1 \ldots e_{blocksize}]$ is stored along with *blockid* and *blocksize* for the purpose of *decoding* packets on receivers. When generating a $\mathbf{c}$, each e_k is chosen randomly from $\mathbb{F}$, which is in general referred to as the random linear coding.

The reliable delivery service agent (or layer) residing on the multimedia data source generates and transmits code packets to the receivers. As a block of frames is required to generate a coded packet, the agent residing on the video source collects frames generated by the application and buffers them. On reception of a coded packet $\mathbf{c}_{(blockid, blocksize)}$, every node stores the packet in its local memory for later decoding and forwarding. To recover *blocksize* original frames belonging to (*blockid, blocksize*), a node should collect a *blocksize* number of coded packets tagged with (*blockid, blocksize*) and encoding vectors that are linearly independent of each other. Once collected, the reliable delivery service agent recovers the *blocksize* original data frames and deliver them to the upper layer. Let $\mathbf{c}_k$ be a coded packet labeled (*blockid, blocksize*) in a node's local memory, $\mathbf{e}_k$ be the encoding vector prefixed to $\mathbf{c}_k$, and $\mathbf{p}_{blockid+k-1}$ be an original data frame to be recovered where $k = 1, \ldots, blocksize$. Further, let $\mathbf{E}^{\mathrm{T}} = [\mathbf{e}_1^{\mathrm{T}} \ldots \mathbf{e}_{blocksize}^{\mathrm{T}}]$, $\mathbf{C}^{\mathrm{T}} = [\mathbf{c}_1^{\mathrm{T}} \ldots \mathbf{c}_{blocksize}^{\mathrm{T}}]$, and $\mathbf{P}^{\mathrm{T}} = [\mathbf{p}_{blockid}^{\mathrm{T}} \ldots \mathbf{p}_{blockid+blocksize-1}^{\mathrm{T}}]$, then conceptually $\mathbf{P} = \mathbf{E}^{-1}\mathbf{C}$, which corresponds to the original data frames where superscripts T denote the transpose operation. Note that all $\mathbf{e}_k$s must be linearly independent to be able to invert $\mathbf{E}$.

When a node receives a coded packet with a new tuple (*blockid, blocksize*), it sets up a timer for the tuple (*blockid, blocksize*) expiring in *blocktimeout* seconds. When the timer expires it broadcasts one coded packet $\mathbf{c}'$ (*blockid, blocksize*) after local re-encoding to its neighbors. The local re-encoding is carried out through the same process that the data source has undergone to generate a coded packet, that is, a random linear combination of packets with the same (*blockid, blocksize*) available in local memory as shown in Figure 6.7. The timer for (*blockid, blocksize*) is reset on expiration unless a decodable set of packets is collected for the tuple (*blockid, blocksize*). On the expiration of the timer for (*blockid,*

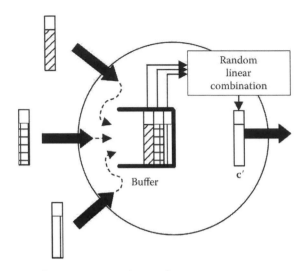

FIGURE 6.7   Re-encoding at an intermediate node.

blocksize), even though there are less than *blocksize* number of packets of (*blockid, block-size*) in the local memory, a node has to generate and transmit a coded packet using packets available in memory. The number of frames/packets that are combined to yield a coded packet is recorded in the field *rank* in the header of the coded packet. As a coded packet $c_{(blockid, blocksize)}$ with *rank* smaller than *blocksize* indicates that the sender of **c** is in need of more coded packets tagged with (*blockid, blocksize*) to recover original frames, on reception of such packets, a node transmits another coded packet to help the sender of **c** collecting more coded packets.

Owing to this recovery process in combination with buffering of packets, the protocol can deliver packets efficiently and reliably across partitions. Suppose that a vehicle encounters a platoon of other vehicles carrying data that the vehicle does not have. The vehicle runs the recovery process and it collects data from the platoon. In the recovery process, a vehicle sends out a coded data packet tagged with (*blockid, blocksize*), and *rank* where *rank* < *blocksize* and in response to the *help request* packet, and neighbors of the vehicles send appropriate coded data packets. If a vehicle in search of help has no data, then it just sends out header-only packets with *rank* = 0. In fact, the help request and responses handshaking is not necessarily to be done block by block (or generation by generation) if a vehicle wants to collect consecutive blocks of frames.

6.6.1.3 Tavarua: Video Streaming over 3G Services

Qureshi et al.[5] proposed Tavarua, a novel real-time multimedia communications subsystem designed to support mobile telemedicine applications that require high band-width (e.g., >500 kb/sec) and QoS. As shown in Section 6.2.1, they evaluated the behavior of 3G services (i.e., 1xEvDO) in a vehicular network and found that 1xEvDO has relatively high latency (~600 msec), and low upload bandwidth (e.g., <140 kb/sec). To support adequate bandwidth and QoS, Tavarua uses multiple simultaneous 3G connections. Tavarua has the following components: Tribe, which provides the lowest-level connection between Tavarua and the network interfaces; Horde, which provides the network-striping layer, including congestion control; and a video services subsystem. Prototype results show that Tavarua significantly mitigates the impact of packet loss on video quality and provides sufficient upstream bandwidth to transmit high-quality video data.

6.6.2 FleaNet: A Virtual Marketplace in VANETs

Wireless communications in vehicles will guide us into a new era of pervasive computing in which seamless access to information sources is provided. When traveling or shopping, for instance, one can search the web to get directions or locate specific products. In fact, not only do such devices empower us with ubiquitous Internet access, but they also create a new environment of vehicular social networking where opportunistic cooperation can emerge among users with shared interests/goals, such as drivers exchanging safety-related information, shoppers/sellers trading goods, and so on.[60]

Following this model, Lee et al.[37] considered a "virtual flea market" in urban vehicular networks called FleaNet. FleaNet operates on the vehicular "ad hoc grid" without any infrastructure support and provides an excellent method for people to communicate with each other as buyers and sellers of goods (or information) and to efficiently find matches of interest, potentially leading to transactions. Figure 6.8 shows an illustration of FleaNet scenarios. Vehicles as well as static roadside Advertisement Stations (or Adstations) generate and propagate queries. Adstations can be stores advertising their products. For example, a pizzeria could advertise its special pizza offer to vehicles passing by, and a driver who received the advertisement could place an order.

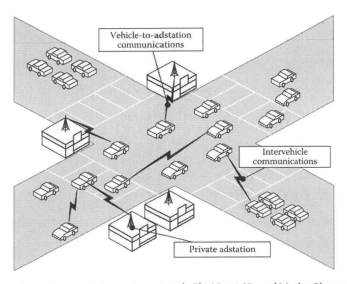

FIGURE 6.8 FleaNet scenario. (From Lee, U. et al., FleaNet: A Virtual Market Place on Vehicular Networks, in V2VCOM'06, San Jose, CA, July 2006. With permission.)

To illustrate FleaNet, consider the following examples. One day Joe Bruin wants to sell some of his items, but he is too busy with his work to do a garage sale. In this situation, FleaNet helps him to sell the items while he is behind the wheel (i.e., a mobile garage sale). He inputs details of the items using FleaNet software to create queries of items, for example, "Consumer Electronics/MP3 Players/Apple iPod Mini, 4G." As he is commuting between downtown LA and west LA, he wants to find buyers near that area. Using a digital map provided by FleaNet software, he can easily set the area of interest to which his queries will be disseminated. For some items, he wants to see multiple matches, say five, to make the best deal by simply setting the "number of matches" field. He also wants to sell the items while he is commuting, which takes about half an hour, and thus, he sets the expiration time accordingly. As a result, this query will be advertised and is spreading near his commuting path through vehicular networks using the FleaNet query dissemination protocol. Some time later, the query will be responded to with a match message (i.e., a sell query of a ticket). Joe Bruin will then send a transaction request message to sell his item, and in the end, he will receive a transaction response from the originator of the matched query.

FleaNet uses mobility-assisted query dissemination where the query "originator" periodically advertises his query only to one-hop neighbors. Each neighbor then stores the advertisement (i.e., query) in its local database without any further relaying; thus, the query spreads only because of vehicle motion. Upon receiving a query, a node tries to resolve it locally in its database; in the case of success, the originator will be automatically informed. A match only happens in its neighbors and, thus, there is no redundant match notification. As this match could lead to an actual transaction, FleaNet provides a mechanism that routes the transaction request/reply. A user could see multiple matches for a given query. Based on his own criterion (either on distance from his current location or on the offered price), he selects the best one and sends the transaction request. Then, the target user responds with the transaction reply after seeing the request. For this purpose, FleaNet uses Last Encounter Routing (LER).[61] LER is based on georouting and combines a location service and a routing service. In FleaNet the query packet includes the originator geocoordinates, and thus, LER does not incur any initial flood search routing cost.

Figure 6.9 shows an example of match notification. Let us assume that node *B* and node *S* are advertising sell and buy queries respectively, and *N1* is carrying the sell query of node *S* because node *N1* has already met *S*. In Figure 6.9a, node *B* advertises its buy query to its neighbors. Node *N1* then finds a local match and sends LOCALMATCH to node *B* as shown in Figure 6.9b. As a result, node *B* makes its final decision by sending TRANXREQ to node *S* and, thus, *S* will respond with TRANXREQ to node *B* as shown in Figure 6.9c and 6.9d.

Readers can find an extensive evaluation of the FleaNet protocols via both analysis and simulation.[37] A random query can be resolved, in most cases, within a tolerable amount of time and with minimal bandwidth, storage, and processing overhead; if the advertiser is stationary (e.g., Adstation), the query resolution time is critically dependent on its location.

The use of FleaNet is not limited to vehicular ad hoc networks. It can be extended to other networks such as personal area networks formed by pedestrians with PDAs or SmartPhones. Also, FleaNet can be associated with infrastructure. For instance, mobile

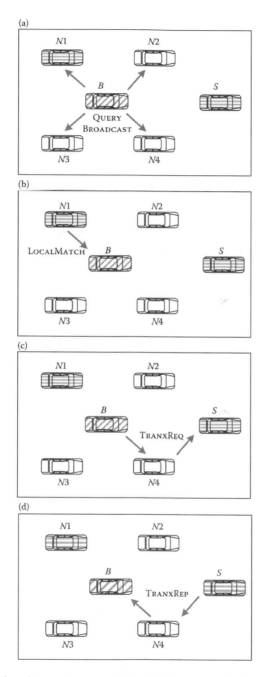

FIGURE 6.9 Match and transaction notification: (a) Buy query advertisement; (b) local match from N1; (c) transaction request; (d) transaction response. (From Lee, U. et al., FleaNet: A Virtual Market Place on Vehicular Networks, in V2VCOM'06, San Jose, CA, July 2007. With permission.)

users' advertisements can be uploaded in the Internet (e.g., Craigslist), and similarly, Internet users' advertisements can be posted in the vehicular networks. It is very important to provide incentives and security mechanisms to deal with noncooperative and/or malicious users. Lee et al.[62] proposed Signature-Seeking Drive (SSD), a secure incentive framework for commercial ad dissemination in VANETs where a PKI (public key infrastructure) is leveraged to provide secure incentives for cooperative nodes.

6.6.3 Vehicular Information Transfer Protocol (VITP)

VITP[35] aims to support various on-demand location-aware services such as traffic conditions (e.g., congestion and traffic flows), traffic alerts (e.g., accidents), and roadside service directories (e.g., location/menu of a local restaurant). VITP is an application-layer, stateless communication protocol that specifies the syntax and semantics of messages carrying location-sensitive queries and replies between the nodes of a VANET. One of the key features of VITP is that it allows nodes to aggregate (or summarize) location-sensitive information and to report the summarized results to the requester.

Figure 6.10 shows the illustration of VITP operations. Node V wants to know the average vehicle speed nearby the gas station, and it sends the query Q to the destination location (i.e., dispatch phase). VITP-enabled peers in that destination location cooperatively resolve the query (i.e., computation phase) and return a reply R to the requester (i.e., reply-delivery phase).

In VITP, locations are represented as a (*roadID*, *segmentID*) tuple where roadID is a unique key representing a road, and segmentID is a number representing a specific segment of the road. The VITP-enabled peers in a road segment become virtual ad hoc servers (VAHS), and these peers cooperatively answer the query. This can be best

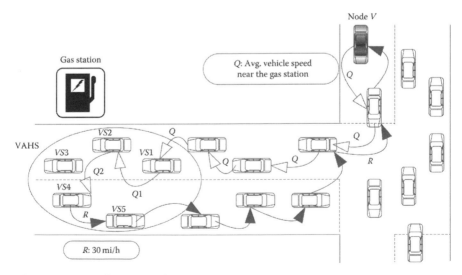

FIGURE 6.10 An illustration of VITP.

explained using the above example. In other words, node *V* wants to know the average speed of the vehicles near the gas station. The query reaches to node *VS1* in the destination zone (i.e., an oval area). After node *VS1* performs a local computation of average speed and updates the query with its partial result, it sends the updated query *Q1* to node *VS2*. Similarly, node *VS2* performs the updates, and passes the updated query *Q2* to *VS4*. In this way, on-the-fly updates continue until the query resolution process satisfies a return condition specified by the requester (i.e., the average value of at least three vehicles). VITP provides best-effort query resolution: any available VITP-enabled peers in a target road segment can participate in query resolution, and it is possible that peers can move out of the area before the query resolution completes.

6.6.4 Infrastructure-Based P2P Traffic Information Systems

Rybicki et al.[10] designed a distributed traffic information system (TIS) that allows vehicles to exchange information about the current traffic status, say in a city. The major shortcoming of the VANET is its low penetration ratio which makes long-distance communications difficult (or opportunistic delay-tolerant data exchange at best). Under current circumstances, they claimed that infrastructure-based mobile wireless access methods such as 3G, GPRS, and WiMAX are more amenable. For instance, UMTS/HSDPA is becoming popular in many countries, and WiMAX has already been deployed and used commercially.

To realize such a service, vehicles must be able to store and retrieve traffic information. Infrastructure allows using a "structured" overlay network, or distributed hash table (DHT) such as Chord and CAN. Recall that DHTs provide a lookup service similar to a hash table such that (name, value) pairs, and any participating node, can efficiently retrieve the value associated with a given name. However, there are still several challenges. Unlike Internet-based DHT where data tends to be static, sensed data on the roads are highly dynamic. Frequent handoffs could incur a short intermittent connectivity. Moreover, the system load is much heavier as it can be potentially used by tens of thousands of vehicles at the same time. Thus, the system must be scalable/reliable and must balance the system load well. Some of the issues could be efficiently handled using conventional techniques such as locality-based load balancing and redundancy/coding-based reliability solutions.

Given that this traffic information system shares the same concept of the publish/subscribe paradigm, Rybicki et al.[10] proposed to build a DHT-based publish/subscribe system for TIS. Vehicles publish traffic information to DHT and it can be delivered to the subscribers based on subscription information. For instance, a user who is interested in accessing traffic conditions during his commuting hours can automatically receive such information. However, there are still several "open" challenging issues. First, the road condition must be constantly monitored and reported in a timely manner. Second, significant updates of driving routes or other changes must be updated in a timely manner, so that users can receive information without any interruption. Third, the system must be able to reliably deliver information (e.g., have multiple distribution trees). Finally, the system must be able to preserve privacy, coordinate traffic patterns, and serve specialized requests (e.g., customized data delivery).

6.6.5 Interactive Applications

6.6.5.1 RoadSpeak: Voice Chatting in Vehicular Social Networks

Smaldon et al.[63] presented a framework for building virtual mobile communities they call vehicular social networks (VSNs), to allow commuters to communicate with each other. RoadSpeak, their VSN-based system, allows drivers to form voice chat groups (VCGs) and communicate on the road via voice chat messages. RoadSpeak uses a centralized server to manage users, and clients connect to the server using infrastructure that provides Internet connections (e.g., 3G services, WiFi access points, and WiMAX).

Users need to first download the client and create an account in the system, which is required to log in and access groups. During the account creation process, the client will generate an asymmetric key pair (using PKI) and register its public key (K_{PUB}) with the server. At this point, they are free to join or create their own groups. To join a group, a user submits their profile, and if the group admission manager validates and accepts the user, the group key (K_G) is transmitted back to the user.

Once on the road, the user chooses one group to listen to and connects to the voice chat server. The connection handler spawns an admission control handler (ACH) to perform admission control. The client will then send the user's name, profile, and SHA1 hash of the group key to the server, which is encrypted using the client's private key. After validating the information, the ACH passes the connection to the channel handler for the VCG. The channel handler maintains one thread per client connected and handles delivery of messages to clients. The client itself keeps three queues: send, receive, and control. The first two buffer outgoing and incoming voice messages, respectively, and the third is used for flow control, allowing the server to send pause and resume messages.

To illustrate RoadSpeak, consider the following example. Joe Bruin downloads the RoadSpeak client onto his smart phone, and joins a politics discussion group that is active between 8 and 10 A.M. on the 10 Freeway between west LA and downtown LA, consistent with his daily commute. The next day, as Joe merges onto the 10, his GPS-enabled smart phone detects his position and contacts the RoadSpeak server. Joe receives an alert notifying him that he has joined the group, and begins to listen to the ongoing discussions. When Joe speaks, his RoadSpeak client transmits his message to the server, which is distributed to all other members of the group. As Joe nears the end of his commute, his smart phone detects his position and the client notifies him of his imminent departure from the group.

6.6.5.2 Online "Passenger" Games

Interactive online games have been mainly based on the Internet infrastructure. As the wireless networking technology becomes ubiquitous in mobile platforms (e.g., Smart Phones and vehicles), wireless and mobile online gaming will soon emerge. In particular, vehicles will provide an ideal platform for wireless and mobile gaming. Vehicular entertainment systems are ever gaining popularity—many millions of TV/DVD systems are sold every year. The convergence of VANETs and entertainment will bring online passenger gaming to reality. Given this, a family or a group of friends driving multiple vehicles in a caravan to a distant place can spend their time engaging in online game sessions with each other through wireless P2P communications. Internet connectivity via access points or 3G services can also support this application, yet these alternatives are less attractive given that wireless access points cannot guarantee quality of service due to intermittent connectivity, and 3G services require a subscription fee.

One of the key challenges is to broadcast game events with high bandwidth and low latency to all the players, because high interactivity is a fundamental feature for online games.[64] For instance, the tolerable delay of fast-paced games is ~150 to 200 msec for ordinary players and 50 to 100 msec for professional players, respectively. Moreover, these events must be delivered via the shared wireless medium. Fairness among online players is another key issue.

To this end, Palazzi et al.[64] designed an efficient multihop broadcast scheme called Fast Multi-Broadcast Protocol (FMBP). Most broadcasting protocols use a random jitter (or backoff) to prevent collision. Also, those protocols force the nodes that are located furthest away to forward packets in order to minimize the hop-count and, thus, the average delay. This can be implemented by setting one's backoff window to be inversely proportional to the distance from the originator: that is, $\tau(1 - d/R)$, where d is the distance from the sender, R is the maximum transmission range, and τ is a fixed time slot value. For instance, if a node's distance is R, it will immediately send out a packet as so on as one receives a packet. In reality, however, the transmission range is not fixed due to road layouts and obstacles in vehicular networks. In Figure 6.11, vehicle A has a shorter communication range due to road layouts and obstacles. Thus, in FMBP, a node estimates its maximum distance in both forward and backward directions. In the example, vehicle B set the backward distance as $|P_B - P_A|$ and the forward distance as $|P_D - P_B|$. A node selects an R value based on the direction of the broadcast message. The simulation results in Ref. 64 show that FMBP can effectively reduce the delay compared to the other schemes.

6.6.6 Other Applications and Related Works

TrafficView[36] disseminates, or pushes (through flooding), information about the vehicles on the road, thus providing real-time road-traffic information such as speed of vehicles

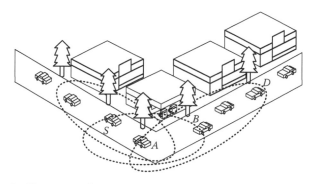

FIGURE 6.11 An Illustration of FMBP.

to drivers. To alleviate broadcast storms, this work focuses on data aggregation/fusion based on distance from the source. Gradinescu et al.[65] proposed adaptive traffic lights where a wireless controller installed in an intersection determines the optimum values for the traffic light phases. The wireless controller collects the volume of vehicles using TrafficView. Zhou et al.[43] proposed EZCab, which discovers and books free cabs using V2V communications. A client (e.g., a node located at a taxi stand) first sends a BookSM packet to find free cabs using flooding or probabilistic forwarding, and free cabs send back ReportSM packets to the client. Given this, the client chooses a cab and makes a reservation by sending a ConfirmSM packet to the cab. Similarly, Huang et al.[66] investigated financial and technical feasibility of a taxi dispatching application.

6.7 Conclusion

In this chapter we have surveyed the emerging vehicular applications, ranging from vehicular sensors to entertainment. We have outlined a potential vehicular network architecture based on various wireless protocols and access methods such as DSRC, 3G, and WiMAX. We have described these protocols and discussed the unique characteristics of vehicular communications. Given this, we proceeded to classify a representative set of VANET P2P applications based on the vehicle's role in managing data: as source, consumer, source/consumer, or intermediary. For the data source scenario, we reported the state-of-the-art vehicular sensing applications; for the consumer scenario, we summarized the contention distribution applications such as multimedia files or advertisements; and for the source/consumer scenarios, we reviewed (emergency) P2P video streaming, virtual flea markets, vehicular information services, and interactive applications (e.g., chatting/games).

References

1. Standard Specification for Telecommunications and Information Exchange Between Roadside and Vehicle Systems—5 GHz Band Dedicated Short Range Communications (DSRC) Medium Access Control (MAC) and Physical Layer (PHY) Specifications, September 2003.
2. Frenkiel, R., Badrinath, B.R., Borras, J., and Yates, R.D., The infostations challenge: Balancing cost and ubiquity in delivering wireless data, *IEEE Personal Communications*, 7-2, 66–71, April 2002.
3. Small, T. and Haas, Z.J., The Shared Wireless Infostation Model—A New Ad Hoc Networking Paradigm (or Where There is a Whale, There is a Way), in ACM MOBIHOC, Annapolis, Maryland, USA, June 2003.
4. Jiang, D., Taliwal, V., Meier, A., Holfelder, W., and Herrtwich, R., Design of 5.9 GHz DSRC-based vehicular safety communication, *IEEE Wireless Communications*, 13(5); 36–43, 2006.
5. Qureshi, A., Carlisle, J., and Guttag, J., Tavarua: Video Streaming with WWAN Striping, in ACM Multimedia 2006, Santa Barbara, CA, October 2006.
6. Han, M., Moon, S., Lee, Y., Jang, K., and Lee, D., Evaluation of VoIP Quality over WiBro, in Passive and Active Measurement Conference (PAM), Cleveland, OH, April 2008.

7. Tan, W. L. and Yue, O.C., Measurement-based Performance Model of IP Traffic over 3G Networks, in TENCON 2005 IEEE Region 10, Melbourne, Australia, November 2005.

8. Hadaller, D., Keshav, S., Brecht, T., and Agarwal, S., Vehicular Opportunistic Communication Under the Microscope, in MobiSys'07, San Juan, Puerto Rico, June 2007.

9. Gerla, M. et al., Vehicular Grid Communications: The Role of the Internet Infrastructure, in WICON'06, MA, USA, August 2006.

10. Rybicki, J. et al., Challenge: Peers on Wheels—A Road to New Traffic Information Systems, in MobiCom'07, Quebec, Canada, September 2007.

11. Torrent Moreno, M., Jiang, D., and Hartenstein, H., Broadcast Reception Rates and Effects of Priority Access in 802.11-based Vehicular Ad-hoc Networks, in ACM VANET, Philadelphia, PA, USA, October 2004.

12. Xu, Q., Mak, T., Ko, J., and Sengupta, R., Vehicle-to-Vehicle Safety Messaging in DSRC, in ACM VANET, Philadelphia, PA, USA, October 2004.

13. Yin, J., ElBatt, T., Yeung, G., Ryu, B., and Habermas, S., Performance Evaluation of Safety Applications Over DSRC Vehicular Ad Hoc Networks, in ACM VANET'04, Philadelphia, PA, USA, October 2004.

14. ElBatt, T., Goel, S.K., Holland, G., Krishnan, H., and Parikh, A., Cooperative Collision Warning Using Dedicated Short Range Wireless Communications, in ACM VANET'06, Los Angeles, CA, USA, September 2006.

15. Korkmaz, G., Ekici, E., Ozguner, F., and Ozguner, U., Urban Multi-Hop Broadcast Protocols for Inter-Vehicle Communication Systems, in ACM VANET, Philadelphia, PA, USA, October 2004.

16. Tang, K. and Gerla, M., MAC Reliable Broadcast in Ad Hoc Networks, in IEEE MILCOM'01, Washington, DC, October 2001.

17. Sun, M.T., Huang, L., Arora, A., and Lai, T.H., Reliable MAC Layer Multicast in IEEE Wireless Networks, in ICCP'02, Vancouver, August 2002.

18. Yadumurthy, R.M., Adithya, C., Sadashivaiah, M., and Makanaboyina, R., Reliable MAC Broadcast Protocol in Directional and Omni-directional Transmissions for Vehicular Ad Hoc Networks, in VANET'05, September 2005.

19. Chen, Z.D., Kung, H.T., and Vlah, D., Ad Hoc Relay Wireless Networks over Moving Vehicles on Highways, in ACM MOBIHOC, Long Beach, CA, USA, October 2001.

20. Burgess, J., Gallagher, B., Jensen, D., and Levine, B.N., MaxProp: Routing for Vehicle-Based Disruption-Tolerant Networks, in IEEE INFOCOM, Barcelona, Spain, April 2006.

21. Zhao, J. and Cao, G., VADD: Vehicle-Assisted Data Delivery in Vehicular Ad Hoc Networks, in IEEE INFOCOM, Barcelona, Spain, April 2006.

22. Naumov, V. and Gross, T., Connectivity-Aware Routing (CAR) in Vehicular Ad Hoc Networks, in INFOCOM'07, Anchorage, AK, May 2007.

23. Wu, H., Fujimoto, R., Guensler, R., and Hunter, M., MDDV: A Mobility-centric Data Dissemination Algorithm for Vehicular Networks, in ACM VANET, Philadelphia, PA, USA, October 2004.

24. Sormani, D. et al., Towards Lightweight Information Dissemination in Inter-vehicular Networks, in ACM VANET'06, Los Angeles, CA, USA, September 2006.

25. Maihöfer, C., Leinm¨uller, T., and Schoch, E., Abiding Geocast: Time-Stable Geocast for Ad Hoc Networks, in ACM VANET'05, Cologne, Germany, September 2005.
26. Lee, U. et al., MobEyes: Smart mobs for urban monitoring with vehicular sensor networks, *IEEE Wireless Communications*, 13(5); 51–57, 2006.
27. Lee, U., Magistretti, E., Gerla, M., Bellavista, P., and Corradi, A., Dissemination and harvesting of urban data using vehicular sensor platforms, *IEEE Transaction on Vehicular Technology*, 2008 (to appear).
28. Lee, U. et al., Bio-inspired multi-agent data harvesting in a proactive urban monitoring environment, *Elsevier Ad Hoc Networks Journal, Special Issue on Bio-Inspired Computing and Communication in Wireless Ad Hoc and Sensor Networks*, 2008.
29. Hull, B. et al., CarTel: A Distributed Mobile Sensor Computing System, in ACM SenSys, Boulder, CO, USA, October–November 2006.
30. Eriksson, J. et al., The Pothole Patrol: Using a Mobile Sensor Network for Road Surface Monitoring, in MobiSys'08, Breckenridge, Colorado, June 2008.
31. Nandan, A. Das, S., Pau, G., Gerla, M., and Sanadidi, M.Y., Cooperative Downloading in Vehicular Ad-Hoc Wireless Networks, in IEEE/IFIP WONS, St. Moritz, Swiss, January 2005.
32. Lee, U., Park, J.-S., Yeh, J., Pau, G., and Gerla, M., CodeTorrent: Content Distribution using Network Coding in VANETs, in MobiShare'06, Los Angeles, CA, September 2006.
33. Nandan, A. et al., AdTorrent: Delivering Location Cognizant Advertisements to Car Networks, in IEEE/IFIP WONS, Les Menuires, France, January 2006.
34. Caliskan, M., Graupner, D., and Mauve, M., Decentralized Discovery of Free Parking Places, in ACM VANET, Los Angeles, CA, USA, September 2006.
35. Dikaiakos, D.M., Iqbal, S., Nadeem, T., and Iftode, L., VITP: An Information Transfer Protocol for Vehicular Computing, in ACM VANET, Cologne, Germany, September 2005.
36. Nadeem, T., Dashtinezhad, S., Liao, C., and Iftode, L., TrafficView: Traffic data dissemination using car-to-car communication, *ACM Mobile Computing and Communications Review (MC2R)*, 8(3); 6–19, 2003.
37. Lee, U., Park, J.-S., Amir, E., and Gerla, M., FleaNet: A Virtual Market Place on Vehicular Networks, in V2VCOM'06, San Jose, CA, July 2006.
38. Guo, M., Ammar, M.H., and Zegura, E.W., V3: A Vehicle-to-Vehicle Live Video Streaming Architecture, in PerCom'05, March 2005.
39. Park, J.-S., Lee, U., Oh, S.Y., Gerla, M., and Lun, D., Emergency Related Video Streaming in VANETs using Network Coding, in ACM VANET'06, Los Angeles, CA, USA, September 2006.
40. Ratnasamy, S. et al., GHT: A Geographic Hash Table for Data-Centric Storage, in WSNA'02, Atlanta, GA, USA, September 2002.
41. Caesar, M., Castro1, M., Nightingale, E.B., O'Shea, G., and Rowstron, A., Virtual Ring Routing: Network Routing Inspired by DHTs, in SIGCOMM'06, Pisa, Italy, September 2006.
42. Vahdat A. and Becker, D., Epidemic Routing for Partially-Connected Ad Hoc Networks, Technical Report CS-200006, Duke University, April 2000.

43. Zhou, P., Nadeem, T., Kang, P., Borcea, C., and Iftod, L., EZCab: A Cab Booking Application Using Short-Range Wireless Communication, in IEEE PerCom'05, Kauai Island, HI, USA, March 2005.

44. Intanagonwiwat, C., Govindan, R., and Estrin, D., Directed Diffusion: A Scalable and Robust Communication Paradigm for Sensor Networks, in ACM MOBICOM'00, Boston, MA, USA, 2000.

45. Fan, L., Cao, P., and Almeida, J., Summary Cache: A Scalable Wide-Area Web Cache Sharing Protocol, in ACM SIGCOMM, Vancouver, British Columbia, Canada, August–September 1998.

46. Yoon, J., Noble, B., and Liu, M., Surface Street Traffic Estimation, in MobiSys'07, San Juan, Puerto Rico, June 2007.

47. Juang, P. et al., Energy-Efficient Computing for Wildlife Tracking: Design Tradeoffs and Early Experiences with ZebraNet, in ACM ASPLOS-X, San Jose, CA, USA, October 2002.

48. Eisenman, S.B. et al., MetroSense Project: People-Centric Sensing at Scale, in ACM WSW, Boulder, CO, USA, October–November 2006.

49. Shah, R.C., Roy, S., Jain, S., and Brunette, W., Data MULEs: Modeling a three-tier architecture for sparse sensor networks, *Elsevier Ad Hoc Networks Journal*, 1(2–3); 215–233, 2003.

50. Wang Y., and Wu, H., DFT-MSN: The Delay/Fault-Tolerant Mobile Sensor Network for Pervasive Information Gathering, in INFOCOM'06, Barcelona, Spain, April 2006.

51. Burke, J. et al., Participatory Sensing, in ACM WSW, Boulder, CO, USA, October–November 2006.

52. Gibbons, P.B., Karp, B., Ke, Y., Nath, S., and Seshan, S., IrisNet: An architecture for a worldwide sensor web, *IEEE Pervasive Computing*, 2(4); 22–33, 2003.

53. Nath, S., Liu, J., and Zhao, F., Challenges in Building a Portal for Sensors WorldWide, in ACM WSW, Boulder, CO, USA, October–November 2006.

54. Riva, O. and Borcea, C., The Urbanet revolution: Sensor power to the people! *IEEE Pervasive Computing*, 6(2); 41–49, 2007.

55. Lee, K.C., Lee, S.H., Cheung, R., Lee, U., and Gerla, M., First Experience with CarTorrent in a Real Vehicular Ad Hoc Network Testbed, in MOVE'07, Anchorage, AK, May 2007.

56. Gkantsidis, C. and Rodriguez, P., Network Coding for Large Scale Content Distribution, in INFOCOM'05, Miami, FL, USA, March 2005.

57. Chiu, D.M., Yeung, R.W., Huang, J., and Fan, B., Can Network Coding Help in P2P Networks? in NetCod'06, Boston, MA, April 2006.

58. Eriksson, J., Balakrishnan, H., and Madden, S., Cabernet: A Content Delivery Network for Moving Vehicles, Technical Report TR-2008-003, MIT-CSAIL, 2008.

59. Yoon, H., Kim, J.W., Tan, F., and Hsieh, R., On-demand Video Streaming in Mobile Opportunistic Networks, in PerCom'08, Hong Kong, China, March 2008.

60. Rheingold, H., *Smart Mobs: The Next Social Revolution*, Basic Books, 2003.

61. Grossglauser, M. and Vetterli, M., Locating Nodes with EASE: Mobility Diffusion of Last Encounters in Ad Hoc Networks, in IEEE INFOCOM, San Francisco, CA, USA, March–April 2003.

62. Lee, S.-B., Pan, G., Park, J.-S., Gerla, M., and Lu, S., Secure Incentives for Commercial Ad Dissemination in Vehicular Networks, in MobiHoc'07, Quebec, Canada, September 2007.

63. Smaldone, S., Han, L., Shankar, P., and Iftode, L., RoadSpeak: Enabling Voice Chat on Roadways using Vehicular Social Networks, in SocialNets'08, Glasgow, Scotland, UK, April 2008.

64. Palazzi, C.E., Roccetti, M., Pau, G., and Gerla, M., Online Games on Wheels: Fast Game Event Delivery in Vehicular Ad-hoc Networks, in V2VCOM'07, Istanbul, Turkey, June 2007.

65. Gradinescu, V., Gorgorin, C., Diaconescu, R., Cristea, V., and Iftode, L., Adaptive Traffic Lights Using Car-to-Car Communication, in IEEE Vehicular Technology Conference, Dublin, Ireland, April 2007.

66. Huang, E., Hu, W., Crowcroft, J., and Wassell, I., Towards Commercial Mobile Ad Hoc Network Applications: A Radio Dispatch System, in MobiHoc'05, Urbana-Champaign, IL, USA, May 2005.

67. Lee, U., Park, J.-S., Amir, E., and Gerla, M., FleaNet: A Virtual Market Place on Vehicular Networks, in V2VCOM'06, San Jose, CA, July 2007.

7

Use of Infrastructure in VANETs

Michele C. Weigle,
Stephan Olariu,
Mahmoud Abuelela,
and Gongjun Yan
*Department of
Computer Science
Old Dominion University*

7.1 Introduction

Vehicular networks are an exciting platform for developing new and useful applications. As driving is an almost universal experience, just about anyone can think of some application that would be useful to them as a driver. Many of the applications would involve driver safety (as described in Chapter 5), but just as many would involve information delivery. These applications would provide drivers with information about upcoming traffic conditions, weather forecasts, services available at interstate exits, special events in the area, and so on. For some of these applications, vehicle-to-vehicle (V2V) communication is sufficient and even preferred, for example, collision avoidance warning systems. But, for many of these, especially nonurgent information delivery applications, a pervasive roadside infrastructure is the key enabling factor.

The need for this pervasive roadside infrastructure is reflected in the U.S. Department of Transportation's (U.S. DOT) Vehicle–Infrastructure Integration (VII) program, tasked with developing plans and prototypes. As described in Chapter 3, the VII proposes that vehicles be equipped with on-board equipment (OBE) that can communicate using dedicated short-range communication (DSRC) with OBEs in other vehicles and with communications devices in roadside units (RSU). Figure 7.1 depicts an infrastructure-based architecture with RSUs, IEEE 802.11 access points, and satellites connecting vehicles to each other, specialized servers, and the Internet.

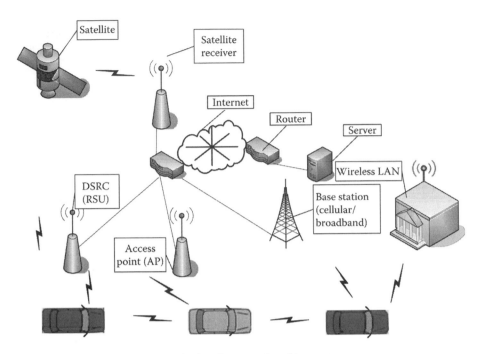

FIGURE 7.1 Infrastructure-based vehicular network architecture.

This chapter presents an overview of how infrastructure is used in VANETs. The manner in which infrastructure facilitates providing security and privacy is described in Sections 7.2 and 7.3, respectively. Section 7.4 outlines how several commercial and infotainment VANET applications take advantage of infrastructure. Note that the focus is on their use of infrastructure, rather than on the details of the applications, as some are covered in Chapter 6. Section 7.5 describes in detail NOTICE, an infrastructure-based system for the notification of traffic incidents and congestion. NOTICE is unique because it embeds infrastructure in the highway itself, rendering the infrastructure an integral part of the system.

7.2 Infrastructure-Based Security

Security in vehicular networks, or vehicular ad-hoc networks (VANETs), has been a popular topic in recent years.[1-7] The consensus appears to be providing security through the use of public key infrastructure (PKI) techniques.[8] As vehicles are not exchanging sensitive information, there is less of a need for data privacy than there is for authentication. In such a system, every vehicle is assigned a unique public/private key pair and an associated certificate (or, many public/private keys and certificates to help achieve some anonymity). When a vehicle sends a message, it first creates a digital signature by signing the message with its private key. Like an actual signature, this signifies that this particular vehicle was the sender of the information. Along with the message and the vehicle's signature, the vehicle would also include its certificate, issued by a certificate authority (CA). This certificate contains the vehicle's public key signed with the CA's private key. Any vehicle receiving a message would first use the CA's public key (assumed to be known to all vehicles) to obtain the sending vehicle's public key. Then, the vehicle would use that public key to verify that the message matches the signed copy of the message that was sent.

As explained by Raya et al.,[3] there are essentially two ways that public keys can be provided to vehicles: by government authorities or by car manufacturers. Each of the public keys must be accompanied by a certificate that is generated by the CA that is assigning the key. In the United States, each of the 50 states is individually responsible for vehicle registration. Assuming that certification is tied to vehicle registration, this would mean that there would be 50 different CAs, one for each state. If a vehicle only knew the CA for its state, it would not be able to communicate with vehicles registered in other states. Raya et al. propose having infrastructure at state borders that recertifies keys once the information has been verified by the original CA. By using roadside infrastructure, all vehicles in a state (that have had their keys recertified at the border) can communicate with each other no matter their state of origin.

In the case of malicious or malfunctioning vehicles, certificates and keys may need to be revoked. In this way, the vehicle would no longer be authenticated, and other vehicles would discard its messages. The preferred method to achieve key revocation is to distribute certificate revocation lists (CRLs).[9] These CRLs would be broadcast by roadside infrastructure and would inform vehicles about the most recently revoked certificates.

Because of the need for vehicles to access CAs, or recertifiers, and be informed about CRLs while on the road, roadside infrastructure plays a large role in VANET security.

7.3 Infrastructure-Based Privacy

With the prevalence of PKI systems as solutions to VANET security, privacy concerns have been raised.[10-14] If each vehicle is given a public key, then each vehicle can be identified by that public key. More importantly, each vehicle and its location, heading, and speed at any time can be tracked just by saving the vehicle's broadcasted reports. The issue here is not to make driving totally anonymous, because driving is an inherently public activity, but to make it difficult for malicious users or overzealous governments to use the power of V2V communication to track and record the movements of particular vehicles. Thus, there have been several approaches to achieving anonymity within the structures of PKI security for VANETs.

Parno and Perrig[6] introduced the idea of roadside infrastructure as reanonymizers, placed at certain intervals on the roadway. In their system, vehicles would be issued temporary certificates and identities. If a vehicle could prove that it already had an authenticated temporary identity, it could request and be issued a new identity at a reanonymizer, thus preventing the vehicle from being tracked. These identities would be designed to be short-lived, to prevent vehicles from obtaining many authenticated identities and launching a Sybil attack,[15] where one vehicle impersonates many nonexisting vehicles. In particular, the reanonymizer would broadcast a random nonce. The vehicle would respond with its public key and the signed nonce. If the signature was accepted, the reanonymizer would broadcast a new certificate for the vehicle. This certificate would be encrypted with the vehicle's old public key, so that only this vehicle could access the certificate.

RSUs also play a large role in providing anonymity in the CARAVAN[7] system by providing vehicles access to CAs that assign vehicles a set of pseudonyms and associated certificates. In CARAVAN, vehicles change their pseudonyms during random-length silent periods during which no messages are sent. Vehicles also form into groups and are represented by a single group leader who communicates with RSUs on behalf of all members of the group. Further, pseudonyms and group leaders can be used when communicating with RSUs to obtain location-based services. This allows a vehicle to anonymously access a particular service.

Crescenzo et al.[16] proposed a vehicular-network public key infrastructure (V-PKI) that uses trusted-party CAs (tpCAs) and vehicle manufacturer CAs (vmCAs), as pictured in Figure 7.2. The tpCA is responsible for revoking certificates (i.e., managing CRLs) and updating keys. To achieve anonymity, the vmCA provides additional cryptographic information, acting as a pseudonym, to each vehicle and the tpCA. When a vehicle uses this pseudonym, the tpCA is able to determine if the vehicle is authenticated without having to contact the issuing vmCA and, more importantly, without having to know the identity of the vehicle. As with other PKI techniques for VANETs, roadside infrastructure plays a large role by allowing the distribution of CRLs and allowing vehicles to access the vmCAs and tpCAs as necessary.

VIPER[17] was designed to protect vehicle privacy when communicating with RSUs. Instead of sending a message directly to an RSU, a vehicle could send through other vehicles in its group, using them as mix points. Vehicles are considered to be in the same group if they are registered with the same RSU. Using the universal re-encryption public-key cryptosystem,[18] the sender of a message would first encrypt its message with

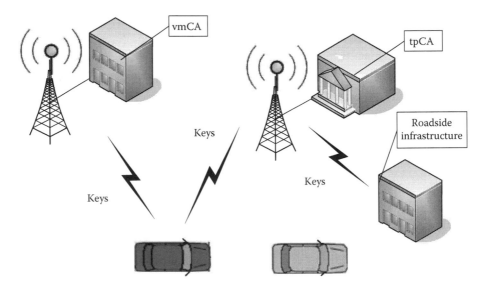

FIGURE 7.2 Vehicular network public key infrastructure.

the public key of the registered RSU, which the RSU periodically broadcasts. This ensures that only that particular RSU can read the message. As a message is passed around by group members, each relay re-encrypts the message before sending.

7.4 Commercial and Infotainment Applications

Infrastructure can assist in commercial and infotainment VANET applications as well as helping to provide security and privacy. In many systems, RSUs facilitate information transfer between commercial servers and vehicles (Figure 7.3) and between information publishers and vehicles (Figure 7.4). This section describes a few of these applications, focusing on how infrastructure contributes. Some of these applications are described further in Chapter 6.

One of the first uses of infrastructure for commercial applications was the electronic toll collection (ETC) system.[19] The main components of ETC are a transponder and a communicator. The transponder is located in the customer's vehicle, and the communicator is located to the side or above the roadway and is used to identify the transponder and charge the customer's account. ETC systems, such as E-ZPass, are widely used on toll roads and bridges in the United States.

Another practical use of infrastructure is in helping drivers find available parking spots. Caliskan et al.[20] proposed a system where parking automats produce and broadcast resource reports containing information on available parking. These reports serve vehicles that are within the automat's broadcast range, but the vehicles also aggregate the reports and disseminate the information to other vehicles.

AdTorrent[21,22] proposes to use RSUs as commercial advertisement dissemination points. These ads should be relevant to the region where the RSU is located (e.g., ads for

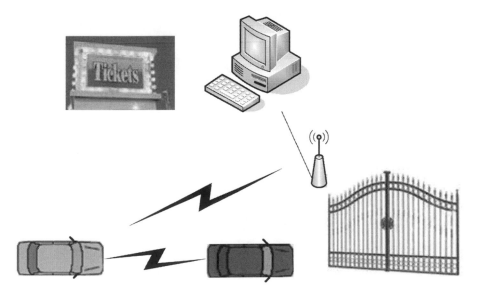

FIGURE 7.3   Infrastructure facilitating a service center.

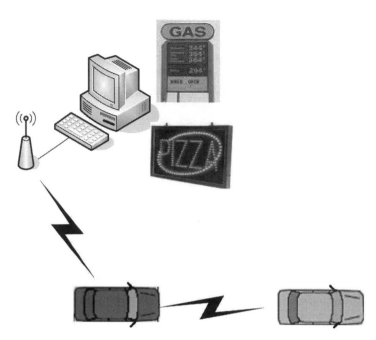

FIGURE 7.4   Infrastructure facilitating an information publisher.

nearby restaurants and hotels). The RSUs periodically broadcast the ads, allowing passing vehicles to download the information. These ads can be more than just digital billboards, including also audio and video. In a separate work, Lee et al.[23] present a framework by which users are encouraged and rewarded for helping to disseminate advertisements. As with AdTorrent, infrastructure plays a role in helping to distribute the advertisements.

Both VANETCODE[24] and CarTorrent[25] are content distribution applications that use RSUs as distribution points. In both systems, the RSUs contain complete file information, and passing vehicles download certain blocks of the file from the RSU and certain blocks from neighbors, much like the BitTorrent file distribution system for the wired Internet. These systems allow vehicles to download complete files faster and more efficiently than if they only downloaded pieces when in the presence of an RSU.

The focus of the vehicular grid (V-Grid)[26] is on exploiting the infrastructure and its connection to the wired Internet to enhance vehicular communications. The authors point out that combining the V-Grid and infrastructure allows V2V communications to take place via the Internet when possible (e.g., for file transfer, email, and voice communications) and allows for monitoring and reporting network conditions. In addition, the V-Grid is able to take over when infrastructure fails (e.g., upon the occurrence of a natural disaster).

7.5 NOTICE—Highway as Infrastructure

The remainder of this chapter is focused on a particular infrastructure-based architecture that embeds the infrastructure in the highway itself. In NOTICE (An Architecture for the Notification of Traffic Incidents and Congestion),[27,28] sensor belts are embedded in the road at regular intervals (e.g., every mile or so). Each belt consists of a collection of pressure sensors, a simple aggregation and fusion engine, and a few small transceivers. The pressure sensors in each belt allow every message to be associated with a physical vehicle passing over the belt. Thus, no one vehicle can pretend to be multiple vehicles and there is no need for an ID to be assigned to vehicles. The belts are better placed than roadside infrastructure to detect passing vehicles and to interact with them in a simple and secure fashion.

By using their pressure sensors, the belts can detect passing vehicles and can initiate interaction with them. In this arrangement, each passing vehicle will exchange data with the belt. It will drop off sensory messages originating with the vehicle or encrypted messages uploaded by the previous belt. The vehicle will then pick up information intended for its own use, along with encrypted messages to be propagated to the next belt. Each pair of belts shares a common encryption key, which allows them to communicate securely. In addition, because the messages carried by vehicles from one belt to another are encrypted, they cannot be tampered with by the vehicles. Thus, passing vehicles act as *data mules*, passing encrypted information between the belts and contributing to the overall traffic-related knowledge in the system.

Note that the security in NOTICE is based on proximity—belts can associate a message received with a physical vehicle. In addition, the encryption used is symmetric, so no PKI is needed. This removes the requirement for a pervasive roadside infrastructure to be used to communicate with certificate authorities as required for PKI.

NOTICE has the following key characteristics:

1. Every message can be associated with a physical vehicle, so there is no need for an ID or pseudonym to be assigned to vehicles.
2. A belt spans the entire roadway, allowing vehicles in oncoming lanes to carry information about congestion to belts that are in a better position to inform vehicles that are approaching the congestion.
3. NOTICE can be used to allow traffic management personnel to disseminate important information to drivers. This feature of the system would be vital in times of large-scale evacuations, allowing emergency managers to alert drivers of estimated travel times, available resources, and contraflow roadways.
4. The belts can use their collective knowledge to detect traffic slowdowns. A belt can monitor the speed and number of vehicles passing at any time and communicate that information to other belts. Thus, a later belt detecting a much lower count or speed of vehicles than the previous belt could imply that congestion is occurring.

7.5.1 Wireless Communications Model

The wireless communications model of NOTICE includes belt-to-vehicle, belt-to-belt, and limited vehicle-to-vehicle communications.

7.5.1.1 Vehicle Model

In NOTICE, it is assumed that vehicles will be fitted with a tamper-resistant *event data recorder* (EDR), much like the well-known black-boxes on-board commercial aircraft. The use of this device has been suggested elsewhere[4,5,29] and is a common device required in many VANET applications. The EDR provides tamper-resistant storage of statistical and private data. The EDR is also responsible for recording essential mobility attributes. For this purpose, all of the vehicle's subassemblies, including the GPS unit, speedometer, gas tank reading, tire pressure sensors, and sensors for outside temperature, feed their own readings into the EDR. The EDR could also be fitted with a cell phone programmed to call predefined numbers (including E-911) in the case of an emergency. The idea here is that the driver may be incapacitated as a result of the accident and may be unable to place the call. This feature exists already on some vehicles and is useful for reporting, upon the deployment of an airbag, that the vehicle was probably involved in a collision. This allows the authorities to be alerted in real time to major traffic events and, ultimately, saves lives.[30]

NOTICE uses DSRC devices for communication. The EDR has a secure connection to two radio transceivers, one placed just behind the front axle and a second transceiver placed in a tamper-proof box at the rear of the vehicle. The front transceiver is essential to ensuring correct handshaking with the belts on the roadway; the second transceiver, placed aft, is responsible for data transfer between the vehicle and the belt.

7.5.1.2 Belt-to-Belt Communications

Each belt consists of several sub-belts, one per lane. The sub-belts are fitted with their own set of sensors and transceivers and are directly connected together for ease of communication between the lanes. This communication can be extended to highways with divided lanes by connecting the sub-belts by wire under the median. Figure 7.5 shows a two-lane roadway where each belt consists of two sub-belts, one for each lane of traffic.

Unlike sub-belts, the belts do not communicate with each other directly. Instead, adjacent belts rely on passing vehicles to communicate. Referring again to Figure 7.5, consider the lane where traffic is moving right to left. If belt C wants to communicate a message m to the next belt, B, it will encrypt m with a time-varying shared key $\mu(C, B, t)$ known only to belts C and B, with t representing the time parameter. The belts switch from one key to the next in a pre-established key chain based on their local time. Tight time synchronization between belts is not essential, given the inherent delays in communications. Given a sufficiently large set of keys in the key chain, use of the belt-to-belt encryption keys appears random to an external observer. To pass the encrypted message

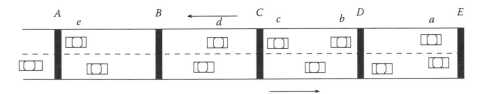

FIGURE 7.5 A collection of belts on a two-lane road. Belts are labeled with upper-case letters, and vehicles are labeled with lower-case letters. Note that as belts will be placed at least 1 mi apart, this figure is not drawn to scale. (From Abuelela, M. et al., NOTICE: An Architecture for Notification of Traffic Incidents, in Proceedings of IEEE Vehicular Technology Conference, Singapore, May 2007, pp. 3001–3005.)

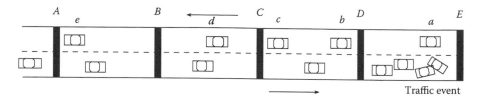

FIGURE 7.6 Information propagation in NOTICE on a two-lane roadway. (From Abuelela, M. et al., NOTICE: An Architecture for Notification of Traffic Incidents, in Proceedings of IEEE Vehicular Technology Conference, Singapore, May 2007, pp. 3001–3005. With permission.)

m to belt B, belt C will upload m onto passing vehicle c. When vehicle c reaches belt B, the message m will be dropped off and decoded by belt B. In turn, belt B may decide to send a message to belt A. This would be done using the symmetric key $\mu(B, A, t)$, known only to belts B and A.

Figure 7.6 is used to illustrate the backwards information propagation in the event of a traffic incident between belts D and E. For now, the issue of how the traffic event is detected is ignored and it is assumed that belt D is aware of the incident and the slow-down in traffic resulting from it. When vehicle a, traveling away from the traffic event, passes over belt D, the belt will upload information about the incident, which will be delivered to belt C. When the information about the traffic incident reaches belt C (and later, belts B and A), the belt will inform vehicles traveling towards the traffic event about upcoming congestion. This information can be processed by the vehicle's navigation system, which might suggest an alternative route. The severity of the traffic incident and the current traffic conditions will affect how far the message is propagated. As mentioned earlier, in a divided highway, belts would be connected via wires under the median, so there would be no change to the way in which message propagation is performed.

7.5.1.3 Belt-to-Vehicle Communications

A vehicle provides two types of messages to a belt. First, the vehicle serves as a data messenger between two belts to further the information propagation described earlier. Second, the vehicle delivers readings stored by its EDR during the time the vehicle was traveling between two belts. For example, in Figure 7.6, vehicle d would report information about its travel between belts C and B. This type of information helps the belt to detect traffic slowdowns and infer when congestion is occurring. Each time a vehicle provides EDR information, the belt updates its model of the traffic conditions between it and the previous belt. The current state of the model is then passed to the belt in the opposite lane (or opposite side of the median, if in a divided highway), so that notifications of traffic slowdowns or incidents can be relayed to traffic approaching the incident. It is important to note that the EDR information that a particular vehicle gives to a belt is not relevant to the conditions ahead of the vehicle. Thus, the belt-to-vehicle communications consists of data exchange only. There is no processing of the data the vehicle gives to the belt while that same vehicle is in communications range.

There are two phases in belt-to-vehicle communications: handshaking and data exchange. Each of these phase will be explained in detail using an example from Figure 7.6.

Consider vehicle c traveling at 100 km/h (~65 mi/h—the legal interstate speed in most U.S. states) towards belt C. Recall that each vehicle will have two transceivers, one behind the front axle for handshaking (T_h) and a second at the rear of the vehicle for data exchange T_d. The very short-range radio transmission that will be used in this communication is deliberate. It renders the communication strictly *local* and, therefore, reduces the chances of eavesdropping by malicious entities positioned by the roadside.

7.5.1.4 Handshaking

Once the pressure sensors in belt C have detected the front wheels of vehicle c, a radio transceiver in the belt will send, at a very low power (range of ~1 m), a "Hello" beacon on a standard control channel to vehicle transceiver T_h. This message contains the ID of the belt, C, and the frequency channel λ on which data is to be exchanged. Once vehicle c receives this information, it will have ~36 msec (time to travel 1m and thus out of communication range) to respond. As the handshaking response will be very short and will not be encrypted, a NOTICE-equipped vehicle will have no problem responding in time. If belt C does not receive a reply to the handshake, it will not communicate further with vehicle c.

7.5.1.5 Data Exchange

Assuming that vehicle c confirms the handshake before it leaves radio range, belt C will initiate the following data exchange with vehicle transceiver T_d on channel λ:

1. Belt C sends a query for information to vehicle c.
2. Vehicle c delivers the encrypted message it carried from belt D.
3. Vehicle c reports the relevant traffic-related data collected by its EDR in the time $T(D,C)$, which is the time spent traveling between belts D and C.
4. Belt C decrypts the message from belt D.
5. Belt C reports traffic information that is relevant to vehicle c.
6. Belt C encrypts a message for belt B with $\mu(C, B, t)$.
7. Belt C sends the encrypted message for belt B to vehicle c.

The belt transceiver used for data exchange has a range of 3 m, giving a total range of 6 m as vehicle c passes over the belt. For vehicle c, traveling at 100 km/h (or, 65 mi/h), there will be 216 msec in which to complete the communication with the belt. Let s be the transmission time for a single message, d the encryption/decryption time for a single message, and p be the processing time for the belt to incorporate new information. There are a total of five messages sent after handshaking and two encryption/decryption events. This results in a total communication time $T = (5s + 2d + p)$ msec. If we set $p = 5$ msec, $d = 20$ msec (measured time to encrypt 1024 bytes of data with DES), and $s = 1$ msec (time to transmit a 750-byte message at 6 Mb/sec, the lower end of the DSRC range[31]), then $T = 95$ msec. Actual messages exchanged should be smaller than 750 bytes, so these are conservative estimates. Even with these conservative estimates, for 95 msec to be too little time for communication, the vehicle would have to be traveling at 277 km/h (141 mi/h), an illegal, not to mention unsafe, speed on U.S. highways.

It is worth noting that the belt-to-vehicle and vehicle-to-belt data exchanges discussed above are perfectly anonymous and do not interfere with vehicle or driver privacy. Indeed, the pressure sensors in the belts allow NOTICE to associate every message with a physical vehicle passing over the belt. A given vehicle cannot interact with a belt more than once because the belt only initiates handshaking after it first detects a vehicle and, consequently, impersonation and Sybil attacks are difficult to perpetrate. In addition, because messages carried by vehicles from one belt to another are encrypted, these messages are secure.

7.5.1.6 Vehicle-to-Vehicle Communications

For reasons of security and privacy, NOTICE minimizes the amount of V2V communications. There are, however, instances where V2V communications are useful and, as such, are supported by NOTICE. Referring again to Figure 7.6, assume that belt D has an emergency message to convey to belt C. Belt D can upload the message onto vehicle a, in which case it will take slightly less than one minute for the message to make it to belt C (assuming that the belts are placed one mile apart and the vehicle is traveling at 100 km/h, or 65 mi/h). In the case of a traffic slowdown, where the traffic moves very slowly or is stopped, it may take considerably longer for the message to reach belt C. In emergency situations, this delay is intolerable. Under such conditions, belt D, having encrypted the message with the key $\mu(D, C, t)$, will upload the message to vehicle a and will also set the "urgent" bit indicating that vehicle a must try to forward the message by radio to vehicles traveling toward C. In Figure 7.6, vehicle a will send a message destined to all vehicles between belts D and C (vehicles b and c in the figure) asking them to drop off the urgent message with belt C. This feature is extremely useful for accident notification and for alerting drivers approaching an accident of the corresponding slowdown.

Thus, with the addition of urgent mode, there are two modes of information dissemination in NOTICE:

1. *Normal.* Vehicle a (and vehicles that follow for a certain amount of time) will carry messages belt C, belt B, and belt A, in succession. The message propagation time depends upon the speed of vehicle a.
2. *Urgent.* Once between belts D and C, vehicle a will transmit the message to any other vehicles that are located between belts C and D, such as vehicles b and c. These vehicles will then drop off the message at belt C before vehicle a would have reached belt C. This speeds up the message propagation time considerably.

7.5.1.7 Role-Based Vehicle-to-Belt Communications

At various times, highway traffic management, local emergency management, and police officers will need to provide drivers with important information. As NOTICE is, at its most basic level, an information dissemination tool, it seems natural to allow these personnel to insert information into the system. Authorized vehicles will be given special encryption keys used to facilitate communication in a role-based fashion with the belts. For example, police vehicles may load messages that ambulances are not allowed to load. This information could be related to planned lane closures, major traffic incidents, suggested detour routes, as well as the availability of resources in the case of a planned evacuation, when finding gas and shelter become critical issues. To input the information, special devices will be installed in authorized vehicles to allow the messages to be encrypted and encoded in a manner that can be later used by civilian vehicles' navigation and in-vehicle information systems. Unfortunately, allowing certain vehicles the ability to insert information into NOTICE also makes the system vulnerable to attack from an adversary that hijacks one of these vehicles. One solution would be to require authorized personnel to provide a password, or even a biometric identifier, before inserting information into the system.

Role-based communication plays a large part in incident detection. Belts may infer a traffic slowdown based on the number of vehicles passing by and the information obtained from those vehicles' EDRs. However, the reason behind the slowdown may not be evident. Many traffic slowdowns are caused by one driver traveling much slower than others, or drivers slowing down to look at an accident in the opposite lane (also known as rubbernecking). In the case of an actual accident, police and other emergency responders are notified (often by witnesses via cell phone).[32] As these responders drive towards the accident, they can provide verification about the reason for the traffic slowdown to the belts. This information can then be relayed to drivers approaching the incident. These messages will also serve to validate and corroborate automatic incident detection performed by NOTICE, as will be described in Section 7.5.3.

7.5.2 Realizing NOTICE

By design, the model for belt-to-vehicle wireless communications in NOTICE allows for anonymous data exchange while preventing impersonation attacks. Here, further approaches to providing security against confidentiality and denial-of-service (DoS) attacks are described, as well as an approach for fault tolerance in NOTICE.

7.5.2.1 Security Solutions in NOTICE

The threat model used here involves an adversary that has not infiltrated the system and has no knowledge of the algorithms for changing the frequency channels employed for belt-to-vehicle communication. It is assumed that the adversary does not know the algorithm by which adjacent belts change their symmetric keys. These solutions are designed to mitigate confidentiality and DoS attacks.

NOTICE provides built-in security for confidentiality attacks. This is because, as discussed in Section 7.5.1, the handshaking protocol between a belt and a passing vehicle involves communication with low transmission range (~1 m) and, consequently, is not intelligible to the adversary, even positioned by the roadside. An intrinsic part of the first handshake is the establishment of a frequency on which subsequent communication between the belt and the passing vehicle will be carried. This, then, cannot be understood by the adversary.

DoS attacks can take several forms, and mitigating their effects is a goal of NOTICE. One type of DoS attack involves inserting false information into a belt, resulting in the information being propagated to many other belts and vehicles. NOTICE has a mechanism for handling false positives. A belt will not react to a single incident reported by a vehicle. Instead it will wait for subsequent corroborations of the reported incident before deciding to propagate the information to the other participants in the traffic. In addition, because NOTICE belts can directly observe traffic conditions by counting the numbers of vehicles passing over, the belts have a method for independently validating some of the information given by passing vehicles. This is one of the main advantages of NOTICE over systems that cannot independently observe events.

Another form of DoS attack involves an adversary laying down a rogue belt on top of a valid NOTICE belt. The effect of this is that instead of dropping off a message with the valid belt, the vehicle will drop off the message with the rogue belt and will pick up a message from the rogue belt. Assuming that the adversary does not know the symmetric key between the belts concerned, it cannot decode the messages, and the message uploaded from the rogue belt will not be decoded correctly by the next belt. If this is repeated by several passing vehicles, this belt will conclude that the previous belt is faulty and this information will be propagated in the reverse direction. As a result, vehicles passing the rogue belt will not drop off or pick up messages, essentially thwarting the adversary's attack.

7.5.2.2 Making NOTICE Fault-Tolerant

An important issue in NOTICE is fault tolerance and the consideration of what would happen if a belt was disabled as a result of sensor malfunction, power failure, or lightning. Figure 7.7 illustrates one approach to providing fault tolerance in NOTICE. Assume vehicle a is traveling between belts D and E. Recall that belts D and E share a symmetric key $\mu(D, E, t)$ used to encrypt, at time t, all the messages carried between belts D and E. If belt E is disabled, the messages uploaded by belt D are lost, because they cannot be decoded by the next active belt, F. This has the effect of interrupting the flow of traffic-related information, drastically curtailing the usefulness of NOTICE.

To address this problem, a simple and natural cluster structure is induced on the set of belts by grouping k adjacent belts into a cluster, where k is the *clustering factor*. In this clustering scheme, every belt belongs to exactly one cluster and, therefore, the resulting clusters are disjoint. The partitioning of the belts into clusters can be static or dynamic. In either case, each belt is informed about the identity of the cluster to which it belongs. Two adjacent clusters, with a clustering factor of $k = 4$, are illustrated in Figure 7.7. Here, belts $A, B, C,$ and D belong to cluster $C1$ while belts $E, F, G,$ and H belong to cluster $C2$.

In the clustered version of NOTICE, all of the belts in a cluster C_i know two keys, $\mu(C_{i-1}, C_i, t)$ and $\mu(C_i, C_{i+1}, t)$. Symmetric key $\mu(C_{i-1}, C_i, t)$ is used to encrypt, at time t, messages originating in cluster C_{i-1} and destined for some belt in cluster C_{i-1} or C_i. Symmetric key $\mu(C_i, C_{i+1}, t)$ is used to encrypt, at time t, messages originating in cluster C_i and destined for some belt in cluster C_i or C_{i+1}.

Referring again to the example in Figure 7.7, vehicle a carries a message encrypted by belt D with key $\mu(C_1, C_2, t)$. If belt E is active, then upon reaching this belt, the message

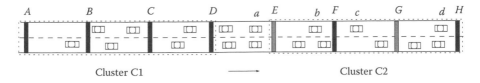

FIGURE 7.7   Clustering and fault tolerance in NOTICE.

is properly encrypted and a new (or the same) message encrypted with the key $\mu(C_2, C_3, t)$ is uploaded onto vehicle a. However, if belt E is disabled, then a fails to establish communication with E and carries the message to the next belt F. Assuming that belt F is active, it correctly decrypts the message and uploads to vehicle a a new message encrypted with key $\mu(C_2, C_3, t)$ informing subsequent belts in the same cluster that belt E is disabled. Observe that key $\mu(C_2, C_3, t)$ must be used because belt E does not know the number of remaining active belts in its own cluster. Determining the number of active belts in a cluster is an avenue for future work.

The above setup guarantees that as long as a cluster contains active belts, messages can be correctly propagated according to the NOTICE semantics. The amount of fault tolerance conferred by this clustering scheme needs to be considered. For this purpose, consider a NOTICE system consisting of n consecutive belts with a clustering factor of k $(k > 1)$. A good measure of the resulting fault tolerance is the probability of the event X that, given that k belts out of the n fail, all of these k belts belong to one of the $\lfloor n/k \rfloor$ clusters. This is, precisely, the probability that NOTICE will fail to propagate messages end to end.

It is assumed that belts fail uniformly at random. To evaluate the probability that NOTICE will fail to propagate messages, the sequence of n belts is considered as a string σ of n symbols A and D, with A denoting an active belt, while D represents a disabled belt. Assuming, as above, that exactly k belts are disabled, NOTICE can fail only if the k belts form a *run* of k consecutive Ds in σ. It is not hard to show (see for example, Ref. 33, p. 62) that the probability $\Pr[X]$ of event X satisfies

$$\Pr[X] < \frac{\lfloor n/k \rfloor}{n-k+1} \cdot \frac{\binom{k-1}{0} \cdot \binom{n-k+1}{1}}{\binom{n}{k}} = \frac{\lfloor n/k \rfloor}{\binom{n}{k}} \leq \frac{n}{k} \cdot \binom{n}{k}^{-1} = \binom{n-1}{k-1}^{-1}.$$

To get a better feel for $\Pr[X]$, consider the case of $n = 100$ and $k = 4$. In this case, given that four belts in the system fail, the probability that some cluster is wiped out is bounded above by $\binom{99}{3}^{-1}$, which works out to be $< 6.4 \times 10^{-6}$, an extremely small probability.

7.5.2.3 Determining Alternative Route Travel Times

The wealth of information collected by NOTICE can be used to help the police and authorized personnel with the important task of incident management. Figure 7.8 shows a divided highway where a multivehicle accident has occurred and blocks both lanes in one direction. The first responders have been alerted and are heading to the scene of the accident. It is anticipated that the accident will take quite a while to clear and, consequently, it is important to alert the drivers approaching the accident area to various detour options. We now outline a simple way in which NOTICE can identify and propagate information about a detour while attempting to minimize the overall detour time.

In Figure 7.8 there are three possible exits, namely, $X1$, $X2$, and $X3$, allowing traffic to detour around the accident area. Assume that the detours end at exit Xk, past the

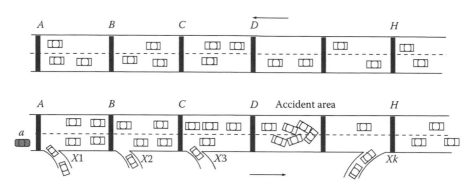

FIGURE 7.8 Incident management in NOTICE.

accident area, just before belt *H*. Belt *H* plays a crucial role in informing the vehicles approaching the accident area of the average duration of the various detour options. Indeed, by querying the vehicles re-entering the roadway at exit *Xk*, belt *H* calculates the average duration for each possible detour option. Every few minutes these averages are updated and propagated to the drivers approaching the accident area. Assume that a vehicle can record where and at what time it left the highway. Also, assume that the vehicle would ignore the time when the engine was turned off (e.g., while refueling) and that vehicles having shutdown times over a certain threshold will not respond to the detour query from the belt.

As expected, the average detour times are highly dynamic (time-dependent) and may change with every update. As an example, suppose that at time t_1, exit *X2* has the shortest average detour time (Table 7.1). As this information is propagated back to the vehicles behind the accident, the drivers will tend to prefer exit *X2*. In turn, due to increased congestion on the secondary roads, this will result in longer and longer detour times for the roads served by exit *X2*. This will be reflected in the averages computed later at time t_2 (Table 7.2). At a later time, exit *X1* may be the most promising one and will be preferred by an increasing number of drivers. This is, then, continued until the accident has been cleared. This tendency towards equilibrium is well known in the traffic management field[34] and works best when drivers have good information about recent congestion levels on secondary roads.

TABLE 7.1 Illustrating the Detour Tables at Time t_1

Exit	Average Detour (min)	Itinerary
X1	17	Hwy 239 to Hwy 244
X2	13	Hwy 233 to Hwy 264
X2	15	Hwy 233 to Hwy 244
X3	21	Hwy 229 to Hwy 244

TABLE 7.2 Illustrating the Detour Tables at Time t_2

Exit	Average Detour (min)	Itinerary
X1	15	Hwy 239 to Hwy 244
X2	22	Hwy 233 to Hwy 264
X2	21	Hwy 233 to Hwy 244
X3	23	Hwy 229 to Hwy 244

In spite of the simplicity of this solution, a number of important issues remain open. Perhaps the most important among them is to determine how far to propagate information about the accident in the absence of topological information about the area. If such topological information were available, belt H could determine the most suitable exits, for example exits X1, X2, and X3, and only propagate information to drivers approaching them. It would be of interest to develop a solution that does not rely on topological information and that adapts dynamically to the situation. One approach could be to use knowledge of standard traffic models in times of congestion.[35]

7.5.2.4 Using NOTICE in Support of Evacuations

In cases of predicted disasters, such as hurricanes, massive evacuations are often necessary in order to minimize the impact of the disaster on human lives. However, there are several issues involved in a large-scale evacuation. For example, once an evacuation is under way, finding available resources, such as gasoline, drinking water, and shelter, quickly becomes an issue.[36,37] In its recent report on hurricane evacuations,[38] the U.S. DOT found that emergency evacuation plans often do not even consider availability of such resources. U.S. DOT also determined that emergency managers need a method for communicating with evacuees during the evacuation in order to provide updated information. The report suggested that traffic monitoring equipment should be deployed to provide real-time traffic information along evacuation routes. NOTICE can be used to provide travel time estimates, notification of available resources, such as gasoline, food, and shelter, and notification of contraflow roadways.

To facilitate this type of monitoring and information dissemination using NOTICE, emergency officials can place temporary hardware in the form of support vehicles, sturdy tamperproof devices, or minitowers in the median, about every 10 mi in advance of an evacuation. It is important to note that these devices are temporary and would only be deployed in an emergency. The temporary devices would plug into the belt system and connect NOTICE to the emergency management center overseeing the evacuation. This would provide the belts with important information that cannot be determined solely by NOTICE. In addition, these devices can use powerful radios to send backward messages in the case of contraflow traffic, where all lanes travel in the same direction, away from the evacuation zone.

Using the temporary hardware, NOTICE can query vehicles for travel times and upload that information to the emergency management center. This vital travel time information could then be released to the public via the news media (TV and radio) and the Internet, which would allow potential evacuees to make informed decisions about if and when to leave the area. In addition, this aggregated information would be fed back into the belt system to give drivers the same information as they pass over the belts.

To determine stations with available gasoline, the EDR in a vehicle can monitor its gas tank and determine when gas was added. In addition, the vehicle will know how far and for how long it has been driven since the gas tank was last filled. The belt closest to each highway on-ramp can query traffic to see when and where they last added gasoline. This is similar to the query by the belts regarding detour information, as described in Section 7.5.2. NOTICE could then use this information to inform travelers of nearby available gas stations. To facilitate the dissemination of additional information, state emergency agencies could require that gas station operators, hotel operators, and restaurants who remain open during the evacuation provide accurate information to the central server, which would then provide this information to drivers via NOTICE. In this same way, emergency managers could upload information about open shelters to the central server. It is important to note that this system would be used to facilitate an evacuation before a disaster strikes, so it is assumed that electricity and network connections are available.

In addition to having state authorities send information to the belts about evacuations or contraflow lanes, using role-based communication as described earlier, the belts themselves could determine the direction and speed that traffic is flowing. Drivers entering entrance ramps onto contraflow roadways (these ramps would likely have been used as exit ramps previously) could be alerted to the direction that traffic is moving. Belts on one roadway could also alert drivers to upcoming entrance ramps that were previously used as exit ramps during noncontraflow travel. As an additional feature, as the belt system can monitor traffic flow, NOTICE could offer recommendations for which roadways are being heavily traveled in only one direction. These roadways are likely good candidates for contraflow.

7.5.3 Incident Detection

One of the issues of any traffic incident notification system is how to input information about incidents into the system. Role-based communications can allow authorized agencies to insert notices into the system, and the NOTICE belts can independently observe traffic conditions by directly monitoring how many vehicles pass over them, whether or not the vehicles are equipped with NOTICE communications devices. The goal of NOTICE is to detect incidents and slowdowns without driver intervention. Drivers should not have to interact with a device to report an incident. Driver distraction is an important issue with the development of new in-vehicle devices that compete for a driver's attention. Allowing unauthorized drivers to directly input warnings would also make the system susceptible to false alerts.

Current traffic-monitoring techniques perform some amount of automated incident detection (AID) using well-known algorithms. Most of the algorithms, however, are threshold-based, meaning that traffic must be monitored for some amount of time to determine what "normal" is. Then, conditions that are different from "normal" are flagged as incidents. The unique communications between the monitoring points (belts) in NOTICE can be used to develop AID algorithms that are self-calibrating. Standard traffic models that describe normal traffic intensity variations can be used to detect significant changes in traffic intensity. In addition, the belts could also use the *rubber-neck* effect that often afflicts drivers in lanes opposite an accident. This occurs when drivers who are not hindered slow down to look at the accident. As belts in opposite lanes are connected, one belt seeing very little traffic and the opposite belt seeing a traffic slowdown could infer that there has been an incident.

7.6 Conclusion

Infrastructure, whether roadside or embedded in the highway, is an important part of vehicular network systems. This chapter has provided an overview of how infrastructure can be used to help provide security and privacy for VANET applications, such as in the CARAVAN and VIPER systems. The chapter has also described how some commercial and infotainment systems, such as AdTorrent and VANETCODE, take advantage of roadside infrastructure. A large portion of the chapter was dedicated to describing the NOTICE system, in which infrastructure plays an integral role. Sensor belts that are embedded in the roadway allow NOTICE to provide essential traffic monitoring and information dissemination functions.

References

1. Golle, P., Greene, D., and Staddon, J., Detecting and Correcting Malicious Data in VANETs, in Proceedings of the ACM Workshop on Vehicular Ad Hoc Networks (VANET), 2004, pp. 29–37.
2. Leinmüller, T., Maihofer, C., Schoch, E., and Kargl, F. Improved Security in Geographic Ad Hoc Routing Through Autonomous Position Verification, in Proceedings of the ACM Workshop on Vehicular Ad Hoc Networks (VANET), Los Angeles, CA, September 2006, pp. 57–66.
3. Raya, M., and Hubaux, J.-P., The Security of Vehicular Ad Hoc Networks, in Proceedings of the ACM Workshop on Security of Ad Hoc and Sensor Networks, November 2005.
4. Raya, M., Papadimitratos, P., and Hubaux, J.-P., Securing vehicular communications, *IEEE Wireless Communications Magazine*, 13(5), 8–15, 2006.
5. Hubaux, J.-P., Capkun, S., and Luo, J., The security and privacy of smart vehicles, *IEEE Security and Privacy Magazine*, 2(3), 49–55, 2004.
6. Parno, B. and Perrig, A., Challenges in Securing Vehicular Networks, in Proceedings of ACM HotNets, 2005.
7. Sampigethaya, K. et al., CARAVAN: Providing Location Privacy for VANET, in Proceedings of the Workshop on Embedded Security in Cars (ESCAR), Cologne, Germany, 2005.
8. Chokhani, S. and Ford, W., Internet X.509 Public Key Infrastructure Certificate Policy and Certifications Practices Framework, RFC 2527, March 1999.
9. IEEE, IEEE P1609.2/D2—Draft Standard for Wireless Access in Vehicular Environments—Security Services for Applications and Management Messages, November 2005.

10. Choi, J.Y., Jakobsson, M., and Wetzel, S. Balancing Auditability and Privacy in Vehicular Networks, in Proceedings of the 1st ACM International Workshop on Quality of Service and Security in Wireless and Mobile Networks (Q2SWinet), Montreal, Quebec, Canada, October 2005, pp. 79–87.

11. Armknecht, F., Festag, A., Westhoff, D., and Zeng, K., Cross-layer Privacy Enhancement and Non-repudiation in Vehicular Communication, in Proceedings of the 4th Workshop on Mobile Ad-Hoc Networks (WMAN), Bern, Switzerland, March 2007.

12. Fonseca, E., Festag, A., Baldessari, R., and Aguiar, R. Support of Anonymity in VANETs—Putting Pseudonymity into Practice, in Proceedings of the IEEE Wireless Communications and Networking Conference (WCNC), Hong Kong, March 2007.

13. Dötzer, F., Privacy Issues in Vehicular Ad Hoc Networks, in Workshop on Privacy Enhancing Technologies, Cavtat, Croatia, May 2005.

14. Papadimitratos, P., Kung, A., Hubaux, J.-P., and Kargl, F., Privacy and Identity Management for Vehicular Communication Systems: A Position Paper, in Proceedings of the Workshop on Standards for Privacy in User-Centric Identity Management, July 2006.

15. Douceur, J., The Sybil Attack, *Lecture Notes in Computer Science: Revised Papers from the First International Workshop on Peer-to-Peer Systems*, 2429, 251–260, 2002.

16. Di Crescenzo, G., Zhang, T., and Pietrowicz, S., Anonymity Notions for Public-Key Infrastructures in Mobile Vehicular Networks, in Proceedings of the IEEE International Workshop on Mobile Vehicular Networks (MoVeNet), Pisa, Italy, October 2007.

17. Cencioni, P., and Di Pietro, R., VIPER: A Vehicle-to-Infrastructure Communication Privacy Enforcement Protocol, in Proceedings of the IEEE International Workshop on Mobile Vehicular Networks (MoVeNet), Pisa, Italy, October 2007.

18. Golle, P., Jakobsson, M., Juels, A., and Syverson, P., Universal Re-encryption for Mixnets, in Proceedings of Topics in Cryptology (CT-RSA), 2004, pp. 163–178.

19. Wiggins, A.E., A Review of Cashless Electronic Toll Collection Technology, in Proceedings of Transport Session, ICAP Expo, 1994.

20. Caliskan, M., Graupner, D., and Mauve, M., Decentralized Discovery of Free Parking Places, in Proceedings of the ACM Workshop on Vehicular Ad Hoc Networks (VANET), 2006, pp. 30–39.

21. Nandan, A., Das, S., Zhou, B., Pau, G., and Gerla, M., AdTorrent: Digital Billboards for Vehicular Networks, in Proceedings of Vehicle-to-Vehicle Communication (V2VCom), San Diego, CA, July 2005.

22. Nandan, A., Tewari, S., Das, S., Gerla, M., and Kleinrock, L. AdTorrent: Delivering Location Cognizant Advertisements to Car Networks, in Proceedings of the International Conference on Wireless On Demand Network Systems and Services (WONS), 2006.

23. Lee, S.-B., Pan, G., Park, J.-S., Gerla, M., and Lu, S., Secure Incentives for Commercial Ad Dissemination in Vehicular Networks, in Proceedings of the 8th ACM International Symposium on Mobile Ad Hoc Networking and Computing (Mobihoc), Montreal, Quebec, Canada, 2007, pp. 150–159.

Applications

24. Ahmed, S. and Kanhere, S.S., VANETCODE: Network Coding to Enhance Cooperative Downloading in Vehicular Ad-Hoc Networks, in Proceedings of the International Conference on Wireless Communications and Mobile Computing (IWCMC), Vancouver, British Columbia, Canada, 2006, pp. 527–532.
25. Lee, K.C., Lee, S.-H., Cheung, R., Lee, U., and Gerla, M., First Experience with CarTorrent in a Real Vehicular Ad Hoc Network, in Proceedings of the IEEE Workshop on Mobile Networking for Vehicular Environments (MOVE), May 2007.
26. Gerla, M. et al., Vehicular Grid Communications: The Role of the Internet Infrastructure, in Proceedings of the International Workshop on Wireless Internet (WICON), Boston, MA, 2006.
27. Weigle, M.C., and Olariu, S., Intelligent Highway Infrastructure for Planned Evacuations, in Proceedings of the IEEE International Workshop on Research Challenges in Next Generation Networks for First Responders and Critical Infrastructures (NetCri), New Orleans, LA, April 2007.
28. Abuelela, M., Olariu, S., and Weigle, M.C., NOTICE: An Architecture for Notification of Traffic Incidents, in Proceedings of the IEEE Vehicular Technology Conference—Spring, Singapore, May 2007, pp. 3001–3005.
29. Plößl, K., Nowey, T., and Mletzko, C., Towards a Security Architecture for Vehicular Ad Hoc Networks, in Proceedings of The First International Conference on Availability, Reliability, and Security (ARES 2006), 2006, pp. 374–381.
30. Virginia Department of Transportation, Commonwealth of Virginia's Strategic Highway Safety Plan, 2006–2010, available at http://virginiadot.org/info/resources/Strat_Hway_Safety_Plan_FREPT.pdf, 2006.
31. U.S. Department of Transportation, Standard Specification for Telecommunications and Information Exchange Between Roadside and Vehicle Systems—5 GHz Band Dedicated Short Range Communications (DSRC) Medium Access Control (MAC) and Physical Layer (PHY) Specifications, ASTM E2213-03, August 2003.
32. U.S. Federal Highway Administration, Traffic Incident Management Handbook, available at http://www.itsdocs.fhwa.dot.gov/JPODOCS/REPT_MIS/13286.pdf, November 2000.
33. Feller, W., *An Introduction to Probability Theory and Its Applications*, 3rd ed., Vol. I, John Wiley & Sons, 2000.
34. Juan de Dios Ortúzar and Willumsen, L.G., *Modelling Transport*, Wiley, 2002.
35. May, A.D., *Traffic Flow Fundamentals*, Prentice-Hall, 1990.
36. Harden, B. and Moreno, S., Thousands fleeing Rita jam roads from coast, *Washington Post*, September 23, 2005.
37. Feldstein, D. and Stiles, M., Too many people and no way out, *Houston Chronicle*, September 25, 2005.
38. U.S. Department of Transportation, Catastrophic Hurricane Evacuation Plan Evaluation: A Report to Congress, available at http://www.fhwa.dot.gov/reports/hurricanevacuation/, June 2006.

8

Content Delivery in Zero-Infrastructure VANETs

Mahmoud Abuelela
and Stephan Olariu
*Department of
Computer Science
Old Dominion University*

8.1 Introduction and Motivation

Since the late 1990s, vehicular ad hoc networks (VANETs), mobile ad hoc networks (MANETs), vehicle-to-vehicle (V2V), and vehicle-to-infrastructure (V2I) communications have received a great deal of attention in the research community.[1] There is good reason for this. Vehicular communications promise to integrate driving into a ubiquitous and pervasive network that is already redefining the way we live and work. The potential societal impact of VANET has been confirmed by the proliferation of consortia and initiatives involving car manufacturers, government agencies, and academia, including, among others, the Car-2-Car Communication Consortium, the Vehicle Safety Consortium, the Networks-on-Wheels Project, the Vehicle Infrastructure Integration Program, and the Advanced Safety Vehicle Program.[2-6]

The original impetus for VANET was traffic safety,[2,3] but more recent concerns involve privacy and security,[7,8] peer-to-peer (P2P) networking,[9-11] as well as integrating VANET into the fast-growing *infotainment* industry.[1,12] The allocation of the 75 MHz spectrum in the 5.9 GHz band for dedicate short-range communications (DSRC)[2] in North America makes it possible to deliver high-data-rate multimedia applications via vehicle-to-roadside (V2R) and V2V wireless links.

In spite of their close resemblance to MANET networks, with which they share the same underlying philosophy, VANET networks have a number of specific characteristics that set them apart from MANET. First, although most MANET networks are deployed in support of special-purpose operations including disaster relief, search-and-rescue, law enforcement, and multimedia classrooms, all of which are intrinsically short-lived and involve a small number of nodes, VANET networks may involve thousands of fast-moving vehicles over tens of miles of roadways and streets. Second, it has been recently noticed that although MANET networks may experience transient periods of loss of connectivity, in VANETs, especially under sparse traffic conditions, extended periods of disconnection are the norm rather than the exception.[13-15] This state of affairs has a significant implication for routing, rendering traditional MANET routing protocols unsuitable for VANETs. To address the specific needs of VANETs, several routing protocols specifically designed for VANETs have been proposed in the literature.[14]

Most P2P file-sharing systems (e.g., Gnutella, BitTorrent) were developed for wired IP-based networks and do not work in MANETs and VANETs without considerable modifications. Recently, because of a potentially growing consumer market, ushered in by the proliferation of VANETs, several P2P schemes targeting VANETs have been proposed.[9-11,16]

Somewhat surprisingly, none of the P2P systems proposed for VANETs is *zero-infrastructure* in form. Instead, they all rely on various instances of V2I communications. In fact, all of the P2P systems rely on pre-existing roadside infrastructure installed every two miles or so. The cost of installing and maintaining this infrastructure, which by all accounts is likely to be exorbitant, casts a long shadow of doubt on the feasibility and scalability of such systems—especially so in developing countries. Also, most of the existing P2P techniques developed for VANETs restrict communications to physical neighbors or they allow a multihop routing. Restricting communication to physical neighbors makes it hard to find data of interest in these neighbors. On the other hand, routing between vehicles far away from each other is very difficult in VANETs that may scale up to tens of thousands of nodes, and theoretically, all of the nodes can be users running P2P protocols. Even with crosslayer optimization, no conventional MANET routing protocols are expected to support such large networks.

The remainder of this chapter is organized as follows. In Section 8.2 we discuss previous work in the area of P2P in VANETs. In Section 8.3 we introduce TAPR,[17] a recently proposed packet-relaying protocol for VANET. An application scenario to illustrate the idea of ZIPPER is discussed in Section 8.4. Section 8.5 introduces ZIPPER in detail. Section 8.6 offers a detailed analytical performance analysis of ZIPPER. Section 8.6.1 discusses an extension of the classic Coupons Collector Problem[18] that arises naturally in the analysis of ZIPPER. Section 8.6.2 offers a simulation-based evaluation of ZIPPER. Finally, Section 8.7 contains concluding remarks and directions for further work.

8.2 State of the Art

The CodeTorrent architecture[11] uses single-hop communication; multihop routes are never used and thus are not required to be maintained explicitly by any layer in the protocol stack. Logical peers, nodes exchanging file pieces, are restricted to physical neighbors. In CodeTorrent, a node that intends to share a file, a seed node, creates and broadcasts to its 1-hop neighbor the description of that file, where each file is stored as a set of n pieces. If any of the 1-hop neighbors is interested in a file, nodes will apply the idea of network coding and will exchange coded frames instead of the file pieces, where a frame is a linear combination of file pieces. When the interested node collects n independent frames, it can decode and recover the entire file.

In spite of its appeal, CodeTorrent has some disadvantages. First, each seed node should broadcast its file information even if no driver is interested in it, which creates unnecessary messages and consumption of resources. Second, if many nodes possess more data to share, the wireless medium will become saturated, because every node will broadcast information about every single file it possesses, requiring a perfect MAC protocol to manage this heavy load. Finally, although network coding is efficient in maximizing the throughput, it is not practical for multimedia streaming, where a user cannot wait to collect the entire file before he can watch or listen to it. By contrast, nodes in the proposed protocol (TAPR) exchange file blocks instead of coded frames. So, a driver may enjoy watching or listening to the first block while downloading subsequent blocks.

In CarTorrent,[16] a work that extends the BitTorrent protocol to the vehicular network scenarios, addressing issues such as intelligent peer and piece selection, given the intermittent connectivity to a preinstalled access point, was proposed. In this work, Lee et al. have implemented and deployed CarTorrent on a real VANET, is the first implementation of a content-sharing application on a real vehicular ad hoc testbed. However, given the hundreds of highway miles for which there is hardly enough budget to carry out maintenance, installing gateways every 2 to 10 mi will be very expensive and not therefore a practical solution.

In VANETCODE,[9] some stationary gateways are supposed to be installed along the road at regular intervals of about 2 to 10 mi as in Ref. 19, and all files are present in their entirety at these static gateways. A gateway splits the file into k blocks, codes all pieces with some random coefficient, and broadcasts the coded block to all interested vehicles covered by its area. Decoding the data requires a node to capture a sufficient number of blocks from its neighbors so that instead of waiting for the next gateway, they can cooperate with each other to collect the entire file.

VANETCODE has a number of problems. First, the gateways are the only sources of data, and we should expect all data of interest to be existing and replicated at all gateways. Second, there is no sharing of the content stored in different vehicles. Third, the system does not scale, and adding or modifying data to all gateways is an issue; again the idea of network coding is not suitable for multimedia streaming, as shown with CodeTorrent.

In PAVAN,[10] the existing infrastructure of a cellular network is used to broadcast a file description to all vehicles in the current area. If a driver is interested in the file, a route should be discovered and maintained between it and the owner of the file. However, scalability is an issue in PAVAN. As the number of vehicles transmitting their file descriptions increases, the cellular network, which supports only a few tens of kilobits per second,[19] will not be able to carry such a load. By contrast, the proposed protocol (TAPR) relies only on vehicles on the road to request and download a file.

8.3 TAPR: A Traffic-Adaptive Packet-Relaying Protocol

TAPR[17] is a traffic-adaptive packet-relaying protocol that is designed specifically for VANETs.

8.3.1 TAPR Motivation

VANETs have a number of specific characteristics that set them apart from MANETs. First, although MANETs are deployed in support of special-purpose operations, all of which are short-lived and involve a small number of players, VANETs involve thousands of fast-moving vehicles over tens of miles of streets and roadways. Second, MANETs may experience transient periods of loss of connectivity, but in VANETs, especially under sparse traffic conditions, extended periods of disconnection are the norm. Directional propagation protocol (DPP)[14] was the first protocol to consider traffic scenarios in which vehicles form disconnected clusters on the road. For routing purposes, DPP prefers clusters codirectional with the packet; when disconnection occurs between two codirectional clusters, clusters in the other direction may be used as bridges to the next codirectional cluster.

8.3.2 Expected Cluster Size

Consider m cars on a single-lane road. The m cars determine $m-1$ intercar segments, populated by (unit-length) objects. Altogether there are n of these objects. Assuming that the m cars and n objects are distributed uniformly at random, the probability $p(m, n, d)$ that between two adjacent cars there are at least $d+1$ objects is $\binom{m+n-d-3}{m-2}$ $\binom{m+n-2}{n}^{-1}$. As we wish to eliminate boundary effects, we assume a very long mighway, where both m and n are very large. However, we assume a uniform traffic density in the sense that for some constant $0 < \lambda < 1$,

$$\lim_{n,m \to \infty} \frac{m}{n} = \lambda. \tag{8.1}$$

If we write $p(d) = \lim_{n,m \to \infty} p(m, n, d)$. Given Equation 8.1, it is easy to confirm that

$$p(d) = (1 + \lambda)^{-(d+1)}. \tag{8.2}$$

In TAPR, two cars are out of communication range if the corresponding intercar segment contains at least $d+1$ objects. Let X be the random variable that counts the number of "gaps" (i.e., the number of intercar segments containing at least $d+1$ objects). It is clear that $E[X] = (m-1)(1+\lambda)^{-(d+1)}$ and, consequently, the expected size of a cluster is

$$\lim_{m \to \infty} \frac{m}{1 + (m-1)(1+\lambda)^{-(d+1)}} = (1+\lambda)^{d+1}. \tag{8.3}$$

8.3.3 Technical Details of TAPR

Cars are assumed to be GPS-enabled and communicate using DSRC operating in the 5.9 GHz band.[2] Being GPS-enabled, the cars know their geographic position and their clocks are synchronized. As mandated by DSRC, every 300 msec each vehicle sends a beacon message with a range of about 200 to 300 m. This beacon contains information that allows vehicles to handshake and synchronize. TAPR uses these beacons for cluster formation and cluster maintenance as well. Mindful of their original intent we shall, nonetheless, refer to these beacons as cluster management beacons (CMB). Using a simple time-out-based strategy, a vehicle that detects no other vehicles in front of (resp. behind) it, declares itself head (resp. tail) of the cluster and piggybacks the information on CMBs.

TAPR accommodates disconnected VANETs and, accordingly, operates in *disconnected* and *connected* modes. Referring to Figure 8.1, assume that the source vehicle has a packet to send to the destination as shown in the figure. In disconnected mode, the source cannot send the packet to the next oncoming cluster through a codirectional cluster. In this case, the source will simply route the packet to the head of its cluster. The head vehicle will wait until it either meets the destination, thus performing the job of a data mule, or it will send the packet to the next oncoming cluster through codirectional clusters. Connected mode occurs when there exists a codirectional cluster that overlaps the source cluster and the next oncoming cluster. In this case, routing proceeds along a shortest path to the farthest oncoming cluster that can be reached. Thus, the basic idea of TAPR is to switch between muling and routing according to current traffic conditions.

8.3.4 Performance Analysis

A 2 km stretch of undivided road was assumed, with one lane of traffic in each direction. In each lane, vehicles were deployed uniformly at random, and for simplicity the size of a vehicle was ignored. Figure 8.2 shows a comparison between TAPR and DPP in terms of the number of messages sent by both protocols to route a packet over the 2 km stretch of road. For low traffic densities, we expect to have clusters that may not overlap. So, TAPR sends fewer messages than DPP because TAPR chooses to mule whenever there is no connection to the next oncoming cluster on the road. By contrast, DPP does not detect

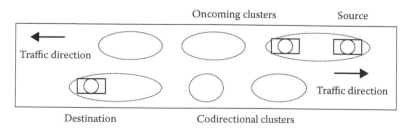

FIGURE 8.1   Illustrating TAPR.

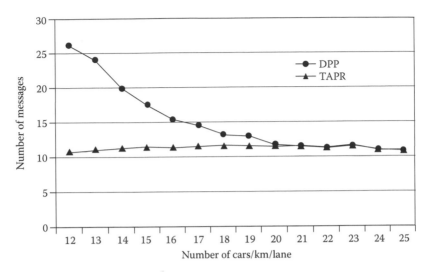

FIGURE 8.2   A Comparison of TAPR and DPP.

this fact and sends the message over a codirectional cluster that ends up at the same oncoming cluster again, which wastes resources.

As the traffic becomes denser, clusters will overlap with high probability and both DPP and TAPR choose to route over the connected clusters. Thus, for high densities, DPP and TAPR behave almost in the same way.

8.4   ZIPPER: An Application Scenario

ZIPPER[20] is a zero-infrastructure P2P system for VANETs that is implemented as a multithreaded architecture that uses TAPR as the underlying packet-relaying protocol. To set the stage for describing ZIPPER contributions, imagine a car cruising down an undivided highway. It is clear that, per time unit, our car will meet far more oncoming cars than codirectional vehicles. This is due to the fact that the relative speed of cars moving in opposite directions is the sum of their absolute speeds, whereas codirectional cars appear to be moving, relative to one another, at the difference of their absolute speeds. In particular, if the majority of codirectional cars are keeping to the legal speed, the relative speed of our car with respect to the traffic going in the same direction is almost nil. As a result, routes established between codirectional cars tend to be stable. This observation has motivated a number of workers to propose establishing routes consisting entirely of codirectional cars. However, it was recently noticed[13–15] that in many highway scenarios, codirectional traffic consists of disjoint clusters with no connectivity between them. This, in turn, implies that end-to-end connectivity between codirectional cars is not guaranteed to exist.

In some situations, a route may be easily established between two codirectional cars, for example when the traffic between them is not sparse and stable. On the other hand when the traffic is sparse or it is highly dynamic, it may be impossible to establish and

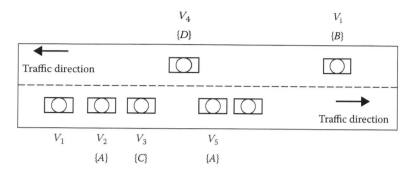

FIGURE 8.3 ZIPPER—an application scenario.

maintain such a route. In the latter case, using oncoming cars as data mules will be the only way to allow communication between two codirectional cars.

The application scenario illustrated in Figure 8.3 is used to introduce the basic idea of ZIPPER. Suppose that the driver of vehicle V_1 initiates a request for a certain movie that consists of four blocks A, B, C, and D. Let us also suppose that vehicle V_2 possesses block A and vehicle V_3 possesses block C, where V_2 and V_3 are within communication range of V_1. Figure 8.4 shows a sequence diagram illustrating the basic control messages (dotted lines) and data messages (solid lines) exchanged between V_1, V_2, and V_3, which can be described as follows. First V_1 broadcasts its request for A, B, C, and D to both V_2 and V_3, as both are 1-hop away from V_1. Both V_2 and V_3 start searching their databases for the required blocks. So, V_2 replies with a control message that it possesses block A, while V_3 replies that it possesses block C. Then V_1 sends the third control message $Ack\{B, D\}$ that contains the new

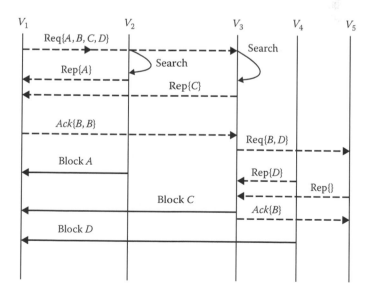

FIGURE 8.4 Application scenario sequence diagram.

remaining blocks to V_3, which is the farthest vehicle from V_1. Now V_1 can download blocks A and C from vehicles V_2 and V_3, respectively. Upon receiving $Ack\{B, D\}$, V_3 can impersonate V_1 in propagating the request and receiving control messages from V_4 and V_5 before sending Ack again to the farthest vehicle and so on. If any vehicle that is not a physical neighbor of V_1 possesses any block, TAPR will be used to relay that block to V_1.

To illustrate the importance of these control messages, let us imagine that they do not exist. So, the application scenario in Figure 8.3 could be explained as follows. V_1 broadcasts its request for blocks A, B, C, and D to both V_2 and V_3. After searching their databases, V_2 may send block A with low adjusted power to V_1, and V_3 sends block C to V_1. In order to propagate the request to the next vehicles on the road, V_3 will update the remaining blocks to be $\{A, B, D\}$ as it did not hear V_2, and V_2 would propagate the new request for blocks $\{B, C, D\}$ to the next vehicles on the road. Obviously, many redundant blocks will be routed to V_1, and the vehicles may not even be able to decide when to stop propagating requests once all blocks have been collected.

8.5 ZIPPER: The Details

In ZIPPER, files are stored as a collection of blocks, and a vehicle might not possess the entire set of blocks for a certain file, as some may be in the process of being downloaded. The typical block size is 256 kB as in most P2P protocols. It is worth mentioning that many blocks could be sent at once because the bandwidth of DSRC is up to 27Mb/sec.[2]

ZIPPER requires no preinstalled infrastructure along the road, so it can be implemented at no extra cost. The following are assumed about vehicles:

1. Power and space are not problems for vehicles. Infinite power supply and memory space are assumed.
2. Each vehicle has a powerful on-board computer.
3. As mandated by DSRC, every 300 msec each vehicle sends a beacon message with a range of about 200 to 300 m. This beacon contains information that allows vehicles to handshake and synchronize.
4. Each vehicle is equipped with a GPS system and a digital map that helps estimate its position accurately.

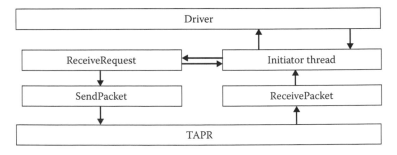

FIGURE 8.5 Illustrating the ZIPPER architecture.

Similar to CarTorrent,[11] Figure 8.5 shows the basic architecture of ZIPPER. ZIPPER operations are performed by four main threads—the Initiator, the ReceivePacket, the ReceiveRequest, and the SendPacket threads—while implementing TAPR[17] as the underlying packet-relaying protocol. We chose multithreaded implementation because it enables a vehicle to perform many operations at the same time. So, a vehicle may initiate a request while searching for and/or relaying some blocks for some other vehicles, which means that at any moment of time, zero or more instances of any thread may be in execution. The various threads may be described as in the following.

8.5.1 Initiator Thread

This thread is created when a driver initiates a search request for a movie or a song of interest. The initiator thread will send a request to every new physical neighbor running on the road. New neighbors can be identified easily because every vehicle is broadcasting periodically about its existence. The request should contain some information about missing blocks, because the initiating vehicle may already possess some blocks. Then the initiator thread waits to collect replies from physical neighbors before sending an *Ack* control message that contains the new remaining blocks to the farthest vehicle on the road. The lifetime of the initiator thread ends when it collects all blocks or after a preset time-out. Whenever the initiator thread receives a new block, it delivers it to the application at which the driver may play it out while still downloading the remaining blocks. It is worth mentioning that the request has a limited time to live (TTL) value; consequently, the request will not flood the network indiscriminately. Instead, it is highly expected that the initiator thread will collect all blocks, either from vehicles in the front or vehicles in the opposite direction, using that TTL value.

8.5.2 ReceivePacket Thread

The ReceivePacket thread is responsible for receiving packets at the initiating car from any other vehicle that possesses or relays a block. Multiple instances of this thread may exist at the same time to be able to receive from many vehicles in parallel. When the ReceivePacket thread receives a block, it notifies the initiator thread so that the latter can update its list of missing blocks.

8.5.3 ReceiveRequest Thread

The ReceiveRequest thread is created when a vehicle receives a request for a certain file. Once created, the ReceiveRequest thread will search in its car for the missing blocks. Then it will reply to the vehicle that sent the request with the search result. If any is found, the ReceiveRequest thread will invoke the SendPacket thread. If the ReceiveRequest thread receives an *Ack* message from the current requester vehicle, it means that it is the farthest vehicle from the current requester, and it should act now as the new requester that will take the responsibility of propagating the request for the remaining blocks to the next vehicles on the road.

8.5.4 SendPacket Thread

Discovering and maintaining routes in VANET is a very complex task because of high mobility and frequent disconnection on the road. Opportunistic packet relaying is a promising way to relay a packet from one vehicle to another over a disconnected network. When the ReceiveRequest thread finds any of the requested blocks, it creates a SendPacket thread that in turn relies on TAPR to relay these blocks to the initiating vehicle.

8.6 Performance Analysis

The main goal of this section is to offer an analytical evaluation of ZIPPER in terms of the number of probes generated in order to collect all the blocks of the file of interest.

8.6.1 Coupon Collector's Problem

The analysis of the efficiency of ZIPPER relies, in part, on an extension of the classic Coupon Collector's Problem,[18] which can be stated as follows. Suppose that there are n distinct types of coupons. At each trial, we draw a coupon whose type is uniformly distributed among the n possible types. It is well known that the expected number of steps needed to collect the entire set of coupons is nH_n, where H_n is the nth harmonic number. As $H_n = \ln n + \gamma + \mathcal{O}(1/n)$, it follows that nH_n is close to $n \ln n + \gamma n + \mathcal{O}(1)$, where $\gamma = 0.57721\ldots$ is Euler's well-known constant.

There is an obvious connection between the Coupon Collector's Problem and the interest in evaluating the number of cars that need to be probed to ensure that we collect all blocks of interest. However, the problem at hand involves an additional difficulty, as not every car needs to have a block of the desired file. For example, suppose we are interested in collecting the various blocks of the popular TV show "Sesame Street." As this is a show of immense interest to children but seldom to adults, we expect that most cars will not store any of its blocks.

For definiteness, let m be the number of blocks of the file of interest, and let p denote the probability that an arbitrary car stores a block of the desired file. For ($1 \leq i \leq m$), let X_i be the random variable that counts the number of vehicles to meet until the initiator thread finds a new block, given that it has already collected $i - 1$ blocks. The probability, p_i, of finding a new block given that we have $i - 1$ blocks in hand is given by

$$p_i = p \cdot \left(1 - \frac{i-1}{m}\right).$$

It is fairly easy to see that X_i has a geometric distribution and that its expected value satisfies

$$E(X_i) = \frac{m}{p \cdot (m - i + 1)}.$$

Let X be a random variable that counts the number of cars to probe in order to find the entire set of blocks. As $X = \sum_{i=1}^{m} X_i$, the linearity of expectation guarantees that

$$
\begin{aligned}
E(x) &= \sum_{i=1}^{m} E[X_i] \\
&= \sum_{i=1}^{m} \frac{m}{p(m-i+1)} \\
&= \frac{m}{p} \sum_{i=1}^{m} \frac{1}{m-i+1} \\
&= \frac{mH_m}{p} \\
&= \frac{m}{p} \ln m + \frac{m\gamma}{p} + O(1).
\end{aligned}
\tag{8.4}
$$

It is, of course, of considerable interest to determine to what extent the values of X are clustered around $E[X]$. This information is invaluable to assessing the performance of ZIPPER because it can help understand the extent to which requests need to be flooded. It is well known[18] that in the classical Coupon Collector's Problem there is a sharp concentration around the expected value. We show that the same behavior is displayed by the expected problem that models the collection of file blocks in ZIPPER. For this purpose, imagine that we are interested in collecting the m blocks $B_1, B_2, \ldots, B_m$ of a file and for all i ($1 \leq i \leq m$), let A_i be the event that block B_i has not been obtained in the first t probes. It is easy to see that the probability $\Pr\{A_i\}$ of event A_i satisfies

$$
\begin{aligned}
\Pr\{A_i\} &= \left(1 - \frac{p}{m}\right)^t \\
&= \left[\left(1 - \frac{1}{m/p}\right)^{(p/m)(m/p)}\right]^t \\
&= \left[\left(1 - \frac{1}{m/p}\right)^{m/p}\right]^{(p/m)t} \\
&\leq e^{-(p/m)t} \quad \left[\text{because for all } x \in \Re, \left(1 - \frac{1}{x}\right)^x \leq e^{-1}\right]
\end{aligned}
\tag{8.5}
$$

Write $A = \bigcup_{i=1}^{m} A_i$. It is clear that the event A occurs if some of the m blocks are still not available at the end of t probes. As the A_is are not necessarily pairwise independent, we have

$$
\begin{aligned}
\Pr[A] &\leq \sum_{i=1}^{m} \Pr[A_i] \\
&\leq \sum_{i=1}^{m} e^{-(p/m)t} \\
&= m \cdot e^{-(p/m)t}
\end{aligned}
\tag{8.6}
$$

LEMMA 8.1 With probability exceeding $1 - e^{-c}$ at the end of $m \ln m + cm$ probes, all the m blocks have been collected.

PROOF 8.1 Write $t = (m/p) \ln m + (m/p)c$. By (8.6), the probability that all blocks have been collected at the end of t probes reads

$$\Pr[\overline{A}] = 1 - \Pr[A]$$
$$= 1 - m \cdot e^{-(p/m)[(m/p)\ln m + (m/p)c]}$$
$$= 1 - e^{-c}$$

as desired.

To put Lemma 8.1 into perspective, assume that we are probing $(m/p) \ln m + 5(m/p)$ vehicles. How close are we to having collected all the blocks? As $e^{-5} = 0.00673 \ldots$, Lemma 8.1 guarantees that the probability of having obtained all the blocks exceeds 99.36%.

8.6.2 Simulation Results

To obtain a precise evaluation of ZIPPER, it was implemented on the top of a realistic traffic simulator[21] that incorporates all traffic situations such as acceleration, deceleration, traffic jam, changing lanes, and maintaining a safe distance between vehicles.

8.6.2.1 Impact of Traffic Flow and Probability on Collecting the Entire File

In this experiment, road length was set to 15 km and the number of blocks to 10 ($m = 10$). The impact of changing the traffic flow (vehicles/hour), and the probability of finding a block in a number of blocks that could be collected in that distance, was studied.

Figure 8.6 shows the number of blocks collected for different values of traffic flow and the block-finding probability. For sparse traffic, with low traffic flow, connectivity with vehicles in the same direction is not expected, and few vehicles will be met in the opposite direction. So, for low probabilities, the whole file could not be collected. As the block-finding probability increases, we expect all cars in the opposite direction to have blocks, and the whole file may be downloaded. For probability = 0.9 in a very sparse traffic flow (730 vehicle/hour), the entire file could be downloaded.

As the traffic flow increases, the initiating vehicle will have connectivity with vehicles in the same direction, and the forward agent will be propagated to many vehicles in front. So, even with low probability (0.01), most of the blocks will be collected. Although the traffic length is small, for high density and large probability values the entire file will be collected.

8.6.2.2 Impact of Probability and Number of Blocks on Number of Vehicles in Which to Search

Figure 8.7 shows how many vehicles need to be searched in order to collect the entire file while changing the probability from 0.01 to 1, and changing the number of blocks from 1 to 30. As expected, the graph is very close to Equation 8.4, which validates the

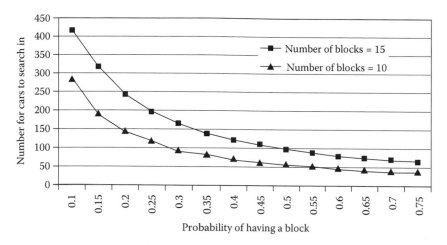

FIGURE 8.6 Impact of changing probability and traffic flow on collecting blocks.

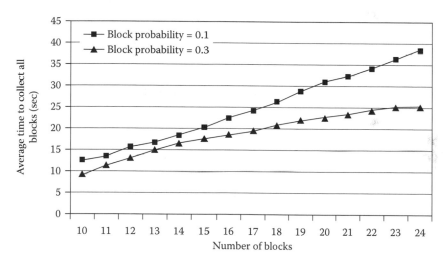

FIGURE 8.7 Impact of changing probability and number of blocks on number of vehicles to search.

simulator. Figures 8.6 and 8.7 shows that as either the probability increases or the number of blocks decrease, a smaller number of vehicles will be searched to collect all blocks.

8.7 Conclusion

In this chapter, we presented an overview of the most noticeable P2P systems proposed in the literature for VANETs. Most of the presented protocols rely on some infrastructure either in the form of roadside units installed every few miles or on the cellular network. We also covered ZIPPER, a zero-infrastructure P2P protocol for VANETs that does not require any form of infrastructure to exist. Through analytical analysis, this chapter shows that a search request does not need to flood the network in order to collect all blocks of a file, and only a limited number of cars needs to be searched.

References

1. Fussler, H., Schnaufer, S., Transier, M., and Effelberg, W., Vehicular Ad Hoc Networks: From Vision to Reality and Back, Proceedings IEEE Conference on Wireless on Demand Network Systems and Services, (WONS '07), 2007, 80–83.
2. Dedicated Short Range Communications (DSRC), available at http://grouper.ieee. org/groups/scc32/dsrc/, 2007.
3. The FleetNet Project Homepage, available at http://www.et2.tu-hamburg.de/ fleetnet, accessed June 22, 2008.
4. Car2Car Communication Consortium, available at http://www.car-2-car.org/, accessed June 22, 2008.
5. Japanese Ministry of Land, Infrastructure, and Transportation, VICS: Vehicle Information and Communication System, available at http://www.its.go.jp/etcvics/ vics/index_e.html, accessed May 12, 2007.
6. NoW: Networks-on-Wheels, available at http://www.network-on-wheels.de/ documents.html, 2007.
7. Raya, M. and Hubaux, J.-P., The Security of VANETs, Proceedings 2nd ACM International Workshop on Vehicular Ad Hoc Networks (VANET '05), 2005, 93–94.
8. Dötzer, F., Privacy Issues in Vehicular Ad Hoc Networks, Workshop on Privacy Enhancing Technologies, May 2005, 541–546.
9. Ahmed, S. and Kanhere, S.S., VANETCODE: Network Coding to Enhance Cooperative Downloading in Vehicular Ad-Hoc Networks, Proceedings ACM International Conference on Communications and Mobile Computing, Vancouver, British Columbia, Canada, 2006, 527–532.
10. Ghandeharizadeh, S.K.S. and Krishnamachari, B., PAVAN: A Policy Framework for Content Availability in Vehicular Ad-Hoc Networks, Proceedings First ACM Workshop on Vehicular Ad Hoc Networks (VANET '04), Philadelphia, PA, 2004, 57–65.
11. Lee, U., Park, J.-S., and Gerla, M., CodeTorrent: Content Distribution Using Network Coding in VANET, Proceedings First ACM Workshop on Decentralized Resource Sharing in Mobile Computing and Networking, 2006, 1–5.
12. Ott, J. and Kutscher, D., Drive-thru internet: IEEE 802.11b for Automobile Users, Proc. IEEE INFOCOM, 2004, 367–373.
13. Agarwal, A., Starobinski, D., and Little, T.D.C., Exploiting Downstream Mobility to Achieve Fast Upstream Message Propagation in Vehicular Ad Hoc Networks, Proceedings IEEE Mobile Networking for Mobile Environments (INFOCOM/ MOVE), May 2007, 13–18.
14. Little, T.D.C. and Agarwal, An Information Propagation Scheme for VANETs, *Proc. IEEE Intelligent Transportation Systems*, 155–160, 2005.

15. Wisitpongphan, N., Bai, F., Mudalige, P., and Tonguz, O.K., On the Routing Problem in Disconnected Vehicular Ad Hoc Networks, Proceedings IEEE Infocom 2007, May 2007, 2291–2295.

16. Lee, K.E., Lee, S.-H., Cheung, R., Lee, U., and Gerla, M., First Experience with CarTorrent in a Real Vehicular Ad Hoc Network Testbed, Proceedings IEEE Mobile Networking for Mobile Environments (INFOCOM/MOVE), May 2007, 109–114.

17. Abuelela, M. and Olariu, S., Traffic-Adaptive Packet Relaying in VANET, Proceedings ACM Workshop on Vehicular Ad Hoc Networks (VANET '07), Montreal, Quebec, Canada, 2007, 77–78.

18. Feller, W., *An Introduction to Probability Theory and Its Applications*, 3rd ed. Vol. I, John Wiley & Sons, 2000.

19. Zang, Y., Stibor, L., and Reumerman, H.-J., Vehicular Wireless Media Network (VWMN): A Distributed Broadband MAC for Inter-Vehicle Communications, Proceedings 2nd ACM Workshop on Vehicular Ad Hoc Networks (VANET '05), 2005, 95–96.

20. Abuelela, M. and Olariu, S., ZIPPER: A Zero-Infrastructure Peer-to-Peer System for VANET, in WMuNeP '07: Proceedings of the 3rd ACM Workshop on Wireless Multimedia Networking and Performance Modeling, Chania, Crete Island, Greece, ACM, 2007, 2–8.

21. Treiber, M., Microsimulation of Road Traffic, available at http://www.traffic-simulation.de, 2005.

IV

Networking Issues

9

Mobile Ad Hoc Routing in the Context of Vehicular Networks

Francisco J. Ros,
Pedro M. Ruiz, and
Juan A. Sánchez
*Department of Information
and Communications
Engineering
University of Murcia*

Ivan Stojmenovic
*School of Information
Technology and Engineering
University of Ottawa*

9.1 Introduction

Recent advances in short-range wireless technologies have enabled a new wealth of networking possibilities for user devices. In particular, ad hoc networks have emerged as one of the most researched areas in the networking community over the last few years. An ad hoc network consists of a set of nodes equipped with wireless interfaces, which are able to communicate among themselves in the absence of any kind of network infrastructure. One of the most salient features of ad hoc networks is the concept of wireless multihop communications. Unlike traditional wireless networks, mobile nodes are allowed to send messages to destinations that are not within the sender's radio range. When the destination is several hops away, intermediate ad hoc nodes act as relays to forward data packets to their intended destinations. Therefore, nodes need to use a routing protocol to find paths to deliver data messages from sources to destinations. In general, ad hoc nodes can be mobile, which makes the design of routing protocols a very challenging task. However, it is one of the research lines that has attracted most attention within the research community.

During the last few years, this general concept of ad hoc multihop communications has evolved into a number of specialized arrangements for different kinds of networks. These specializations differ regarding the mobility of the nodes, their processing power, the criticality of their energy efficiency, and so on. Although they all retain the same basic principles of multihop communications, they also have enough peculiarities to prevent a one-size-fits-all solution from being practical. These evolved wireless networks include, among others, the following:

1. *Mobile ad hoc networks (MANET).* These consist of a collection of wireless nodes with arbitrary mobility patterns. Nodes are usually battery-operated, making energy efficiency one of the important design issues. Computation and memory resources can also be scarce, so routing protocols designed for MANETs should not be too complex. In addition, a particular mobility pattern cannot be assumed, nor the existence of additional valuable information such as the position or trajectory of the nodes. Routing protocols assume that the network is fully connected. That is, if the destination of a data packet lies on a different part of the network, such a packet is simply discarded.

2. *Wireless mesh networks (WMN).* These are a particular case of ad hoc network in which nodes are like static base stations that are able to communicate using multihop routes. Client devices are mobile and switch among mesh nodes as they move around. Mesh nodes can be equipped with multiple radio interfaces for higher efficiency. In this case, energy, computation, and memory resources are not a concern. Mesh routers are, commonly, dedicated devices with a continuous power supply. Routing protocols are required to find the best possible routes for the aggregated user traffic.

3. *Wireless sensor networks (WSN).* These consist of a set of generally tiny wireless devices with very limited energy, computation power, and memory. Therefore, energy efficiency and simple algorithms are the factors of paramount importance in these networks. They are mainly used to monitor the environment. So, a WSN may consist of hundreds or even thousands of devices. Sensors are usually assumed to have knowledge about their own positions and those of their neighborhood (commonly by including position information in periodic beacons). In many applications, the destination is a sink device that processes the data sensed by the nodes, so that its position is known a priori. When this is not the case, the destination's position is unknown and must be discovered. The process by which a node retrieves the current position of another one is commonly referred to as the "location service." Scalable location services are hard to develop.

4. *Vehicular ad hoc networks (VANET).* These are a particular case of MANET in which nodes are vehicles able to move at very high speeds. In addition, they may consist of a very large number of nodes, and their mobility patterns are constrained by the topology of the roads, streets, speed limits, and so on. Cars do not usually have energy constraints, and can be equipped with high computing and communication capabilities. Thanks to the on-board navigator, they often know

their own position and the street maps of the surrounding area. Nevertheless, those maps might not be fully accurate if they are out of date or if a special event is happening (e.g., a road is closed for repairs). Going on, it is expected that vehicles will issue periodic beacons to support collision avoidance applications (a control message about every 300 msec[1]). Position information, as well as velocity and trajectory, can be carried within these messages to learn the neighborhood topology and its evolution in time. The destination's position might be known in some cases, although this cannot be assumed in general. Finally, under many scenarios the network is expected to be highly disconnected, as vehicles tend to travel around forming groups, but the distance between groups can be much longer than the communication range.

We summarize the main differences between these different kinds of networks in Table 9.1. As can be seen, vehicular ad hoc networks (VANET) have emerged as a specialization of MANETs in which participating nodes are expected to be vehicles. Of course, this raises a number of issues such as the need for high scalability, the importance of making use of valuable information (e.g., position of nodes and street maps) to enhance the performance of the protocols, and so on.

As in MANETs, multihop communications are one of the most important building blocks of VANETs. The possibility of distributing information very efficiently using neighboring vehicles becomes a very important enabler for many vehicular applications. Safety applications such as collision avoidance, safety warnings, and the like are probably the most popular examples, in which the importance of multihop message dissemination becomes prevalent. However, there are many other applications, some of them yet to be discovered, which can greatly benefit from multihop relaying.

Many proposed solutions exist for routing in MANETs; these were the natural starting point for routing protocols in VANETs. For this reason we will devote this chapter to describe the most relevant routing solutions for MANETs, and note which of their features can be carried over to VANET routing.

In general, MANET routing algorithms can be classified according to a number of criteria. The most used one classifies routing protocols as (1) proactive, (2) reactive, and (3) hybrid. Proactive protocols follow a very similar approach to those used in wired networks such as the Internet. Nodes taking part in proactive routing maintain a routing table, which is built by exchanging messages with other nodes of the network. Thus,

TABLE 9.1 Properties of Different Types of Ad Hoc Networks

Property	MANET	WMN	WSN	VANET
Network size	Medium	Moderate	Large	Large
Node's mobility	Random	Static	Mostly static	High, nonrandom
Energy limitations	High	Very low	Very high	Very low
Node's computation power	—	High	Very low	High
Node's memory capacity	—	High	Very low	High
Location dependency	Low	Very low	High	Very high

routes are computed and maintained in advance, even if no data traffic is present in the network. A reactive routing protocol (also known as "on-demand") only searches for a route when it is needed, in other words, if the sending node does not know of a route to reach the destination. Unlike proactive routing protocols, route discovery only takes place when needed, but this usually increases to a small extent the end-to-end delay as the data source must wait some time before the routing path is established. Hybrid routing protocols are those that cannot be fully classified into the previous categories. Typical examples include those protocols that behave proactively for some destinations and reactively for others. In addition to those categories, we add for clarity an additional one: (4) geographic routing. Geographic routing generally works "on-demand." However, it is very different from traditional reactive routing protocols. Rather than routing based on the topology of the network, geographic routing protocols take routing decisions hop-by-hop in a per-packet basis. Each relay selects its next hop based on its position, its neighbors' positions, and the position of the destination.

MANET protocols rely heavily on flooding control messages throughout the whole network. Proactive approaches employ this primitive technique to acquire network-wide topology information and compute the shortest paths (e.g., using Dijkstra). It is also needed in reactive solutions in order to find the destination and set up a route towards it. Obviously because hybrid routing proposals are a mixture of the previous two, they also need flooding to some extent. The exception occurs in geographic routing when the destination's position is known. In such a case, no more global knowledge is necessary and the packet can be routed by taking local decisions. However, if the source has to discover the position of the destination, it will likely be forced to use some sort of restricted flooding.

We shall elaborate a little more on the particular operation of existing MANET routing protocols in the next sections. However, one of the issues we need to discuss is the validity of those routing protocols for the particular case of VANETs. As we studied in Table 9.1, the main differences between a VANET and a general MANET are related to the mobility patterns of the nodes and the scalability requirements. However, VANET nodes also have access to very relevant information (e.g., their position, speed, direction, road maps, and so on), which may be helpful to routing protocols. Putting it all together, we can see that, although MANET routing solutions should be able to work in VANETs, their performance is expected to be low compared to other solutions that are able to exploit all the available information. Thus, we can summarize the main technical limitations for MANET routing solutions in VANET scenarios as follows:

1. *Scalability.* Most of the routing protocols designed for MANETs are only able to support limited number of mobile nodes (about one or two hundred). The path computation mechanisms used by those protocols are very costly for very large networks such as VANETs. For instance, proactive protocols store routes to all other nodes in the network within their routing tables. In the case of a VANET, storing routing tables for all vehicles is really impractical.

2. *Full connectivity.* This assumption is not realistic in vehicular networks. Although the destination is not reachable at the moment of sending a packet, there could be a noncontemporaneous path between source and destination. This means that

vehicles can move, carrying the packet until the destination is eventually reached. This paradigm is called delay (and disruption) tolerant networking (DTN)[2], and is more appealing than MANET routing for delay-insensitive packets.

3. *Mobility prediction.* Most MANET routing protocols have not made any assumption about particular mobility patterns of mobile nodes. They assume arbitrary mobility patterns. Although that approach favors flexibility regarding the scenarios in which those protocols can be deployed, they are also inefficient in the cases where mobility of the nodes can be somehow predicted. This is the case for VANETs, where node movements are restricted by the topology of the streets, speed limits, traffic signals, and the like. Thus, traditional MANET routing solutions neglect the advantages that can be obtained by considering a constrained mobility pattern.

4. *Anticipation of path breakages.* MANET routing protocols deal with the mobility of the nodes (i.e., path breakages) either by periodic control messages or by periodic path creations. The timers used by MANET routing protocols are adjusted so that the protocol can react after a route breaks. However, in many VANET scenarios the knowledge about the mobility patterns of neighboring nodes can help prevent path breakages before they happen.

5. *Extensive use of flooding.* Most MANET routing protocols are based on flooding. In reactive routing protocols the data source uses flooding to find a route to a destination. In the case of proactive protocols, every node sends periodic control messages to either its neighborhood (one message issued by each node is as costly as a flooding initiated by one node) or the entire network. That kind of operation consumes a lot of bandwidth with control messages and limits very much the performance in large networks such as VANETs. Because in vehicular environments the number of nodes is very high (and unknown beforehand), any devised flooding mechanism must be scoped, that is, restricted to a limited area.

6. *Nonlocal operation.* MANET routing protocols are distributed algorithms used to compute routing paths. However, the creation and maintenance of routing paths usually requires the effort of all nodes in the network. In proactive routing all nodes take part in building routing tables. In reactive protocols all nodes participate in the initial flooding required to find a route towards the destination. In VANETs with a potentially large number of nodes, localized routing solutions in which nodes only need information from their neighborhood are more appealing in terms of scalability, control overhead, and adaptation to different network conditions. However, in order for this to work, the destination of the communication must be known or an efficient location service must be designed.

7. *Exploitation of existing knowledge.* VANET nodes can be assumed to be equipped with on-board units providing relevant information about expected trajectories, current speed and direction, topological map of streets or roads, and so on. All the information is extremely valuable for enhancing the performance of routing protocols. Unfortunately, MANET routing solutions, by trying to be effective regardless of the mobility of the nodes, usually just neglect all that information.

We shall describe in detail all these issues in the remainder of this chapter, which is organized as follows. Section 9.2 classifies and describes the most well-known routing protocols for mobile ad hoc networks. We describe their operation, strengths, and drawbacks. In Section 9.3, we study the particular properties of VANET networks and their influence in the design of routing solutions. In particular, we discuss the validity of MANET routing solutions as well as which of their features can be taken over for VANET networks. Section 9.4 describes the main ideas of the most recent VANET-specific routing solutions. Finally, Section 9.5 summarizes all the conclusions and discusses some open issues for further research.

9.2 Routing in Mobile Ad Hoc Networks

The issue of routing messages from a source to a destination in generic ad hoc networks with unknown or random mobility patterns has been heavily investigated over the last ten years. Both academia and the IETF (by means of the MANET Working Group[3]) have developed protocols that deal with routing in such unpredictable networks. In addition, some of these proposals have been deployed in real networks and testbeds (e.g., Ref. 4), demonstrating their ability to perform well in static or low mobile environments.

As we saw in the previous section, we can classify ad hoc routing protocols according to their behavior into proactive, reactive, hybrid, and geographic protocols.

Like traditional protocols used within the Internet, proactive algorithms issue periodic messages in order to learn the network topology and create routes to every other node present in the network (Figure 9.1). In this way, routes are established regardless of the data traffic pattern. Therefore, network resources are wasted setting up routes that might never be used. This problem is particularly dramatic when the number of nodes is high, provoking a big control overhead. Furthermore, all proactive protocols must face a trade-off between the freshness of the routing information and the generated signaling overhead. In order for the protocol to compute effective routes from source to destination (avoiding routing loops and long paths), the nodes have to exchange frequent messages with topology information. However, this consumes a high percentage of bandwidth, which decreases the communication resources left to user data packets. On the other hand, as paths between every possible pair are created beforehand, there is no associated delay in the process of finding a route when a connection is about to be established. In addition, the protocol can immediately know if a destination is reachable within the MANET just by looking up the routing table.

On-demand or reactive routing protocols follow the opposite direction of proactive ones. This approach remains silent until a data flow is about to be sent. At this moment, the protocol floods the network with a query that tries to find the intended destination. Once the message arrives at this node (or maybe another one that knows a route to it), it replies to the source in order to establish a path between them. This generic process is illustrated in Figure 9.2. The main benefit of this approach is that it does not incur any signaling overhead to create or maintain routes that are not used. Therefore, scarce network resources are just used when needed. However, when there are many flows between different pairs, the protocols need to issue many requests, reducing the scalability of the solution. Additionally, the search process (commonly referred to as

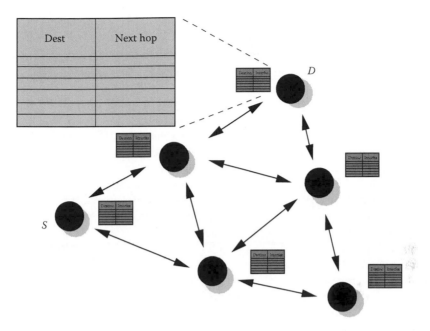

FIGURE 9.1 Proactive routing protocols maintain a global view of the network topology via the exchange of periodic messages.

"route discovery") introduces a nonnegligible delay before the node can actually inject packets into the network. The route discovery process also generates unnecessary overhead when a destination is not reachable (perhaps because it is not a MANET member or there is no available path at the moment), initiating a network-wide flooding that is not replied to by any node.

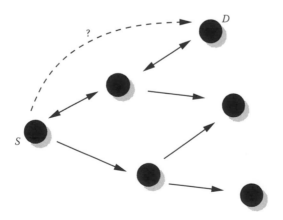

FIGURE 9.2 Reactive routing protocols ask the network to find a route up to the destination when there is a packet to send.

We have just seen that both proactive and reactive routing protocols have their own advantages and drawbacks. As a trade-off between them, hybrid protocols propose to proactively set up routes to the nodes inside a given zone, while letting the process of acquiring routes outside the zone operate on-demand. This idea is based on the assumption that, in a MANET, the communication pattern will likely involve nearby nodes. Thus, for most of the communications the routes are computed beforehand. For those few cases in which the route is outside the proactive zone, the route can still be established but with a delay cost. The use of zones also solves the scalability issues present in proactive protocols. Unfortunately, it is not possible to find a constant zone size that performs well for a variety of network conditions. Therefore, it should be dynamically adapted depending on the network density, mobility, communication patterns, and so on.

All the former kinds of protocols assume that the destination's identifier (usually the IP address) is known and exchange nonlocal information between peers in order to find it and establish a route. When the destination's position is known, a different approach is taken by geographic routing protocols. They can exploit location information (see Figure 9.3) to make forwarding decisions solely based on local information. In order to route a packet, a node must know its own position, the destination's position, and the 1-hop neighbors' position. Global positions can be learnt by means of geographic positioning systems (such as GPS and Galileo), and are exchanged between neighbors. Alternatively, relative positions can be estimated based on the received signal strength of messages issued with fixed signal power.[5] As protocol messages are not exchanged between distant nodes, geographic routing is scalable and can be used in really large networks.

However, the destination's position may be unknown and the former situation dramatically changes. In that case, a location service must be employed, which maps a node identifier with its current position. Given a scenario where the destination is mobile, the service would likely involve the sending of control messages to update and

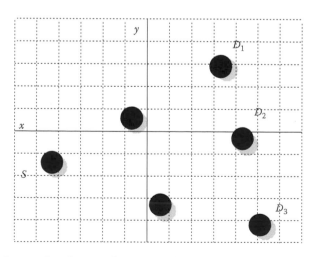

FIGURE 9.3   Position-based geographic routing scheme.

TABLE 9.2 Comparison between the Different Kinds of Ad Hoc Routing Protocols

	Proactive	Reactive	Hybrid	Geographic
Overhead	Very high	Low	Medium	Very low
Connection delay	Very low	Very high	Medium	Very low
Mobility impact	High	High	High	Very high/very low
Scalability (nodes)	Very low	Low	Medium	Very high
Scalability (flows)	Very high	Very low	Medium	Very high

retrieve the position, which extremely reduces the scalability of the solution. Indeed, the location service by itself can be a more difficult problem than routing.[6] Section 9.2.4 provides a summary of location services for MANET after going into details about the routing protocols themselves.

The most salient features of every routing protocol type are summarized in Table 9.2. The impact of mobility on geographic routing protocols varies from veries low to very high depending on whether the destination's position is known or not. This is because the overhead of the location service must be taken into account in order to offer a fair comparison. In the following subsections, we will describe some of the most representative protocols of each type.

9.2.1 Proactive Protocols

9.2.1.1 DSDV

Bellman[7] and Ford[8] independently designed a centralized algorithm to compute shortest paths in weighted graphs. Bertsekas and Gallager extended it to be executed in a distributed fashion, creating the so-called Distributed Bellman–Ford (DBF) algorithm.[9] In this protocol, every node maintains a vector of distances to every known destination. Thus, the routing table consists of a set of tuples *<destination, distance, next-hop>*. Initially, the routing tables are empty, and each node starts issuing periodic broadcast messages to its 1-hop neighborhood. With the first received message, every node sets up routes of distance one to every direct neighbor. These new tuples are inserted in future messages, so that network topology is slowly learnt by every node. To illustrate the convergence of the algorithm, Tables 9.3 through 9.5 show the evolution of the distance vectors over the network topology depicted in Figure 9.4.

TABLE 9.3 Example of Distance Vectors at Time t_0

	S	A	B	C	D
S	0	1	—	—	—
A	1	0	1	—	1
B	—	1	0	1	—
C	—	—	1	0	1
D	—	1	—	1	0

The first message exchange round has happened, so that 1-hop neighborhood information is acquired.

TABLE 9.4   Example of Distance Vectors at Time t_1

	S	A	B	C	D
S	0	1	2	—	2
A	1	0	1	2	1
B	2	1	0	1	2
C	—	2	1	0	1
D	2	1	2	1	0

The second message exchange round has happened, so that 2-hop neighborhood information is acquired.

TABLE 9.5   Example of Distance Vectors at Time t_2

	S	A	B	C	D
S	0	1	2	3	2
A	1	0	1	2	1
B	2	1	0	1	2
C	3	2	1	0	1
D	2	1	2	1	0

The third message exchange round has happened, so that topology information up to three hops is acquired.

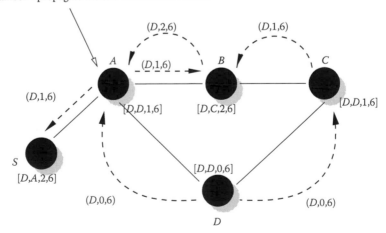

FIGURE 9.4   Example of DSDV operation. The figure shows the exchanged messages (*destination, distance, sequence_number*) and the routing table entries [*destination, next_hop, distance, sequence_number*] related to node *D*.

DBF is known to suffer from a bouncing problem that leads to the *count-to-infinity* and looping issues.[10] DBF might not converge in the presence of link failures, and loops can appear if out-of-date information is used to compute the shortest path.

One of the first proposals specifically designed for mobile ad hoc networks, the highly dynamic destination-sequenced distance-vector (DSDV) routing protocol,[11] modifies DBF in order to prevent the formation of loops in the face of link failures. DSDV augments the routing table and the messages with a sequence number that indicates the freshness of the information. In this way, a route is "better" than another if the sequence number is higher or, in case of a draw, if the distance is lower. Every destination increments its sequence number up to the next even number before sending a routing message. When a link breakage occurs, the destination marks with distance ∞ all the routes that used that neighbor as their next hop. This situation is immediately advertised by a protocol message, and the receiving nodes update those routes with the next odd number. So, whenever the route is reestablished, the destination will generate an even sequence number bigger than the one that indicated a broken link and the routes will be restored.

The basic procedure of DSDV is depicted in Figure 9.4. Please note that the figure only refers to node D related messages and routing entries for the sake of clarity, although information about every known node in the network is exchanged between neighbors. Initially, node D sends a message in which it announces itself to its neighborhood, A and C. These nodes update their routing tables and issue a new message to inform their neighbors that destination D is reachable through them. Next, A receives a message from B, which announces D at distance 2 and sequence number 6. As A already has a routing entry for D with the same sequence number and lower distance, it just ignores this message because both pieces of information are equally fresh, but the first provides a shorter route. Therefore, finally, S can set up a route to D with distance 2 and A as the next hop.

Now suppose that node D moves and is no longer in the neighborhood of A and C, but is within radio range of S (Figure 9.5). D issues a new message with a higher sequence number, 8. This invalidates the routing entries for D in every node of the network, as the new sequence number is higher. The information is propagated until all the network learns a new route to D. In this way, we can see how sequence numbers are used to avoid the use of stale information, which might cause the formation of loops or nonoptimal routes.

Even though DSDV enhances DBF for its use on mobile networks, all nodes still issue frequent broadcast messages intended for their neighborhood (some of them are periodic and others are triggered by topology changes, i.e., when links come up or down). As every message must contain information about each destination in the network, or at least about each destination that is affected by a topology change, this leads to a great control overhead in large networks, especially if they are highly mobile. Even worse, the bigger a message is, the less likely it is that if will be successfully delivered in a wireless link. When routing protocol messages are lost, many nodes will become unreachable for part of the network, and routing loops will be created.

9.2.1.2 OLSR

A different approach is taken by the Optimized Link State Routing (OLSR) protocol.[12] It was designed by the IETF community and is specified in the Request for Comments

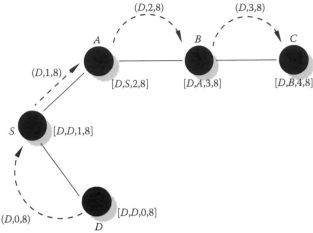

New sequence number is bigger than
the stored—propagate the new distance

FIGURE 9.5 Continuation of the DSDV example. Node *D* moves and the network topology changes accordingly. The figure depicts how sequence numbers avoid the use of stale information.

3626 as an experimental protocol. It implements a shortest path routing algorithm that extends the traditional link state approach to reduce the overhead of link state updates, especially in dense ad hoc networks.

As in every link state algorithm, nodes sense their neighborhood by the periodic exchange of HELLO messages. In this way, nodes learn their local vicinity and the status of the link with each neighbor (i.e., if the link is considered unidirectional or bidirectional). This local information is disseminated throughout the whole network via periodic topology control (TC) messages. With the information acquired via HELLO and TC messages, each node has its own view of the network topology and can run the Dijkstra algorithm[13] to obtain shortest routes to every potential destination. Similar to DSDV, OLSR tags every protocol message with a sequence number to distinguish between stale and fresh information.

The flooding of TC messages is a very expensive operation in terms of network resources. By using regular blind flooding, each node would forward a copy of the message. To limit the cost of forwarding flooded messages, OLSR uses the multipoint relay (MPR) technique,[14] which reduces the number of nodes that need to forward a message, although it still reaches the whole nonpartitioned part of the network. The MPR approach assumes that a node *N* has knowledge of its 2-hop neighborhood, which is accomplished in OLSR by enriching HELLO messages with neighborhood information. Then, *N* selects a subset of relays among its 1-hop neighbors, which covers the same 2-hop nodes as the complete 1-hop neighborhood. This subset is called the MPR set of *N*, and *N* is an MPR selector of each node in the set. If a message is intended to reach the whole 2-hop neighborhood, only those nodes selected as MPRs by the source are needed to forward the message. Figure 9.6 shows an example where only four retransmissions are needed to make the message reach all the nodes, instead of the eight transmissions that would be needed in the case of blind flooding. Applying the same behavior to bigger networks, only those broadcast messages received from a MPR selector are forwarded.

Obviously, the smaller the size of the MPR sets, the lower the number of transmissions needed to flood the network. Unfortunately, the problem of finding the minimal MPR set is known to be NP-complete.[15] A greedy heuristic proposed by Quayyum et al.[14] is described as follows:

1. Insert all 2-hop neighbors of N in a set $MPR'(N)$ of uncovered 2-hop neighbors.
2. Select all 1-hop neighbors of N that are the only possible relays for some 2-hop neighbors. Relays are placed in $MPR(N)$ and covered 2-hop neighbors are removed from $MPR'(N)$.
3. While the set $MPR'(N)$ is not empty, select the 1-hop neighbor not already selected that covers the greatest number of nodes in $MPR'(N)$. Add it to $MPR(N)$ and remove the covered 2-hop neighbors from $MPR'(N)$. In the case of a tie, choose the node with the highest degree (number of neighbors excluding N and N's neighbors).

OLSR introduces two more optimizations based on the use of partial topology information. Only those nodes selected as MPR are needed to generate topology information (TC messages). In addition, only the links between a node and its MPR selectors need to be reported in TC messages in order to obtain optimal routes in terms of number of hops.

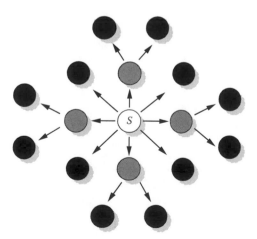

FIGURE 9.6 Illustration of the MultiPoint Relay (MPR) technique. Gray nodes are selected as MPRs by the source. Just four out of eight retransmissions are needed to reach all the nodes.

OLSR assumes that accessing the medium for transmitting several packets is more costly than putting more bytes in a given packet, which is true for most wireless data-link layers. Thus, OLSR messages are aggregated as long as possible into control packets. Obviously, the trade-off between medium accesses and probability of reception is biased to the former, because a large message is more likely to suffer from transmission error.

Let us examine an example of the OLSR functioning. Figure 9.7 shows the HELLO messages issued by the nodes of an ad hoc network. HELLO messages contain the list of neighbors of the sending node and a link code indicating the type of link existing between them: asymmetric, symmetric, or MPR. Consequently, the HELLO mechanism serves as the basis to compute the MPR set and the way to notify a neighbor if it has been chosen as an MPR. In the figure, D computes its optimal MPR which is $\{A\}$. In order to add redundancy, it could have selected the nonoptimal MPR set $\{A, C\}$. Moreover, the optimal MPR need not be unique. For instance, the optimal MPR set for C is either $\{B\}$ or $\{C\}$, as both sets cover the whole 2-hop neighborhood $\{A\}$ with the same number of elements.

Using only the information acquired via HELLO messages, every node can set up optimal routes (in terms of number of hops) for every 1-hop and 2-hop neighbor, even though, for more distant nodes it is necessary to exchange link states with the remaining nodes in the network via TC messages.

Following with the same example, in Figure 9.8 it can be observed that just A and D, the nodes that were selected as MPR by at least one neighbor, are in charge of sending TC messages. Furthermore, they are only forwarded by MPR selectors; that is, A forwards messages from D because the latter selected the former as MPR, and vice versa. And finally, only those links with the MPR selectors are needed to be announced within the TC. This information suffices to compute optimal routes. For instance, node B knows that it has a direct 1-hop route to A and C because of the HELLO exchange. In addition,

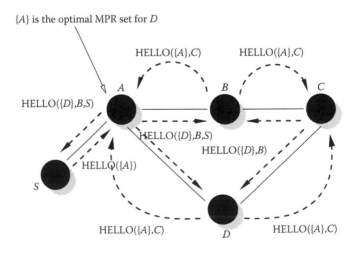

FIGURE 9.7   OLSR example. HELLO messages are used to sense the neighborhood, determine the link status and compute (and notify) the MPR set. MPR sets are enclosed in braces.

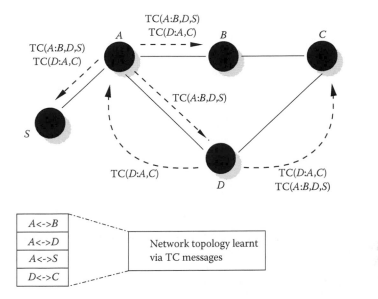

FIGURE 9.8 Continuation of the OLSR example. TC messages (*originator*: <*neighbor_list*>) are only generated and relayed by MPR selectors, only need to advertise the MPR selectors, and are used to learn the network topology.

it also knows that S and D are neighbors of A, so it can configure routes to them of two hops each with A as the next hop.

A node detects that a neighbor is not reachable either when the link layer notifies that it cannot deliver packets to it, or when several periodic HELLO messages from the neighbor should have been received but were not. In that case, the topology change can be immediately propagated by including the new topology in HELLO and TC messages.

Despite the great optimization achieved by means of MPRs, the flooding of TC messages still imposes a big burden in large networks. Furthermore, in highly mobile scenarios with frequent topology changes, the issue of triggered TC messages to prevent loops further increases the protocol overhead. In fact, it has been shown[16] by an analytical study that the OLSR protocol does not meet the scalability bound imposed by the well-known capacity results* of Gupta and Kumar.[17] Therefore, OLSR traffic would use up the whole bandwidth in really large ad hoc networks. However, it is a very good candidate for the deployment of mesh networks[4] and low-mobile medium-scale ad hoc networks.

Currently the IETF is working on the OLSRv2 protocol,[18] an evolution of its predecessor, which keeps the same behavior but adds better protocol extensibility.

---

* The achievable throughput by each node in a random network is $\ominus[W/(N\log N)]$, with N being the number of nodes and W the bits per second transmitted by each one.

9.2.1.3 FSR

However, OLSRv2 does not solve the scalability problem and is still a heavy weight protocol for big networks. Trying to reduce the amount of information exchanged between nodes in link-state algorithms, the Fish eye State Routing (FSR) protocol[19] varies the frequency of topology messages depending on the distance towards the destination. In fact, every node defines a set of scopes around it, as depicted in Figure 9.9. The inner scope is formed by the nearest nodes, which will receive link-state messages at the highest frequency. The update frequency of the remaining scopes is lower because they are farther away. So, to summarize the idea, FSR often exchanges topological information with nearby nodes, but less frequently with nodes which are farther away. The rationale behind this is that nodes do not need up-to-date topology information about far away destinations. They just need a vague idea about where the destinations can be reached, and nearer intermediate nodes that have more information are able to successfully deliver the packet to the intended destination. Obviously, there is a trade-off between the reduction of control overhead and the use of stale link states, as this causes routing loops and creates suboptimal paths. In addition, there is no clear criterion about the process of choosing the number of scopes and their characteristics (nodes inside each one and update intervals).

9.2.1.4 OLSR-Based Scalable Protocols

The fish eye technique has also been tested within OLSR, showing (by means of simulation) a great overhead reduction when compared to the original version of the protocol.[20]

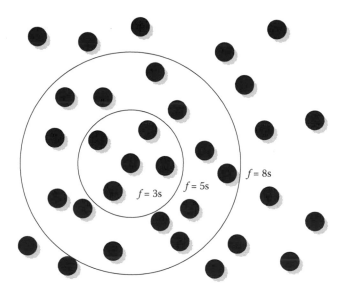

FIGURE 9.9   Three different scopes for the fish eye technique. Every scope has a different update frequency.

In addition, this approach can theoretically scale up to large-scale ad hoc networks.[16] The modification to support the fish eye algorithm in OLSR is straightforward. Every ad hoc node divides the network in scopes, each one being a group of nodes that are at a given range of number of hops away. For example, scope 1 can consist of nodes up to two hops away, scope two of nodes at three to four hops away, and so on. TC messages are propagated at a given interval for the nearest scope, and at higher intervals as the scope increases. Common OLSR features allow for such a kind of modification without altering the core protocol, by just tuning some fields of the TC messages. Unfortunately, the operational troubles associated with fish eye routing are also incorporated in OLSR.

Another approach to improve the scalability of OLSR (indeed, the idea can be applied to other proactive protocols) is taken by the hierarchical OLSR (H-OLSR) protocol.[21] H-OLSR creates a hierarchical topology because it assumes that some nodes are equipped with better communication capabilities. So, nodes at level 1 have a single interface, whereas nodes at level 2 own two different interfaces: one to communicate with nodes at level 1, and another with longer transmission range to communicate with nodes at the same level. The same criteria applies if there are more levels in the hierarchy. The point of the protocol is that nodes with higher capacities automatically become cluster heads of some lower-level nodes. The topology information exchange is restricted to every cluster, and direct communication between the cluster heads of a same level is used to exchange local membership within the cluster. In this way, routing traffic is heavily reduced and nodes can set up routes to other same-level same-cluster nodes. In addition, the MPR technique is used at every level, reducing the forwarding of topology information for both low-capacity and high-capacity nodes. However, to reach any other destination, data packets must traverse the cluster head (which knows enough topology information to route the packet). This leads to suboptimal routes when source and destination are close but belong to different clusters.

Similarly, the clustered OLSR (C-OLSR) protocol[22] makes use of clusters in order to restrict the forwarding of TC messages and to apply the MPR technique at multiple levels. In contrast to H-OLSR, this proposal does not assume nodes with higher capabilities and relies on an underlying clustering mechanism. Therefore, the protocol is more flexible but needs extra signaling to create and maintain the clusters. In addition, cluster heads are no longer neighbors by definition (no long-range communications), which translates into new messages exchanged between cluster heads and traversed along the network to learn the topology at the cluster level. In this way, every node computes a host route to every other node inside its own cluster, but aggregates the routes to the remaining ones into routes to clusters. This provokes suboptimal routes when the destination belongs to a foreign cluster.

9.2.2 Reactive Protocols

9.2.2.1 DSR

A good example of reactive ad hoc routing is the Dynamic Source Routing (DSR) protocol,[23] which in 2007 became an IETF Request for Comments as an experimental protocol.[24] The protocol consists of two fundamental parts, namely the route discovery and the route maintenance processes.

When a node needs a route to a destination, it first checks whether it already knows an appropriate route. If not, it starts the route discovery process by issuing a route request (RREQ) broadcast message, indicating the destination, to its 1-hop neighborhood. The receivers that are not the target of the RREQ and do not know an appropriate route to it, forward an updated copy of the RREQ in which add their own address. In this way, the RREQ records the complete path followed by the message. This feature provides a very simple mechanism to avoid loops, because received messages in which the own node appears in the route record are discarded. If the RREQ arrives at the destination, this returns a route reply (RREP) message back to the source. It can reverse the route record carried in the RREQ and use this information to add IP routing headers (source routing) and eventually deliver the RREP, which also contains the route record. Upon successful reception of the RREP, the source can establish the route to the destination. Subsequent data packets destined to that target will use source routing following the path recorded by the route discovery mechanism. Figure 9.10 shows a simple example of this method.

While forwarding or overhearing any kind of messages, the nodes can cache that routing information in order to avoid future route discoveries. In addition, when forwarding a RREQ message, intermediate nodes that know a route to a destination can return a RREP on behalf of it. So, the delay and overhead associated with a route discovery are decreased.

Despite the fact that the route has already been set up, any of its constituent links may break due to network dynamics. As part of the route maintenance method, every node of an active route is responsible for sensing the link status with its next hop. This can be accomplished by the link layer, for example, IEEE 802.11 acknowledgments (ACKs) or passive ACKs,[25] or via DSR-specific ACKs. If the confirmation of reception does not arrive at the transmitter, it marks the link as broken, and indicates the situation to every node that has sent a packet routed over that link by means of a route error (RERR) message.

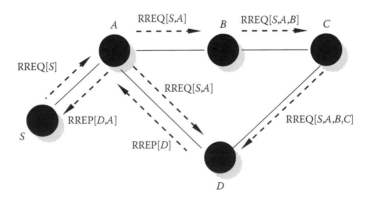

FIGURE 9.10   Example of DSR operation. The path followed is recorded within the protocol messages.

A node that is forwarding a packet might know via route maintenance that the link with the next hop is broken. In such a case, the forwarder can try to salvage the packet if its cache contains an alternative route to the destination. Then, it replaces the original routing headers in the packet with the ones that match its own route. Another interesting feature is the automatic shortening of established routes. When a node overhears a packet with a source route, in which it is not the next intended destination but is listed afterwards, the nodes in between are no longer needed and can be eliminated from the route. To inform the source that it can shorten the route, such a node sends a gratuitous RREP containing the new route.

Finally, we briefly discuss the optional "flow state" extension, by which a sending node establishes a hop-by-hop forwarding state over a known route. Once the route has been set up using the route discovery mechanism, the first data packet is augmented with a header option indicating that this extension is to be used. Intermediate nodes that receive such a header are required to maintain the information of the next hop used for this flow between the source–destination pair. Subsequent packets issued by the source in this flow do not need to use source routing, because the forwarding state is already stored at each next hop.

DSR generates low control overhead and supports several interesting features. The main drawback is the use of source routing in data packets, especially if the average route length is high. This is a problem because the expected number of data packets injected in a network is much higher than that of control ones. Thus, DSR favors the low state maintenance per node in detriment to the data packets overhead. This is partly solved by flow state extension at the cost of higher per node maintenance. In addition, protocol messages used to find a route are usually smaller than the actual data packets that will traverse it. This can cause problems because the route that has been set up might not be able to deliver the data packets. This is due to the nature of the wireless lossy links, which make big messages more prone to suffer from transmission errors. This problem is also exhibited by other reactive protocols like AODV and DYMO.

9.2.2.2 AODV

There is another IETF Request for Comments describing a reactive routing protocol, the Ad hoc On-Demand Distance Vector (AODV) routing protocol.[26] The main difference with respect to DSR is the use of hop-by-hop forwarding instead of source routing. Loop freedom is achieved by the intelligent management of sequence numbers within control messages.

As happened in DSR, the route discovery process is initiated just when a communication is about to begin, but the source does not know any appropriate route to the destination. It then propagates an RREQ, which is forwarded by intermediate nodes until either the destination itself or another node with a fresh enough route to the destination replies in unicast to the source with an RREP. As the RREQ is forwarded, every node adds a routing entry in which the destination is the route discovery originator (data source) and the next hop is the node that retransmitted the message. Similarly, upon processing an RREP, every node creates a route where the destination is the route discovery target (data destination) and the next hop is the previous neighbor, which retransmitted the message. A representation of this process is depicted in Figure 9.11.

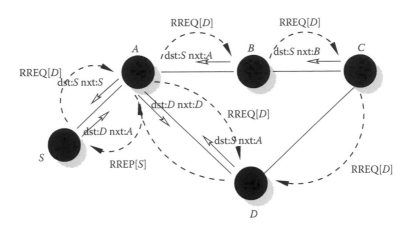

FIGURE 9.11 Example of AODV operation. The route is maintained by each node on a per-hop basis.

The freshness of a given item of routing information (message or routing entry) is given by its sequence number. The bigger the sequence number, the more up-to-date is the information. Every destination maintains a sequence number included within the protocol messages. When a route to the destination is created by any node, it is tagged with the corresponding sequence number. The nodes that receive a protocol message can infer the freshness of the information by comparing the sequence number included in the protocol message with that contained within their routing table. By avoiding stale information, the protocol achieves loop-free operation. A node must increase its own sequence number immediately before it originates a route discovery, and must update it to the maximum of its current sequence number or the destination sequence number of an RREQ message that is about to be answered with an RREP.

An optional feature of AODV is the generation of gratuitous RREPs. When an intermediate node replies an RREQ on behalf of the destination, it also sends an RREP to the destination in order for this to create a route to the source. This avoids the need of future route discoveries to establish the reverse route. AODV also provides guidelines for its operation over unidirectional links. During a route discovery process, it could occur that the transmission of an RREP fails because the link is unidirectional (the RREQ can be transmitted but not the RREP). If no other RREP is received, the source will attempt a new discovery, in which the same situation could be repeated. To avoid this, AODV uses link-layer acknowledgments or RREP-ACK messages to assure the reception of the RREP. If the acknowledgment is not received, the node adds the next hop to its blacklist, so that future RREQs issued by that next hop will be ignored. With this approach, the protocol is able to find a bidirectional route if it exists, because unidirectional paths are no longer explored.

Every node must monitor the link status of its next hops in active routes. Both link-layer mechanisms and periodic HELLO messages are available to detect when a link breaks, and therefore so do all the routes that use such a link. When this occurs, the detecting node issues an RERR message to the nodes in its precursor list. This list includes the neighbors that are using the node under consideration as next hop for any route. By propagating the RERR in this way, all the nodes affected by the broken link are finally notified that it is not available at that moment.

When a link break occurs in an active route the upstream node may try to locally repair the route by issuing an RREQ for the destination. Meanwhile, the upstream node is in charge of buffering the data packets received from the source. If an RREP is not received, it must notify the source via an RERR message, as we have already explained. Otherwise, the route has been repaired and the buffered data packets can be delivered to the destination. In addition, if the new route is longer than the previous one, the node that repaired it informs the source with a special RERR message that does not destroy the route. It just notifies that a longer one is being used, in case the source is interested in starting a new discovery.

A problem that arises with AODV (and is common to DYMO) is that the maximum number of route discovery attempts for a packet is established to three by default. However, if we consider a lossy link with less than 33% success rate, the route discovery will fail. This problem becomes more serious when such a link is acting as a bridge between two otherwise unconnected subnetworks, as both sides will not be able to find routes to destinations on the other side. This could happen often in vehicular scenarios; think, for example, of an actual bridge or two cities linked by a road.

9.2.2.3 DYMO

Currently, the IETF MANET Working Group is working on an evolution of the AODV protocol. Much as OLSRv2 provides easy extensibility on the basis of OLSR, the Dynamic MANET On-demand (DYMO) routing protocol[27] uses the same packet format[28] for the same purpose. The route discovery and maintenance processes of DYMO are like those in AODV, including the possibility of an intermediate node sending an RREP on behalf of a destination. Also, DYMO allows intermediate nodes that are forwarding a routing message (RREQ or RREP) to add their own address to the message. This additional feature is partly inherited from DSR, although the motivation is not the use of source routing, but the incorporation of extra routing information to help avoid future route discoveries. Nonetheless, whether the extra overhead is worthwhile or not is very scenario-dependent (the communication pattern would determine this).

9.2.3 Hybrid Protocols

9.2.3.1 ZRP

The most representative hybrid scheme is the Zone Routing Protocol (ZRP).[29] It defines a proactive zone centered on each node. The radius ρ of the zone, in terms of number of hops, is chosen independently by each node. Those nodes inside the zone whose minimum distance to the central node is exactly ρ are called peripheral nodes, while the others are termed interior nodes. The proactive protocol, which is executed locally, receives the name of IntrA-zone Routing Protocol (IARP), and can be any proactive protocol given that the topology information exchange is restricted within the zone. When a route to an unknown destination is needed, a reactive IntEr-zone Routing Protocol (IERP) is used. Such a protocol uses the topology knowledge provided by IARP to deliver the route request directly to the peripheral nodes. This concept is called bordercasting, and is implemented by the Bordercast Resolution Protocol (BRP). BRP can use multicast to send the route request to the peripheral nodes and thus reduce the overhead of broadcasting the message inside the zone. Whenever the request arrives a node that is aware of a route to the destination, it issues a route reply back to the source. As the reply is propagated, either the path information is recorded in the message (as in DSR) or maintained on a hop-by-hop basis (as in AODV). If the receiving node of a request does not know the destination, then it forwards the packet using bordercasting. An example is illustrated in Figure 9.12.

The main issue with ZRP is the selection of the optimum radius. Even with complete knowledge about network size, node density, and relative mobility, the calculation of the best radius is complicated. The situation becomes even worse in highly dynamic networks whose characteristics are time-dependent. Two dynamic sizing techniques have been proposed:

1. *Min searching.* This scheme tries to achieve a local optimum for the total ZRP traffic. The radius is either incremented or decremented in steps of one hop. The process is repeated in the same direction as long as the new measured traffic is smaller than the previous one.

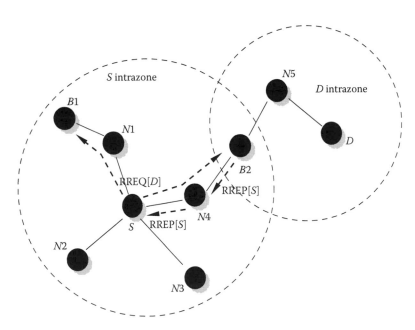

FIGURE 9.12 ZRP example with ρ = 2. Source *S* wants to communicate with destination *D*, which is not inside its intrazone. *S* sends a RREQ to its peripheral nodes, *B1* and *B2*, using multicast. Because *B2* is within *D*'s intrazone, it knows a proactive route to *D* and sends a RREP back to *S* using unicast.

2. *Traffic adaptive.* The ratio of IERP and IARP traffic is compared with a threshold. The radius is increased if *overhead(IERP)/overhead(AIRP)* is larger than the threshold (too many reactive queries), and decreased if it is lower (too much proactive overhead).

9.2.4 Geographic Protocols

The work on geographic routing, also called position-based routing, started in the late 1980s. Almost all the variants proposed to date are based in the notion of progress. As MANETs are normally modeled by graphs, calling nodes and edges to routers/mobile nodes and links respectively is a widely accepted convention. Using this notation, Figure 9.13 shows a node *S* that needs to send a message to node *D*. Some other nodes in the network are located inside *S*'s coverage range. These nodes are called 1-hop neighbors. As *D* is located outside *S*'s coverage range, *S* has to select one of its 1-hop neighbors as next relay for the message. In general, the progress of a 1-hop neighbor is defined as the distance of the segment between *S* and the 1-hop's projection onto the line between *S* and *D*. The progress can be positive as in the case of 1-hop neighbors *A*, *C*, and *F*, or negative as in the case of nodes *B* and *E*.

Unlike in traditional distance vector or link-state-based-routing protocols, in geographic routing algorithms there is no need of interchanging routing tables among 1-hop neighbors. Nodes take routing decisions solely based on position, the position of the destination, and the 1-hop neighbors's positions. These decisions are taken using geometric criteria, therefore nodes' positions must be defined in a common coordinating system. It is common to use coordinates defined in the standard $\mathbb{R}^2$ Cartesian plane, but this is not the only one possible.

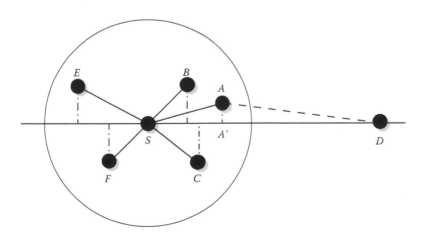

FIGURE 9.13 Different ways of measuring the progress achieved by selecting a 1-hop neighbor as next relay towards the destination, node *D*.

The operation of a standard geographic routing algorithm normally comprises the following four phases:

1. *Determining the destination's coordinates.* This information is determined by the source of the message. The coordinates are included in the header of the message to inform intermediate relays. Here, an open issue is how to determine the coordinates of a given destination node. Nodes cannot easily maintain an accurate list of node–coordinate pairs. Flooding can be used for searching the destination node and then waiting for its response, but this is not a good option because of the unacceptable communication overhead.

2. *Determining 1-hop neighbor's coordinates.* This information is usually obtained through the periodical interchange of short messages called beacons. Beacons contain the identifier of the sender and its coordinates. Alternatively, some beaconless geographic routing protocols have been proposed. In these protocols, nodes use a proactive approach to determine which other nodes are their neighbors, and their respective coordinates.

3. *Determining the next relay.* The next relay is determined using the information about the position of the current node, the destination's coordinates, and the coordinates of the 1-hop neighbors.

4. *Message delivery.* After choosing which 1-hop neighbor will become the next relay, the last step consists of delivering the message to it. Usually, the message header includes the identifier of that node because all 1-hop neighbors receive the same message, but only one of them must continue routing it.

As can be seen there are some important assumptions that every geographic routing protocol makes. First, nodes must be able to determine their own position. This can be accomplished by using GPS devices or through the use of virtual coordinates such as in Refs. 30 and 31.

The idea of virtual coordinates consists of electing some nodes and giving them some coordinates. The rest of the nodes obtain their own virtual coordinates by averaging the ones of their neighbors, or by triangulation. The most difficult task here is to determine which nodes to select as base points, and how to deal with the error propagated through the network due to accumulative imprecision of calculus.

There are also some proposals for deploying a wireless network in which only some nodes use GPS devices and the others determine their coordinates using a distributed localization system[32,33] or even use spatial triangulation.[34] Second, determining the coordinates of the destination is a task not solved satisfactorally to date. Several algorithms have been proposed to deal with the location updates of destinations in MANETs and mobile WSNs.

In some of them, nodes send position updates to multiple zones. This is the case of the *doubling circles* approach, in which updates are issued to all nodes located within circles of radii P, $2P$, $4P$, ..., $2^t P$ centered at the sender. When thenode leaves the circle $2^t P$, it notifies its new position inside the circle $2^{t+1} P$. In this way, as a message is routed towards its destination, more updated position information is found at every hop until it is eventually delivered to the current position of the destination. The main problem of this approach is the occasional flooding of the entire network (or a big part of it).

Quorum-based strategies try to reduce the former overhead by replicating the location information on certain nodes, which act as repositories. Stojmenovic proposes an algorithm in which location updates are performed in a "column" of the network; that is, the location update message is propagated from the sender to the north and south of the network. When another node needs to retrieve location information for a particular destination, it initiates an horizontal search (i.e., to the east and west) until the search message crosses the column with updated information. Once the search message arrives at any node of the column, it replies to the source to communicate the most recent position information of the destination.[35]

Another way to implement the location service is inspired in the way it is performed in cellular and infrastructure networks, where every node notifies its position to a home agent that stores location information for a set of nodes. In the protocol described in Ref. 36, each node initially informs every other one of its current position. The home agent for a given node is defined as the set of nodes located within a circle of predefined radius, centered at its initial position. When a location update is to be sent, the node geocasts such a message to its home agent. That is, the location update is unicasted using geographic routing until it reaches one node inside the home agent. The message is then broadcast inside the home agent's circle. When a destination must be found, the source issues two search messages. The first is directed to the last known position of the destination, while the second travels towards the destination's home agent. When the search message arrives at the destination or a node within the home agent, the source is notified with the current location.

Note that the protocols described make use of geographic routing to deliver search messages. Once the destination's position is known, either because it is a fixed server that has been preconfigured or because any of the previous location protocols have been run, the routing protocol comes into play. Geographic routing algorithms work in two different modes: greedy and perimeter. The principal difference resides in determining the situation where it is appropriate to use each. The greedy mode is used whenever possible, and the perimeter mode is strictly used when the greedy mode cannot be applied.

9.2.4.1 Greedy Mode

Most geographic routing protocols differ only in the way the next relay selection function works. Concretely, the differences reside normally in the the notion of progress used. Nevertheless, there is a common assumption that the next relay selected should be a neighbor closer to the destination than the current node because not doing that can lead to starting a cycle. Thus, choosing a neighbor providing a positive advance is the idea behind the greedy mode, which is also called greedy routing.

Now we use Figure 9.13 to explain the slight differences between the most notable greedy routing proposals in the literature. In figure, node S is the one currently holding a message whose destination is node D. Node S has a set of 1-hop neighbors and D is not inside its coverage radius. Let us now see a brief description of every major greedy routing proposal.

1. *Greedy scheme.* One of the first geographic routing proposals is the Greedy Scheme by Finn.[37] Here, the next relay selection function just chooses the 1-hop

       neighbor whose progress is greatest. Here the notion of progress used is the one we explained above. In this scheme, when there are no 1-hop neighbors providing advance, the message is dropped.

2. *Compass routing (CR)*. Another greedy scheme is the Compass Routing method depicted by Kranakis, Singh, and Urrutia.[38] This protocol determines the next relay by measuring the angle that each candidate relay (1-hop neighbors) forms with the segment between the current node and the destination. The neighbor selected is the one whose angle is lower. This protocol may lead to fails, as shown by Stojmenovic and Lin.[39] Figure 9.14 shows an example where this greedy approach start cycles.

3. *Most forward progress (MFR)*. Takagi and Kleinrock proposed MFR[40] as a way to overcome the problem of cycles suffered by the two previous algorithms. This protocol is the first to correctly use the concept of progress as we introduced it. Nevertheless, a special kind of cycle can appear when there are only two nodes. The trivial solution consists of introducing a state in nodes to remember the next relay selected. In that way, when a node receives a message from the same neighbor it choose as next relay, the message is discarded.

MFR is an improved version of the Random Protocol,[41] which is a first approach from the same authors in which the next relay is selected in a random way among those neighbors providing a positive progress (nodes *A*, *B*, and *C* in Figure 9.13). Obviously, MFR achieves a lower mean hop count than RP.

Finally, all the protocols have the same problem; when there are no neighbors providing positive progress, the protocol fails to deliver the message. Thus, the second routing mode, the perimeter one, is fully justified, as we will show.

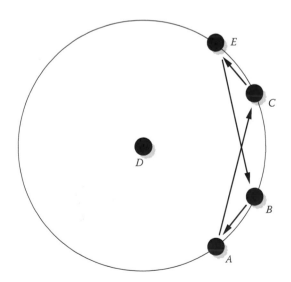

FIGURE 9.14 The Compass Routing Scheme can lead to cycles due to the criteria used in its next relay selection.

9.2.4.2 Perimeter Mode

In the literature, the situations where a node has to route a message towards a destination and it has no neighbors providing positive advance towards that destination are also called local minima. It is very common in scenarios where the mean density of nodes is low, but it can also appear in very dense deployed networks. Thus, geographic routing protocols behaving only in greedy mode cannot guarantee the delivery of messages.

Figure 9.15 shows an example of a local minimum. The node *S* when trying to determine its next relay towards *D*, the destination of the message, cannot find any neighbor among the two known by it (*A* and *B*) providing positive progress towards *D*. In the figure, the coverage range is represented by a circle of radius *r*.

As reaching a local minimum does not mean the nonexistence of a path to the destination, there are some proposals to overcome this issue. Finn considered the possibility of reaching local minimum, and proposed the use of restricted flooding in those situations. Unfortunately, this approaches generates a control packets overhead that is too high, degrading the performance of the protocol to lower levels than traditional nongeographic protocols.

Using well-known algorithms from graph theory, different authors proposed relatively similar protocols with a common goal: trying to surround the area without nodes found on the local minimum to a node where greedy routing can be restarted. The idea is simple. Considering the network as a graph, a local minimum can be seen as the border of a void area, that is, an area without nodes. Void areas are defined by a set of

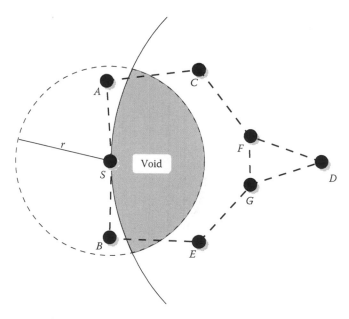

FIGURE 9.15   Greedy routing fails to deliver a message due to a local minimum.

nodes delimiting their perimeter. Thus, applying the right-hand rule[42] it is possible to go through the border of the void area.

To successfully apply the right-hand rule the graph must be a planar one, that is, a graph whose edges do not cross each other. Owing to the localized nature of geographic routing algorithms, the planarization of the graph modeling the whole network is an unaffordable task. However, there some topology control algorithms exist that can be used to obtain a planarized view of the portion of the graph that each node locally knows. These algorithms are designed to be applied in a distributed fashion[43]; that is, they do not need to know the complete topology. Concretely, these algorithms planarize only the subgraph in which the current node and its 1-hop neighbors are embedded.

The planarization algorithms most commonly used in the literature are Gabriel Graph[44] and the Relative Neighborhood Graph.[45] They work by locally removing the edges connecting the current node with some of its neighbors. Other algorithms exist whose planarized subgraphs have subtle differences, but these are used less frequently because of their higher computational costs (e.g., the Localized Delauney Triangulation[46-48] and the Morelia Test[49,50]).

Next we describe the most important geographic routing algorithms developed, taking into account the local minimum problem, that is, considering not only a greedy approach but also a perimeter mode to allow the protocol to guarantee the delivery ratio, at least, in ideal conditions. All of them make use of a localized planarization algorithm, usually the Gabriel Graph.

1. *GFG.* Proposed by Bose et al.,[51] the first version of the algorithm, FACE-1, works as follows. Assuming the whole network is modeled by a graph without edges crossing (a planar graph), the path used to reach the destination must follow the interiors of the adjacent faces intersected by the imaginary line between the source of the message and its destination. For example, starting by following the right-hand rule, the second time the message crosses an edge intersected by the imaginary line, the message continues applying the contrary rule, in this case the left-hand rule.

As this method generates very long paths, the same authors proposed an optimized version, FACE-2. The optimization consists of changing the rule the first time the imaginary line is crossed. FACE-1 needs a maximum of $3|E|$ hops to reach the destination, E being the set of edges of the graph. Although FACE-2 has a higher theoretical superior limit, in practice, the results are better than those of FACE-1.

Both algorithms have a very poor performance because the paths they create always go through a border of faces. Therefore, the authors proposed a third algorithm combining a greedy routing algorithm with the FACE-2 perimeter routing protocol. The new algorithm, called Greedy-Face-Greedy[43,51] (GFG) is thus a greedy approach able to get over void areas through the use of the FACE-2 algorithm. The perimeter routing protocol is only applied when a void area is reached, and then greedy routing is used as soon as possible. It is enough to find a node closer to the one where greedy routing found the previous local minimum, that is, where perimeter routing started.

2. *AFR and GOAFR+.* Kuhn et al. proposed Adaptive Face Routing (AFR) in Ref. 52. AFR is an extension to FACE-1 in which the faces used to reach the destination are

limited by a virtual ellipse whose foci are both the source and the destination of the message. The authors proved that every localized protocol following the geographic routing scheme has a maximum theoretic cost of $O(c^2)$, with c being the length of the shortest possible path between source and destination.

9.2.4.3 Conclusions

Geographic routing algorithms are usually applied to wireless sensor networks due to the special computation, bandwidth, and memory limitation characteristic of these networks. Nevertheless, this routing technique has started to gain momentum in VANET scenarios. Researchers in the field of VANETs have adopted geographic routing almost as a standard base technique because the nodes forming a VANET cannot move through the field with total freedom; instead, nodes (cars) can only move by some concrete path (streets). As the topology of the streets is normally known thanks to maps used by cars in their GPS navigators, the application of geographic routing techniques is almost direct.

9.3 Applicability of MANET Routing to Vehicular Environments

9.3.1 Characteristics of VANET Scenarios

The main nodes involved in VANET communications are the vehicles themselves. However, the ad hoc network is not always isolated but can be attached to an infrastructural deployment. Equipped network interface cards such as IEEE 802.11p can be used to communicate both with neighboring vehicles and roadside units (RSUs). The on-board unit (OBU) can also be provided with other communication technologies (e.g., 2G/3G or WiMax interfaces), which provide direct access to an operator's network when it is available. Indeed, many international consortiums (e.g., CAR2CAR[53]) and standardization bodies (e.g., ISO through the CALM initiative[54]) are defining standards to manage the different communication capabilities expected to be present on future private vehicles. So, a given car could be equipped with multiple network interface cards of different technologies.

In general, a personal vehicle exhibits the following properties:

1. As we have just seen, a vehicle may have high communication capabilities, depending on the interface cards installed in the OBU.
2. Vehicles are equipped with long-lived batteries, so the energy consumption due to communications is almost negligible.
3. Memory and computational resources are high enough to develop complex algorithms. Here, memory and CPU savings are not a critical concern.
4. Position information may be acquired via geographic positioning systems like GPS or Galileo.
5. Vehicles can have digital maps of the geographic zone they are traveling around. Moreover, they might be aware of the route that is to be followed.

However, not only private vehicles take part in communications. Public transport systems are also expected to be part of the network. For instance, buses could participate in the VANET and additionally provide Internet connection by means of a 3G link with the network infrastructure.[55] In such way, public transport vehicles act as mobile gateways that extend the coverage of the VANET up to the Internet.

From a routing protocol designer perspective, the availability of infrastructure (either mobile or fixed) allows for connectivity to external networks (such as the Internet) and, interestingly, can also be used to route packets inside the VANET.[56] On the one hand, if a communication between two far-away vehicles is taking place through a long path, the likelihood of reception at the destination decreases as the number of hops gets higher. To improve reliability, either the source or an intermediate node can route the packet via the infrastructure network. The packet is then routed until it reaches the VANET again, where multihop ad hoc routing is used to deliver the packet to the intended destination. On the other hand, the previous approach can be necessary to enable communications between scattered subnetworks, especially when the degree of market penetration is low.

A possible vehicular scenario is shown in Figure 9.16, where several communication infrastructure deployments are illustrated.

Another salient characteristic of vehicular scenarios is that nodes cannot freely move around an area. They have to respect the road layout, traffic signals, and other vehicles'

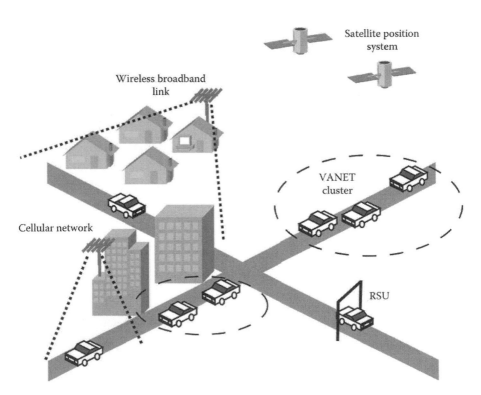

FIGURE 9.16 High-level view of an urban vehicular environment with several communication possibilities.

movements. Also, the distribution of nodes is not uniform. That is, vehicles tend to drive around forming groups, and the distance between groups is usually larger than the radio coverage of the VANET wireless interfaces. In addition, the traffic pattern differs depending on the kind of road on which the vehicles are passing:

1. *Rural road.* In a rural environment, the traffic density is expected to be low and therefore the resulting ad hoc network would be highly disconnected. This means that the network is partitioned into many little *clusters* (or groups of vehicles), which are not close enough to communicate directly. To overcome this, vehicles can route data packets to others by means of any existing communications infrastructure. However, in rural regions there are typically few infrastructure deployments, which dramatically worsens the connectivity problem. Additionally, the average speed of the vehicles is expected to be moderately low.
2. *Urban road.* In this case, there is a moderately high number of vehicles which makes it easier to find a path from source to destination. In addition, a communications infrastructure is likely to be available, increasing the communication capability of the equipped vehicles. These would run at moderately low or high speeds, depending on the specific road.
3. *City road.* Traffic density is expected to be very high, with many vehicles at very low speeds and with long periods stuck in traffic jams or stopped at traffic signals. The availability of network infrastructure would be very high, with several technological choices to establish communications. The large number of cross-overs and roundabouts, as well as traffic signals, determines the possible movements of vehicles. In addition, the city streets where nodes are located and move around are normally surrounded by tall buildings and other obstacles; these shadow the wireless signals and cause multipath fading.
4. *Highway.* The traffic pattern is clearly different in this case, where vehicles are driven at high speeds following a road without crossovers or traffic lights. A special case appears at on-ramps, where new automobiles entering the highway must be aware of those already on it. Traffic density, as well as infrastructure deployment, are very much highway-dependent, although it is expected that RSUs will be encountered, with which the vehicles can communicate.

The traffic pattern followed by vehicles in each of the former scenarios is also affected by geographic, sociological, and external factors. Thus, some regions are more populated and/or developed than others, which supposes a higher number of automobiles. The timeframe is also very important, because a low traffic density can be expected late at night, but it would be extremely high at rush hours. Similarly, traffic conditions are different on weekends, on holidays, when a special event (e.g., a concert or sport match) takes place, and so on. It also differs depending on meteorological conditions and when an unexpected situation occurs, such as accident or a street being repaired. To sum up, many factors come to play in order to determine the mobility pattern of vehicles.

Moreover, the degree of connectivity of a network is dependent on the market penetration of communication technology in the vehicles. In the first few years following products being available for purchase, the networks will be very sparse and partitioned. This must be taken into account by routing protocols and applications designers.

All these issues have long been investigated in traffic engineering studies. There are two main sorts of models to analyze traffic, namely macroscopic and microscopic models. Macroscopic models describe the traffic as a compressible fluid, and deal with parameters such as traffic density, average speed, and traffic flow. In contrast, microscopic models consider the behavior of each individual vehicle in a (set of) road(s). This achieves higher levels of accuracy at the cost of higher computational effort.

Another characteristic of vehicular scenarios is the great variety of services that are to be deployed over the network. Every service has its own requirements, and the network should be able to accommodate them. The following is a brief description of common services used for VANET*:

1. *Safety.* Safety-related services are of paramount importance is minimizing the number of victims on the road. So, they impose the hardest requisites to the network, as safety messages must be relayed through the VANET as fast and as reliably as possible. In the case where the network fails to do so, human casualties and great economic losses may arise. These kinds of services can also use the network infrastructure to alert an emergency center.
2. *Advanced driver assistance systems (ADAS).* This can also improve the security on the road by preventing collisions (e.g., the car can automatically break if it is too close to the preceding one) or automatically adapting speed (e.g., based on messages issued by RSUs or traffic lights). When human lives are at risk, these services also have maximum priority.
3. *Traffic management.* The actual traffic situation in a given location is valuable information for road operators and float management applications. The service can interact with the vehicles in order to direct them to other roads in response to several criteria, for example, to avoid collapsed ways.
4. *Infotainment.* Multimedia and entertainment-related applications have been proposed for use in future vehicular networks. The exchanged information is not of high priority, but is sensitive to delay and jitter. The network must provide an adequate treatment to these messages to make the applications usable.
5. *Legacy internet services.* Vehicle occupants can connect their devices (laptops, PDAs, and so on) to the OBU, either with a wireless or wired link, and make use of traditional Internet applications when external connectivity is available. Thus, the users will likely try to use regular services like email and web browsing.

All these features mean that solutions originally designed for MANETs are not appropriate, in their current arrangements for use in vehicular ad hoc networks.

9.3.2 Problems

The specific characteristics of VANETs make the design of routing protocols a challenging task. As we have seen in the previous subsection, VANETs are a very special case of MANETs, with many differences in the assumptions made when designing routing

---

* There are many other services designed for vehicular environments. This is not an exhaustive list, but on illustration of the wide range of expected applications and the different requisites they impose on the network.

protocols for them. Typical MANET nodes were primarily characterized by low energy, computation, and memory resources, which is not a big concern for VANETs. Furthermore, the information available to a node in a vehicle is enormous, partly due to the navigation of the vehicle, which requires a map of the area where the vehicle is driving, the trajectory suggested by it, traffic information related to congested streets, and so on. There are also some other sources of information that can be used, such as knowledge about recurrent events or the movement pattern of certain vehicles such as buses. As MANET routing protocols do not need to take advantage of all this information, it is clear that more specialized solutions would achieve better performance in VANETs.

When evaluating the performance of MANET protocols with simulation studies, researchers have commonly used synthetic mobility models based on random movements, for example, the Random Walk and Random Waypoint mobility models.[57] These cannot be used in VANET tests, because nodes in VANETs can have very different speeds to the ones in MANETs and the movements are clearly not random. Some works[58] illustrate the problems that traditional mobility models have. They show the low performance achieved by AODV and GPSR protocols when using real logs in contrast with the good marks achieved using the Random Waypoint model. Additionally, depending on the considered scenario, mobility patterns can differ a lot. For instance, in an urban scenario drivers do not normally drive quickly, but in a highway this is obviously not the case. The highway scenario does not normally have buildings disturbing radio signals, but the high speed of nodes can play an important role in the quantity of information that two nodes can exchange, especially when traveling in opposite directions.

Given that, and as some performance evaluations have shown,[59] traditional MANET routing protocols do not successfully work in VANET environments. In the remainder of this subsection we shall elaborate on the causes of this.

One of the most important causes of the poor performance of MANET routing protocols in VANETs is the intensive use of flooding. This primitive method consumes a lot of network resources in terms of bandwidth. This is particularly dramatic in dense vehicular networks, such as big city scenarios, where many nodes are expected to initiate a communication. For each of them, a message must be forwarded to the whole network (or a big portion of it). This has several inconveniences. First, much bandwidth is wasted in routing control messages, leaving little bandwidth for applications or services, and these are the ones that matter to the user. Even worse, because of the nature of wireless communications, such a *broadcast storm*[60] undoubtedly leads to lost messages because of collisions. Moreover, if a CSMA-based MAC layer like IEEE 802.11 is used, flooded messages increase the contention to access the medium, increasing communication latency. These flooded message will interfere with other data flows, and if the message does not reach its intended destination, a route cannot be set up, preventing the communication.

Reactive protocols use the technique of flooding whenever they want to establish a communication, as it is used to find the destination. Although many protocols try to reduce the scope of the flooding by means of the *expanding ring search* approach, this does not always work well and can cause even more overhead. The expanding ring search consists of starting the flooding inside a ring of a few hops away from the source, and incrementally expanding the ring until the destination is found. Obviously, this technique relies on the assumption that the destination is near the source. Otherwise, it provokes more overhead than blind flooding.

Link-state proactive algorithms also need to flood the network with link states. This is the case for OLSR, which issues TC messages that have to be received by every node in the network. In spite of the great overhead reduction achieved by the use of MPRs, OLSR is not able to scale up to large networks like those in vehicular environments. Although Fish eye OLSR has been proven to improve scalability, it also generates a lot of control overhead, which requires high quantities of bandwidth. Moreover, because it uses less fresh information, longer routes and loops are more likely to occur. Other proposals like C-OLSR and H-OLSR are not suitable either, the former because of the overhead to maintain the clusters' structure, and the latter for the extremelyhigh deployment effort needed (in addition to the suboptimal routing achieved in both cases).

Hybrid protocols like ZRP still have to flood a restricted zone of the network, so high contention and likelihood of collisions and interferences occur. Limiting the problem does not solve it.

A problematic consequence of this the lack of scalability of ad hoc routing in VANET. The protocols that use flooding do not perform well when the number of nodes in the network is high, which is the case for VANET. Owing to the specific functioning of on-demand protocols, they also cannot scale when the number of data communications is high, because each (or many) of them requires a route discovery process with its associated network-wide search. Proactive distance vector solutions like DSDV do not issue flooded messages, but cannot scale up to big networks either. The problem is that every node needs to exchange its routing table with its neighbor. In a VANET, the size of such a message would be prohibitive due to the bandwidth waste. Obviously, hybrid approaches need periodic message exchanges inside the proactive zone, and need to flood requests for destinations outside it. This does not scale either.

In summary, reactive, proactive, and hybrid protocols exchange nonlocal information or local information with nonlocal nodes. This is a burden for protocol operation because it translates into big control messages or propagation to far away nodes. In addition, this procedure establishes a route that is probably not valid after a short period of time, because of wireless signal dynamics and node mobility. We would like to emphasize this concept. Under many vehicular scenarios, any existing path between a source and a destination is short-lived and therefore it might not be worth the effort of setting up such a route. A much more appealing approach is the one taken by geographic routing, where only local information (in the case where the destination's position isknown) is applied to route a packet on a hop-by-hop basis. In fact, geographic routing has been chosen as the foundation of most of the routing algorithms used for VANETs, as we will see in Section 9.4.

Unfortunately, geographic routing algorithms developed for MANETs and WSNs cannot be directly applied to VANETs. The rationale is quite simple, because nodes cannot freely move but need to follow the paths allowed by the environment. So, the idea of progress is no longer directly applicable, because there is no physical way to find neighbors in a direction where vehicles are not allowed to be located (think, for example, of a building between two streets). This fact is corroborated by the low performance obtained by protocols like GPSR.[61]

Because of this, vehicular ad hoc routing protocols must use as much information about the surrounding road topology as possible. This helps the protocol when deciding the most appropriate next hop for a packet. Knowing the layout of the streets, the node can decide to follow a given path of streets instead of a path of nodes, given that such a route leads to the destination. An additional problem with geographic routing in many applications shows up when the communication is with a mobile destination. Because the source needs to know the position of the destination to make the forwarding decision, a location service must be deployed in order to acquire such information. The design of a scalable location service can be quite cumbersome, and the adopted solution could limit the scalability of the approach. Clearly, the location algorithms developed for MANETs and WSNs (see Section 9.2.4) are not applicable to VANETs. The doubling circles approach (and similar) introduce too much overhead. The quorum-based approach, which performs vertical updates and horizontal searches, is useless in a VANET, because the message propagation is determined by the environment and road layout. And finally, the home agent-based approach has no point because vehicles travel very long distances, so it does not seem appropriate to send update/search messages to far-away home agents through the VANET. In conclusion, new specific location services adapted to the very distinct characteristics of vehicular environments must be developed.

In addition to exploiting the information of the surrounding area in the forwarding algorithm, information about the speed, direction, or route plan can be exchanged between passing vehicles. With such knowledge, a vehicle can predict link breakages and change its decision accordingly. More interesting, if the message is not delay-sensitive, a DTN (delay- and disruption-tolerant network) approach can be taken. DTN routing[2] is based on the principle of store-and-forward; that is, a message can be buffered at a node until an appropriate next hop appears. Then, the node forwards the message to the next such hop. This sort of operation is very useful for many vehicular applications. For example, think of an alert application that propagates emergency messages when an accident occurs. Just after the accident the application forwards a message that is rapidly propagated in order for the upcoming vehicles to break and stop. Now think of the last vehicle that received the message. If there are no more neighbors to send the message to, it has no trigger to forward and forget the packet. The intelligent decision is to buffer the packet and forward it as soon as a new approaching vehicle is detected. This behavior could save another accident. So, DTN routing can be useful for similar applications or services. This, combined with the ability to know the travel plan of other vehicles, can be a powerful mechanism to deliver a message to a destination by relaying it to a vehicle that is going to pass through the same area. Once the forwarder is near the destination, it can deliver the packet to it.

However, the classical MANET algorithms assume that a path exists between a source and a destination at the moment of sending a message from the former to the latter. If this is not the case, the message is simply discarded. As we have just seen, this is not the best approach for vehicular scenarios.

9.4 Routing Protocols for Vehicular Ad Hoc Networks

Routing protocols designed specifically for VANETs can be classified into two big groups: the source routing protocols and the geographic routing ones. In the former the source node decides the trajectory that the message should take through the network to reach the destination. The path is normally specified to include a list of points, nodes, or subareas. The letter group, based on geographic routing, does not directly use geographic routing over the graph made of nodes and wireless links. Instead, it is used over the graph of streets. Thus, the decisions taken are related to the next street that the packet should take, not the next node. In the following subsections we describe the principal protocols in each group and their internal subclassifications. But first, we briefly summarize some of the approaches that have been studied for the development of a location service protocol suitable for vehicular scenarios.

The European project CarTalk2000[62] defines a local location service in which vehicles announce their position to others that are within a limited number of hops. The update frequency is adapted depending on the distance between the source and the other vehicles, being lower as the node is farther away. In addition, positions are carried within data packets, allowing intermediate nodes to cache others' positions.[63] This scheme is suitable when communications only happen between close vehicles, as in the case of the CarTalk2000 scenario under consideration. However, it does not work as a general solution when the destination is far from the source.

In the case of the German project FleetNet,[64] two different location services have been evaluated. The Grid Location Service (GLS)[32] considers the network area as a grid. Every node notifies its position to certain "location servers," which are vehicles with a similar identifier located in squares of the grid of increasing size. The number of servers decreases logarithmically with the distance from the node. When another vehicle needs the position of a destination D, it issues aquery that is forwarded to the known destination with the closest identifier to Ds, by using geographic routing. This process is repeated hop-by-hop until the query eventually reaches any of Ds servers, which reply to the petitioner. However, some simulation results show that the performance of GLS are not that good,[65] especially with high mobility and node density. On the other hand, the Reactive Location Service (RLS)[66] initiates a scoped flooding to find the destination's position whenever a data packet is to be sent but no position information is available. Although the flooding protocol incorporates a neighbor elimination scheme to avoid the broadcast storm problem, and other nodes' positions can be cached, simulation results show that this approach is not scalable.

A more interesting approach is proposed in V-Grid,[67] which takes advantage of the intrinsic characteristics of vehicular scenarios. Two complementary location services are deployed, one in the network infrastructure and the other in the vehicular network. In this way, the position of each node can easily be acquired when there is infrastructure connectivity, but the service still works (though not so well) when the ad hoc network is isolated.

9.4.1 Source-Routing-Based Protocols

In general, the application of the source routing technique is useful when the intermediate hops determined are not nodes but locations. The three protocols described in the following sections are the most representative ones in this category.

9.4.1.1 Geographic Source Routing

The geographical source routing (GSR) protocol is described in Ref. 61 along with a reactive location service. After evaluating the bad performance results that traditional protocols (concretely DSR, AODV, and GFG/GPSR) achieve in VANET scenarios, the authors present GSR. The key idea is to use the knowledge of the street map of the area where nodes are deployed. As we have already noted, this information can be found in the navigator device of every vehicle. Knowing the area topology, GSR applies Dijkstra's algorithm to discover the shortest path to the destination. The list of crossroads is then included in the header of the message and is transmitted in a greedy way. Concretely, the message is passed to the 1-hop neighbor located closer to the next crossroad. The authors assume that the greedy routing part of the algorithm used between crossroads cannot have problems, but, in fact, the message can be lost if there are very few vehicles, and therefore connectivity is low.

Besides, an on-demand location service protocol is presented as a tool to be used by source nodes wanting to send a message to a destination node. This protocol finds out the destination node and retrieves its location. The main problem is that it uses a global flooding, thus, the scalability is not guaranteed. Moreover, the paper also presents the results of some simulations comparing DSR, AODV, and GSR. GSR's results are better than DSR due mainly to the overhead introduced by DSR. Nevertheless, GSR only achieves similar results to AODV. Obviously, this is due to the initial flooding used by GSR to determine the location of the destination.

9.4.1.2 Spatial Aware Routing

The novelty that spatial aware routing (SAR)[62] introduces consists of a strategy to avoid losing packets when a local maximum is reached. SAR uses GSR as the basis protocol, so a local maximum can be reached due to the lack of 1-hop neighbors providing an advance towards the next intermediate crossroad included in the source routing header. In GSR the packet is discarded in these situations, but here, the idea is to find an alternative route to that previously computed by Dijkstra in the source node.

When a node is a local maximum it computes the shortest path from itself to the next point, determined by the source routing header. To do that, it uses the Dijkstra algorithm again after eliminating the edge representing the current street. Thus, an alternative path can be found. The authors also propose one more strategy to avoid discarding packets. This consists of storing the packet a predefined time while waiting to find an adequate 1-hop neighbor.

When comparing the performance evaluation of SAR against a plain GPSR the results show that SAR achieves a better packet reception ratio, but the mean number of hops of the paths is longer than that of GFG/GPSR. Also, when comparing the overhead introduced by the header, in low-density networks GFG/GPSR is better than SAR, but in high-density networks SAR outperforms GFG/GPSR because its overhead is almost independent of the density.

9.4.1.3 A-STAR

The A-STAR protocol[69] uses the same idea as GSR and Terminode Routing (TRR)[70]; that is, the source node determines the trajectory the message must follow applying Dijkstra's algorithm over the street map. The novel aspect proposed by the authors consists of including information about the traffic density of the streets as edges' weights. The goal is to determine a sequence of streets with a high probability of having enough nodes (vehicles) to allow the transmission of the message. Additionally, when a message reaches a local maximum the recovery strategy used is similar to the one proposed in SAR.

Messages being routed through alternative trajectories are also used to disseminate information about the current and temporary state of streets where routing is having problems. That is, the message contains information about the recently discarded street, so nodes forwarding it can have this information and they update the weights of the graph representing the streets they use to determine new routes. Obviously this information has to be discarded after a predefined time, assuming thatthe state changes over time and that the old information can be outdated.

9.4.1.4 Connectivity-Aware Routing

The authors of connectivity aware routing (CAR)[71] present this algorithm as a solution to two problems present in most of the previous solutions. On the one hand, there is the unsuitability of applying pure geographic routing techniques to VANET scenarios, that is, using geographic routing algorithms over a graph of nodes instead of a graph of streets. On the other hand, there is the assumption of connectivity between nodes inside streets that some protocols make when computing the path to follow to the destination. Assuming this can lead to reaching nodes in which there exists a local maximum.

CAR is also a source-routing-based protocol that applies a greedy scheme to progress through the crossroads determined by the source node. In fact, the greedy routing protocol used is the Advanced Greedy Forwarding (AGF) algorithm.[58] However, the list of intermediate points forming the path towards the destination is not determined applying Dijkstra's algorithm over a graph made of streets. Instead, CAR uses a preliminary flooding-based phase to localize the destination node. The destination's answer travels back to the source following the inverse tree created by the flooding phase. On its way back, nodes relaying the message add their location when they think they are close to a crossroad. In that way, the source receives a response message including a list of anchor points that is used as a source routing header.

The flooding phase uses an optimized version of the Preferred Broadcast Group (PGB) algorithm,[58] which achieves a better delivery ratio. Nodes infer they are near a crossroad using a special field included in the periodic beacons interchanged. Each beacon contains the velocity vector, so that when a node receives beacons from two different 1-hop neighbors whose velocity vectors are not parallel, it identifies itself as being near a crossing. Additionally, the authorsintroduce a new adaptive beaconing mechanism to maintain the overhead of control messages independent of the density of the network.

The performance results of CAR are not exceptionally good. The main problem lies in the use of an initial broadcast that prevents the protocol from scaling with density. Nevertheless, CAR outperforms GPSR when applied in a standard way.

9.4.2 Geographic-Routing-Based Protocols

As we have commented above, the geographic routing scheme is used in most solutions but with some provisos. First, it is only used to transmit messages between nodes in the same street. Next, standard recovery mechanisms are never used directly over the graph of nodes. When applied, they are used to determine the next crossing instead of the next hop.

Now we present some representative examples of protocols not using source routing. These protocols try to avoid increasing the header of messages with source routing headers. To do that, the basic idea is to apply geographic routing techniques directly over the map of streets, that is, deciding the next direction when reaching a crossroad.

9.4.2.1 Greedy Perimeter Coordinator Routing

Most proposals presented at the moment assume each node knows the complete street map of the area. The authors of greedy perimeter coordinator routing (GPCR)[72] eliminate this precondition and, at the same time, their protocol does not use flooding or source routing headers. Thus, the main goal of GPCR is to achieve a reasonably good delivery ratio while reducing the control overhead due to control messages including excessive information in the form of headers. GPCR bases its operation in the fact that the underlying street graph is planar, thus it is possible to apply a geographic routing scheme directly over it.

GPCR models the street map as a graph in which crossroads are a vertex and streets are edges. The nodes located in junctions are called *coordinators* and they include their role in the beacons they periodically broadcast along with their position. Thus, each node can learn which neighbors are coordinators. Coordinators route messages in a different way to noncoordinator nodes. A normal node forwards the packet along the street towards the next junction. That is, the forwarding node selects among those neighbors whose positions approximate an extension of the line between the forwarding node's predecessor and the forwarding node itself, the farthest one.

On the other hand, coordinators take decisions based on the availability of streets starting from the junction in which they are located. Here, a greedy-face-greedy approach is followed. Thus, when the packet is being routed in greedy mode, coordinators select as the next junction the one closer to the destination. However, routing in perimeter mode makes them apply the right-hand rule to determine the next junction.

To allow nodes to determine whether they are coordinators or normal nodes, the authors propose two approaches. The first consists of including information in beacons so that, nodes know information about 2-hop neighbors. Thus, a node determines it is a coordinator when it has two neighbors that are within transmission range with each other but do not list each other as neighbors. However, this approach can lead nodes to wrongly determine they are in a junction, for example when the vehicles are in a curve. The second approach consists of applying a statistical method to determine the relative position of a node with respect to its neighbors. Studying the variation of this parameter over time can determine its role.

Finally, the authors do not explain how coordinator nodes acquire the information about the streets starting from the junction they are placed on. This topic is not covered in the paper.

9.4.2.2 VADD

The VADD[73] includes some specific solutions to alleviate the problems caused by the high-mobility and low-connectivity characteristics of vehicular networks. One solution consists of using not only the street-level map available in most vehicles but also information about the mean traffic density, the maximum allowed speed, and the mean vehicle speed of each street. Using that information, the authors propose a mathematical formula to determine the estimated transmission speed of each street segment, that is, the segment of streets between crossroads. This estimation is then used by nodes carrying a message to decide whether to continue carrying the message or forward it to another node. These decisions are only taken when nodes are near a crossroad because the number of possible routing options is larger at these points.

Once the next street to take has been selected, the next step is to chooe the next relay. At this point, different solutions are proposed because none is totally perfect:

1. *Select the node closest to the next intersection.* This strategy can lead to cycles and prevention of this is not possible. These cycles appear when the selected node is in fact moving away from the destination. Using an additional header including the last *n* previous hops it is possible to avoid cycles of up to *n* nodes, but in that case some good alternative paths are not being considered, for example, when choosing a concrete next relay initiates a short cycle but, in the end, the cycle is automatically resolved thanks to the movement of nodes and the path followed is a good one.

2. *Select the node whose movement direction is aligned with the relative position of the current next intersection selected.* This alternative does not produce cycles, but it can lead to worse paths in terms of number of hops or end-to-end delay.

3. *Select more than one next relay.* This is a multipath approach, and although the authors claim it is the best one in terms of delivery ratio and end-to-end delay, the overall network traffic overhead is too high.

As the connectivity of vehicular networks is unstable over time, the authors propose to use a store-and-forward approach, especially when nodes are not near intersections, thus, assuming a node carrying a packet is in the middle of a street segment between two intersections. If the intersection it has to reach next is the intersection ahead, the packet is forwarded towards that intersection by geographical greedy routing. If there is no vehicle available to forward ahead, the current packet carrier continues to carry it. If the identified target intersection is the one behind, the packet carrier keeps holding the packet, and waits for a vehicle in the opposite direction. Upon meeting one, it immediately forwards the packet.

9.4.3 Trajectory-Based Protocols

As we have already commented, most vehicles today are equipped with car navigators. These devices use GPS technology to determine position and include street-level maps of the most important cities of the world. Moreover, as Internet connection via UMTS has grown over time, these devices also support periodical updates of useful information such as traffic congestion, roads closed for maintenance, density of traffic in some major roads, weather advice, traffic advice, the location of restaurants and, fuel stations, and other services. Indeed, that kind of information is very valuable from the point of view of routing protocols, as has been shown in previous sections.

The principal function of navigators is to determine the optimal way to a certain destination, and, whether or not the driver knows the way to reach a determined point, programing the navigator to give him advice can be of interest. If the driver already knows the way, the up-to-date information available at the navigator can help to avoid a recently closed road or a traffic congestion. On the other hand, when the driver does not know the way the instructions of the navigator are necessary to reach the destination.

In any case, the information about the trajectory that the vehicle is following, and even more importantly, the rest of the path still to cover is an important piece of knowledge that routing protocols can use in the process of taking routing decisions. We next describe some important protocols based on this idea.

9.4.3.1 Trajectory-Based Forwarding

The idea of using a predefined trajectory to guide routing decisions was first introduced in trajectory-based forwarding (TBF).[74] The idea behind TBF is to determine an imaginary curve or trajectory to follow from the source node to the destination one. Ideally, thiscurve should be easily described as some form of parametric equation. Thus, just including that equation in the header of a message is enough to route the message. The routing process consists of selecting as next relay a neighbor closer to a point inthe curve. Obviously, in order to have progress at every step, the selected point in the curve must be more ahead in the curve than the current one.

There are some aspects to carefully consider in applying this technique. First, there are several different characteristics used in defining the best neighbor at each step:

1. The neighbor closest to the curve on a straight line
2. The neighbor whose closest point to the curve provides the greatest advance along the curve
3. The node closer to the centroid of candidate neighbors
4. A neighbor randomly chosen between the best three

Also, when nodes are not stationary, it is possible to know the current directions and speeds of neighbors. In such cases a good approach can be to choose the neighbor whose trajectory fits best with the imaginary curve.

Finally, this approach has a drawback, the local minimum. When the density of nodes is low it is possible to reach a node without available candidate neighbors. TBF does not specify what to do in these cases although there are some possible solutions that are similar to those used by previous protocols.

9.4.3.2 Opportunistic Geographical Routing

Leontiadis et al.[75] proposed an opportunistic forwarding approach for VANETs. This is based on the assumption that the source node already knows the position of the destination by some kind of location database or similar approach. It also assumes that nodes already know in advance their own route along the street map. Finally, it also assumes that cars can interchange their expected routes using beacon messages. These beacons serve two main purposes: they are used to discover neighbors and they report relevant information about the expected route of those neighbors.

Opportunistic geographical routing (GeOpps) nodes, rather than following a predefined trajectory, use the expected trajectory of neighbors to take routing decisions. Nodes perform an opportunistic geographic routing approach for VANETs using a simple forwarding strategy. The general idea is that if a node finds a neighbor whose trajectory goes closer to the destination's position than its own, it forwards the packet to that neighbor. By choosing nodes whose trajectory goes closer and closer to the destination coordinates, it is expected that data packets will eventually make it to the destination.

In practice, given a destination and given the expected route of the current node, it can compute the nearest point of the route to the destination, the expected time of arrival to that point, and the estimated time required to drive from that nearest point to the destination. Based on that information, the current node can compute the minimum estimated time of delivery as the sum of the estimated time to reach the nearest point and the estimated time from there to the destination. Based on that metric, next hops can be selected.

Unfortunately, the proposed approach may have local minima. That is, this forwarding along the most promising nodes, may end up reaching a car such that the following occur:

1. Its route is not close enough to the destination to allow a direct transmission.
2. It does not meet another car with a route whose nearest point is closer than that of this one.

This means that some packets may not reach the destination.

9.5 Conclusions

Throughout this chapter we have described and analyzed routing solutions for MANETs. As mentioned, routing protocols for MANETs face very challenging problems. In addition to having to find efficient routes in scenarios where nodes may experience high mobility, they must be efficient enough to minimize the consumption of network resources. This is particularly important for wireless networks where those resources are very scarce.

The main routing paradigms for MANETs are proactive and reactive protocols. In proactive protocols, nodes flood (generally using optimized approaches) their topology information throughout the network. Nodes aggregate that information to compute their best route to any other destination in the network. Reactive protocols only compute routes when they are needed. When a source has data to send and it does not know a route towards the destination, it floods the network to build a reverse shortest-path tree, which is then used by the destination to report back along its shortest path. The choice of the best approach is highly dependent upon the particular scenarios. In general, proactive protocols have a lower delay than reactive ones because routes are precomputed. On the other hand, reactive protocols usually have a lower control overhead because paths are only created when really needed.

We also studied how MANET routing protocols would perform in vehicular networks. We concluded that MANET routing is definitely not the best approach for efficient routing in VANETs. At first glance, it may seem that vehicular networks are one of the best candidate scenarios to deploy MANET routing solutions. Those routing protocols could be used to find multihop paths across different vehicles. However, as we described, VANET presents some very particular properties which prevent MANET routing solutions from performing well. First, the extensive use of flooding prevents those routing protocols from scaling in VANET scenarios where the number of cars can be very large. Second, MANET routing protocols assume full connectivity, which is certainly not guaranteed in VANET scenarios. That is, in a VANET there is no guarantee that a path from the source to the destination exists. In fact, if it exists, it may only last for a short time. Thus, delay-tolerant solutions are preferred for these scenarios. Finally, vehicles have a lot of mobility information readily available that could be used by routing protocols to improve their performance. Thus, MANET routing, neglecting knowledge about the mobility of the nodes, offers suboptimal performance. That information includes, among others, car positions, trajectories, current headings, and speed.

Given that MANET routing protocols are not suitable for VANET scenarios, we finally analyzed some examples of existing routing solutions for VANETs. As we show, these solutions do a much better job at exploiting existing information. In addition, the scalability achieved is much higher through using localized forwarding schemes such as geographic forwarding, trajectory-based forwarding, and the like.

Although there exist some basic routing solutions for VANETs, this is still a very hot research topic. The problem of efficient geographic routing in VANET scenarios poses a number of unique open research issues. They include among others, the efficient integration of VANET routing with infrastructure-based communications, the design of efficient and adaptive broadcasting and multicasting schemes, and the provision of highly scalable application-specific routing solutions able to adapt their forwarding approach to different scenarios and application requirements.

References

1. Armstrong, L., Dedicated Short Range Communications (DSRC), published online. http://www.leearmstrong.com/DSRC/DSRCHomeset.htm.
2. Jain, S., Fall, K., and Patra, R., Routing in a Delay Tolerant Network, in SIGCOMM '04: Proceedings of the 2004 Conference on Applications, Technologies, Architectures, and Protocols for Computer Communications, ACM Press. ISBN 1-58113-862-8, New York, NY, USA, 2004, pp. 145–158.
3. Macker, J. and Chakeres, I., Mobile ad-hoc networks (MANET). http://www.ietf.org/html.charters/manet-charter.html.
4. MIT Computer Science and Artificial Intelligence Laboratory, MIT roofnet. http://pdos.csail.mit.edu/roofnet/doku.php.
5. Capkun, S., Hamdi, M., and Hubaux, J.P., GPS-free positioning in Mobile Ad-hoc Networks, in Proc. of the Hawaii International Conference on System Sciences (HICSS '01), January 2001.
6. Stojmenovic, I., Location updates for efficient routing in ad hoc wireless networks, in *Handbook of Wireless Networks and Mobile Computing*, Wiley, 2002, pp. 451–471.
7. Bellman, R., On a routing problem, *Quarterly of Applied Mathematics*, 16(1), 87–90, 1958.

8. Ford Jr., L.R., Network flow theory, paper P-923, The RAND Corporation, Santa Monica, CA, USA, August 1956.
9. Bertsekas, D. and Gallager, R., *Data Networks*, Prentice-Hall, 1987, pp. 297–333.
10. Cheng, C., Riley, R., Kumar, S.P.R., and Garcia-Luna-Aceves, J.J., A loop-free extended Bellman-Ford routing protocol without bouncing effect, *SIGCOMM Comput. Commun. Rev.*, 19(4), 224–236, 1989.
11. Perkins, C. and Bhagwat, P., Highly dynamic destination-sequenced distance- vector routing (DSDV) for mobile computers, in ACM SIGCOMM '94 Conference on Communications Architectures, Protocols, and Applications, 1994, pp. 234–244.
12. Clausen, T. and Jacquet, P., Optimized Link State Routing Protocol (OLSR), RFC 3626, Internet Engineering Task Force, October 2003.
13. Dijkstra, E.W., A note on two problems in connexion with graphs, *Numerische Mathematik*, 1, 269–271, 1959.
14. Qayyum, A., Viennot, L., and Laouiti, A., Multipoint Relaying for Flooding Broadcast Messages in Mobile Wireless Networks, in Proc. of the Hawaii International Conference on System Sciences (HICSS '02), Big Island, HI, USA, 2002, pp. 3866–3875.
15. Garey, M. and Johnson, D., *Computers and Intractability: A Guide to the Theory of NP-Completeness*, W.H. Freeman, 1979.
16. Adjih, C., Bacelli, E., Clausen, T.H., Jacquet, P., and Rodolakis, G., Fish Eye OLSR scaling properties, *IEEE Journal of Communications and Networks (JCN)*, Special Issue on Mobile Ad Hoc Wireless Networks, 2004.
17. Gupta, P. and Kumar, P.R., The capacity of wireless networks, *IEEE Transactions on Information Theory*, 46(2), 388–404, 2000.
18. Clausen, T., Dearlove, C., and Jacquet, P., The Optimized Link State Routing Protocol version 2, Internet Draft 04, Internet Engineering Task Force, July 2007.
19. Pei, G., Gerla, M., and Chen, T.-W., Fisheye State Routing in Mobile Ad Hoc Networks, in Workshop on Wireless Networks and Mobile Computing, (ICDCS), 2000, pp. D71–D78.
20. Clausen, T.H. Combining Temporal and Spatial Partial Topology for MANET routing—Merging OLSR and FSR, in Proc. of IEEE WPMC '03, October 2003.
21. Ge, Y., Lamont, L., and Villasenor, L., Hierarchical OLSR—A Scalable Proactive Routing Protocol for Heterogeneous Ad Hoc Networks, in Proc. of IEEE WiMob '05, Vol. 3, August 2005, pp. 17–23.
22. Ros, F.J. and Ruiz, P.M., Cluster-Based Olsr Extensions to Reduce Control Overhead in Mobile Ad Hoc Networks, in IWCMC '07: Proceedings of the 2007 International Conference on Wireless Communications and Mobile Computing, ACM Press, New York, NY, USA, 2007, pp. 202–207.
23. Johnson, D.B., Maltz, D.A., and Broch, J., *DSR The Dynamic Source Routing Protocol for Multihop Wireless Ad Hoc Networks*, Chapter 5, Addison-Wesley, 2001, pp. 139–172.
24. Johnson, D., Hu, Y., and Maltz, D., The Dynamic Source Routing Protocol (DSR) for Mobile Ad Hoc Networks for IPv4, RFC 4728, Internet Engineering Task Force, February 2007.
25. Jubin, J. and Tornow, J.D., The DARPA Packet Radio Network Protocols, in Proceedings of the IEEE, Vol. 75(1), January 1987, pp. 21–32.

26. Perkins, C., Belding-Royer, E., and Das, S., Ad hoc On-Demand Distance Vector (AODV) Routing, RFC 3561, Internet Engineering Task Force, July 2003.
27. Chakeres, I. and Perkins, C., Dynamic MANET On-Demand (DYMO) Routing, Internet draft 11, Internet Engineering Task Force, November 2007.
28. Clausen, T., Dearlove, C., Dean, J., and Adjih, C., Generalized MANET Packet/Message Format, Internet draft 11, Internet Engineering Task Force, November 2007.
29. Haas, Z. and Pearlman, M., ZRP: A Hybrid Framework for Routing in Ad Hoc Networks, Chapter 5, Addison-Wesley, 2001, pp. 221–254.
30. Fonseca, R., Ratnasamy, S., Culler, D., Shenker, S., and Stoica, I., Beacon vector routing: Scalable point-to-point in wireless sensornets, *Intel Research*, IRB-TR-04, 12, 2004.
31. Carus, A., Urpi, A., Chessa, S., and De, S., GPS-free coordinate assignment and routing in wireless sensor networks, in Proc. 24th Annual Joint Conference of the IEEE Computer and Communications Societies (INFOCOM '05), Vol. 1, March 2005, pp. 150–160.
32. Li, J., Jannotti, J., De Couto, D.S.J., Karger, D.R., and Morris, R., A Scalable Location Service for Geographic Ad Hoc Routing, in Proc. 6th Annual ACM/IEEE International Conference on Mobile Computing and Networking (MobiCom '00), ACM Press, New York, NY, USA, 2000, pp. 120–130.
33. Zhang, R., Zhao, H., and Labrador, M.A., The Anchor Location Service (ALS) Protocol for Large-Scale Wireless Sensor Networks, in Proc. First International Conference on Integrated Internet Ad Hoc and Sensor Networks (InterSense '06), ACM Press, New York, USA, 2006, p. 18.
34. Lazos, L. and Poovendran, R., SeRLoc: Robust localization for wireless sensor networks, *ACM Transactions on Sensors Networks*, 1(1), 73–100, 2005.
35. Stojmenovic, I., A Scalable Quorum Based Location Update Scheme for Routing in Ad Hoc Wireless Networks. Technical Report TR-99-09, SITE, University of Ottawa, September 1999.
36. Stojmenovic, I., Home agent based location update and destination search schemes in ad hoc wireless networks, in Advances in Information Science and Soft Computing, Zemliak, A. and Mastorakis, N.E., Eds., WSEAS Press, 2002, pp. 6–11.
37. Finn, G.G., Routing and Addressing Problems in Large Metropolitan-scale Internetworks, technical report ISI/RR-87-180, University of Southern California, Information Sciences Institute, March 1987.
38. Kranakis, E., Singh, H., and Urrutia, J., Compass Routing on Geometric Networks, in 11th Canadian Conference on Computational Geometry (CCCG '99), Vancouver, British Columbia, Canada, August 1999, pp. 51–54.
39. Stojmenovic, I. and Lin, X., Loop-free hybrid single-path/flooding routing algorithms with guaranteed delivery for wireless networks, *IEEE Transactions on Parallel and Distributed Systems*, 12(10), 1023–1032, 2001.
40. Takagi, H. and Kleinrock, L., Optimal transmission ranges for randomly distributed packet radio terminals, *IEEE Transactions on Communications*, 32(3), 246–247, March 1984.
41. Nelson, R. and Kleinrock, L., The spatial capacity of a slotted ALOHA multihop packet radio network with capture, *IEEE Transactions on Communications*, 32(6), 684–694, 1984.

42. Bondy, J.A. and Murty, U.S.R., *Graph Theory with Applications*, Macmillan, London, 1976.

43. Frey, H., and Stojmenovic, I., On Delivery Guarantees of Face and Combined Greedy-Face Routing Algorithms in Ad Hoc and Sensor Networks, in Proc. 12th Annual ACM/IEEE International Conference on Mobile Computing and Networking (MobiCom '06), Los Angeles, CA, USA, September 2006, pp. 390–401.

44. Gabriel, K. and Sokal, R., A new statistical approach to geographic variation analysis, *Systematic Zoology*, 18, 259–278, 1969.

45. Toussaint, G.T., The relative neighborhood graph of a finite planar set, *Pattern Recognition*, 12, 261–268, 1980.

46. Gao, J., Guibas, L., Hershberger, J., Zhang, L., and Zhu, A., Geometric Spanner for Routing in Mobile Networks, in Proc. 2th ACM International Symposium on Mobile Ad Hoc Networking and Computing (MobiHoc '01), ACM Press, New York, NY, USA, 2001, pp. 45–55.

47. Li, X.Y., Calinescu, G., and Wan, P.J., Distributed Construction of a Planar Spanner and Routing for Ad Hoc Wireless Networks, in Proc. 21th Annual Joint Conference of the IEEE Computer and Communications Societies (INFOCOM '02), Vol. 3, 2002, pp. 1268–1277.

48. Li, X.-Y., Stojmenovic, I., and Wang, Y., Partial Delaunay triangulation and degree limited localized Bluetooth scatternet formation, *IEEE Transactions on Parallel Distributed Systems*, 15(4), 350–361, 2004.

49. Boone, P. et al., Morelia Test: Improving the efficiency of the Gabriel Test and face routing in ad-hoc networks, *Lecture Notes in Computer Science*, 3104, 23–34, 2004.

50. Datta, S., Stojmenovic, I., and Wu, J., Internal node and shortcut based routing with guaranteed delivery in wireless networks, *Cluster Computing*, 5(2), 169–178, 2002.

51. Bose, P., Morin, P., Stojmenovic, I., and Urrutia, J., Routing with guaranteed delivery in ad hoc wireless wetworks, *Wireless Networks*, 7(6), 609–616, 2001.

52. Kuhn, F., Wattenhofer, R., and Zollinger, A., Asymptotically Optimal Geometric Mobile Ad-Hoc Routing, in Proc. 6th International Workshop on Discrete Algorithms and Methods for Mobile Computing and Communications, ACM Press, New York, USA, 2002, pp. 24–33.

53. Car 2 Car Communication Consortium. http://www.car-to-car.org/.

54. Continuous Air Interface for Long and Medium Range Communications (CALM) Forum. http://www.isotc204wg16.org/.

55. Namboodiri, V., Agarwal, M., and Gao, L., A Study on the Feasibility of Mobile Gateways for Vehicular Ad-Hoc Networks, in VANET '04: Proceedings of the 1st ACM International Workshop on Vehicular Ad Hoc Networks, ACM Press, New York, NY, USA, 2004, pp. 66–75.

56. Kutzner, K., et al., Connecting Vehicle Scatternets by Internet-Connected Gateways, in Workshop on Multiradio Multimedia Communications (MMC)—Communication Technology for Vehicles, 2003.

57. Camp, T., Boleng, J., and Davies, V., A survey of mobility models for ad hoc network research. *Wireless Communication and Mobile Computing (WCMC)*, Special

Issue on Mobile Ad Hoc Networking: Research, Trends and Applications, 2(5), 483–502, 2002.

58. Naumov, V., Baumann, R., and Gross, T., An Evaluation of Inter-Vehicle Ad Hoc Networks Based on Realistic Vehicular Traces, in Proc. 6th ACM International Symposium on Mobile Ad Hoc Networking and Computing (MOBIHOC '06), May 2006, pp. 108–119.

59. Wong, K.J., Lee, B.S., Seet, B.C., Liu, G., and Zhu, L., BUSNet: Model and Usage of Regular Traffic Patterns in Mobile Ad Hoc Networks for Inter-vehicular Communications, in Proc. 10th International Conference on Telecommunications (ICT '03), April 2003, pp. 102–108.

60. Ni, S., Tseng, Y., Chen, Y., and Sheu, J., The Broadcast Storm Problem in a Mobile Ad Hoc Network, in Proceedings of the International Conference on Mobile Computing and Networking (MobiCom '99), Seattle, WA, USA, 1999, pp. 151–162.

61. Lochert, C., et al., A Routing Strategy for Vehicular Ad Hoc Networks in City Environments, in Proceedings of the IEEE Intelligent Vehicles Symposium 2003, Columbus, OH, USA, June 2003, pp. 156–161.

62. Reichardt, D., Miglietta, M., Moretti, L., Morsink, P., and Schulz, W., CarTALK 2000: Safe and Comfortable Driving Based Upon Inter-Vehicle Communication, in Proc. of the IEEE Intelligent Vehicle Symposium, June 2002, pp. 545–550.

63. Cseh, C., Coletti, L., Moretti, L., Riato, N., and Tian, J. Communication Architecture, Deliverable D6, Project CarTalk2000 (IST-2000-28185), October 2002.

64. FleetNet—Internet on the Road, available at http://www.et2.tu-harburg.de/fleetnet.

65. Kasemann, M., Hartenstein, H., Fubler, H., and Mauve, M., Analysis of a Location Service for Position-Based Routing in Mobile Ad-Hoc Networks, in Proceedings of the 1st German Workshop on Mobile Ad-Hoc Networking (WMAN 2002), March 2002, pp. 121–133.

66. Kasemann, M., Fubler, H., Hartenstein, H., and Mauve, M., A Reactive Location Service for Mobile Ad-Hoc Networks, Technical Report TR-14-2002, Department of Computer Science, University of Mannheim, November 2002.

67. Gerla, M., et al., Vehicular Grid Communications: The Role of the Internet Infrastructure, in WICON '06: Proceedings of the 2nd Annual International Workshop on Wireless Internet, ACM Press, New York, NY, USA, 2006, p. 19.

68. Tian, J., Han, L., Rothermel, K., and Cseh, C., Spatially Aware Packet Routing for Mobile Ad Hoc Inter-Vehicle Radio Networks, in Proc. IEEE Intelligent Transportation System Conference (ITSC '03), October 2003, pp. 1546–1551.

69. Seet, B.C., et al., A-STAR: A Mobile Ad Hoc Routing Strategy for Metropolis Vehicular Communications, in Proc. 3rd International Networking Conference IFIP-TC6 (IFIP '04), Vol. 3042 of Lecture Notes in Computer Science, December 2004, pp. 989–999.

70. Blazevic, L., Giordano, S., and Le Boudec, J.Y., Self-organizing wide-area routing, in Proc. SCI 2000/ISAS 2000, July 2000.

71. Naumov, V. and Gross, T.R., Connectivity-Aware Routing (CAR) in Vehicular Ad-hoc Networks, in Proc. 26th IEEE International Conference on Computer Communications (INFOCOM '07), May 2007, pp. 1919–1927.

72. Lochert, C., Mauve, M., Füßler, H., and Hartenstein, H., Geographic routing in city scenarios, *ACM SIGMOBILE Mobile Computing and Communications Review*, 9(1), 69–72, 2005.
73. Zhao, J. and Cao, G., VADD: Vehicle-Assisted Data Delivery in Vehicular Ad Hoc Networks, in Proc. 25th IEEE International Conference on Computer Communications (INFOCOM '06), April 2006, pp. 1–12.
74. Niculescu, D., and Nath, B., Trajectory Based Forwarding and its Applications, in Proc. 9th International Conference on Mobile Computing and Networking (MOBICOM '03), September 2003, pp. 260–272.
75. Leontiadis, I. and Mascolo, C., GeOpps: Opportunistic Geographical Routing for Vehicular Networks, in Proc. of the IEEE Workshop on Autonomic and Opportunistic Communications, Helsinki, Finland, 2007.

10

Delay-Tolerant Networks in VANETs

Yifeng Shao,
Cong Liu, and Jie Wu
*Department of Computer
Science and Engineering
Florida Atlantic University*

10.1 Introduction

In many commercial applications[1-4] and in road safety systems, vehicular delay-tolerant networks have been envisioned to be useful. For example, a vehicular ad hoc network (VANET) can be used to alert drivers of traffic jams ahead, help balance traffic loads, and reduce traveling time. It can also be used to propagate emergency warnings to drivers behind the vehicles in an accident in order to prevent compounding on accident that has already taken place. Transportation safety issues have been addressed in Refs. 1 and 3, where vehicles communicate with each other and with static network nodes such as traffic lights, bus shelters, and traffic cameras.

The Federal Communications Commission (FCC) has allocated 75 MHz of spectrum for short-range vehicle-to-vehicle or vehicle-to-roadside communications. IEEE is working on standard specifications for intervehicle communication. In the near future, intervehicle communication will be enabled by communication devices equipped in general vehicles and form a large-scale VANET.

The cost of a wireless infrastructure is high and may not be possible when such an infrastructure does not exist or is damaged. Although services can be supported by a wireless infrastructure, from the service provider point of view, setting up a wireless LAN is very cheap, but the cost of connecting it to the Internet or the wireless infrastructure is high. From the user point of view, the cost of accessing data through a wireless carrier is still high and most cellular phone users are limited to voice services. Moreover, in the event of a disaster, the wireless infrastructure may be damaged, whereas wireless LANs and vehicular networks can be used to provide important traffic, rescue, and evacuation information to the users.

Many researchers and industry players believe that the benefit of vehicular networks for traffic safety and many commercial applications[1-3] should be able to justify the cost, although the cost of setting up vehicular networks is high. In the near future, many of the proposed delay-tolerant data delivery applications can be supported with such a vehicular delay-tolerant network already in place.

The fact that vehicular networks are highly mobile and sometimes sparse complicates multihop delay-tolerant data delivery through VANETs. The network density is related to traffic density. Traffic density is affected by location and time. It is low in rural areas and at night time, but very high in largely populated areas and during rush hours.

Finding an end-to-end connection is very difficult for a sparsely connected network. Opportunities for mobile vehicles to connect with each other intermittently while moving is introduced by the high mobility of vehicular networks. There are ample opportunities for moving vehicles to set up a short path with few hops in a highway model, as shown by Namboodiri et al.[5] A moving vehicle can carry a packet and forward it to the next vehicle. The message can be delivered to the destination without an end-to-end connection for delay-tolerant applications through store-carry-and-forward.

This chapter studies the problem of efficient data delivery and dissemination in vehicular delay-tolerant networks.

10.2 Overview

The rest of this chapter is organized as follows. First, we review the most up-to-date research regarding delay-tolerant networks (DTNs), with a focus on the routing problem. After that, we illustrate the car-following vehicle traffic model, which appropriately represents the mobility pattern of VANETs and which has a significant impact on the performance of specific data dissemination algorithms. Based on the traffic model, the problem of data dissemination is studied. We categorize the data dissemination problem into two aspects: vehicle-to-roadside and vehicle-to-vehicle (V2V). In the V2V case,

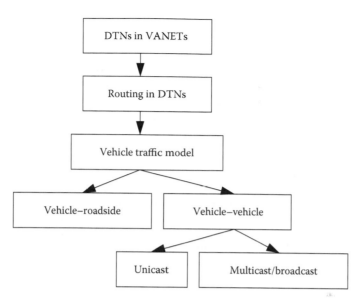

FIGURE 10.1 The structure of this chapter.

we study both the unicast problem and the multicast/broadcast problem. The structure of this chapter is organized as in Figure 10.1.

10.3 Delay-Tolerant Networks

As part of the Internet Research Task Force (IRTF), The Delay-Tolerant Networking Research Group (DTNRG)[6] was created to address the architectural and protocol design principles needed for interconnecting networks operating in environments where continuous end-to-end connectivity is sporadic.

DTNRG members are defining the initial DTN architecture. Kevin Fall was among the first to describe the main challenges facing current IP-based networks.[7] He proposed a DTN communication architecture based on a message-oriented overlay implemented above the transport layer. Messages are aggregated in "bundles" that form the protocol data units in a virtual message-switching architecture. Devices that implement this bundle layer, called DTN nodes, use persistent storage-to-buffer bundles whenever a proper contact is not available for forwarding.

Reliable delivery and optional end-to-end acknowledgment is implemented by the bundle layer. In addition, the bundle layer also implements security services and a flexible naming scheme with late binding. For more details on the DTN architecture, the reader should consult Ref. 7 and the Internet Draft by Vint Cerf et al.[8] Because the bundle layer is implemented above several transport layers, it supports interconnecting, heterogeneous networks using DTN gateways, similar to how Internet gateways route packets between networks with different data links.

Fall et al.[7] point out that routes in a DTN consist of a sequence of time-dependent communication opportunities, called contacts, during which messages are transferred from a source to the destination. Contacts are described by capacity, direction, the two endpoints, and temporal properties such as begin/end time and latency. Routing in a network with time-varying edges involves finding the optimal contact path in both space and time, meaning that the forwarding decision must schedule transmissions considering temporal link availability in addition to the sequence of hops to the destination.

This problem is exacerbated when contact duration and availability are nondeterministic. Contact types are classified in Refs. 7 and 8. Persistent contacts are always available. A scheduled contact is an agreement to establish a contact at a particular time for a particular duration. Opportunistic contacts present themselves unexpectedly. On-demand contacts require some action in order to instantiate, but then function as persistent contacts until terminated. Predicted contacts are based on a history of previously observed contacts or some other information.

Message forwarding requires scheduling in addition to next-hop selection because DTN routing must operate on a time-varying multigraph. To optimize the network performance, DTN routing must select the appropriate contact defined by a next-hop and a transmission time. If a contact is not known when a message is received from the upper layer, the bundle layer will buffer it until a proper contact occurs or until the message is dropped. In conditions of a DTN with sporadic contact opportunities, the main objective of routing is to maximize the probability of delivery at the destination while minimizing the end-to-end delay.

A sketch of the types of DTN routing protocols is illustrated in Figure 10.2. When the routing protocol has better information regarding the current state of the topology and its future evolution, the forwarding decision is more effective. At one end of the spectrum is deterministic DTN routing, where the current topology is known and future changes can be predicted. With deterministic routing, message forwarding can be

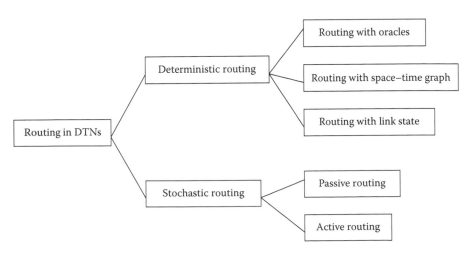

FIGURE 10.2   DTN routing protocols to be covered.

scheduled such that network performance is optimal and resource utilization is reduced by using unicast forwarding.

At the other end of the spectrum, node movement is random or unknown and nodes know very little or nothing about the future evolution of the topology. In this case, stochastic DTN routing forwards messages randomly hop by hop with the expectation of eventual delivery, but with no guarantees. In between, there are routing mechanisms that may predict contacts using prior network state information, or that adjust the trajectory of mobile nodes to serve as message ferries. Stochastic routing techniques rely more on replicating messages and controlled flooding for improving delivery rate, trading off resource utilization against improved routing performance in the absence of accurate current and future network states. The next section describes the principles of operation of representative deterministic and stochastic DTN routing mechanisms.[9]

10.4 Deterministic Delay-Tolerant Routing

In general, deterministic techniques are based on formulating models for time-dependent graphs and finding a space–time shortest path in DTNs by converting the routing problem to classic graph theory or by using optimization techniques for end-to-end delivery metrics. Deterministic routing techniques for networks with intermittent connectivity assume that local or global information on how the network topology evolves in time are available to a certain degree.

Good performance with less resource usage than stochastic routing techniques is provided by deterministic routing protocols using single-copy unicast for messages in transit. Deterministic routing mechanisms are appropriate only for scenarios where networks exhibit predictable topologies. This is true in applications where node trajectory is coordinated or can be predicted with accuracy, as in interplanetary networking.

10.4.1 Deterministic Delay-Tolerant Routing with Oracles

The distribution of network state and mobility information under sporadic connectivity, long delays, and sparse resources is a major problem facing deterministic routing protocols. In Ref. 10, Jain et al. present a deterministic routing framework that takes advantage of increasing levels of information on topology and traffic demand (oracles) when such information is predictable. A DTN multigraph is defined where vertices represent the DTN nodes and edges describe the time-varying link capacity between nodes. It is called a multigraph because multiple directed links between two nodes may exist.

One of the routing objectives is to minimize the end-to-end delay. Reducing the message transit times in the network also reduces contention for limited resources, such as buffer space and transmission time. Four knowledge oracles are defined: contacts summary oracle (for aggregate or summary contact statistics), contact oracle (for the time-varying contact multigraph), queuing oracle (for instantaneous queue state), and the traffic demand oracle (for present and future messages injected in the network). The authors adapt Dijkstra's shortest-path algorithm to support time-varying edge weights defined by the oracles available, and propose six algorithms for finding the optimal contact path.

Time-invariant edge weights is assumed in the first two algorithms in Ref. 10. The First Contact (FC) algorithm is a zero-knowledge approach that chooses a random edge to forward a message among the currently available contacts. If no contact is available, the message will be forwarded on the first edge that comes up. The Minimum Expected Delay (MED) algorithm applies the Dijkstra algorithm where the edge weight is time-invariant and is determined by the sum of the average waiting time (from the Contacts Summary oracle), propagation delay, and transmission delay. MED ignores congestion and does not recompute routes for messages in transit.

A time-varying edge cost, defined as the sum of the waiting, transmission, and propagation delays, is used in the following four proposed partial-knowledge algorithms. The waiting delay includes the time waiting for a contact and the queuing delay. The Earliest Delivery with Local Queuing algorithm (EDLQ) is equal to the local queue size at a particular node, and "0" for all other edges. EDLQ routes around congestion for the first hop and ignores queue occupancy at subsequent hops. Therefore, this algorithm must recompute the route at every hop. Cycles are avoided by using path vectors. Still, EDLQ is prone to message loss due to lack of available buffer space at reception.

The contacts oracle and the queuing oracle are used in the Earliest Delivery with All Queues (EDAQ) algorithm. EDAQ predicts the correct queue space for all edges at all times. In EDAQ, routes are not recomputed for messages in transit because the initial route accurately predicts all delays. EDAQ works only if capacity is reserved for each message along all contact edges. In practice, EDAQ is very difficult to implement in most DTNs with low connectivity, as it requires an accurate global distribution of queuing state. Limited connectivity also severely limits practical implementations of edge capacity reservations.

Simulation results indicate that algorithms that use the knowledge oracles (ED, EDLQ, and EDAQ) outperform the simpler MED and FC algorithms in terms of latency and delivery ratio. The more constrained the network resources are, the better the performance is for the algorithms that are more informed (i.e., use more oracles). A promising result is that routing with EDLQ (using only local queuing information) has a very similar performance to the EDAQ algorithm. This means that similar network performance can be achieved without expensive queue state dissemination and capacity reservations.

10.4.2 Deterministic Delay-Tolerant Routing with Space–Time Graphs

The trajectories and mission objectives of nodes may change. Therefore, in practice, contacts are deterministically predictable for only a finite time horizon. In Ref. 11, Merugu et al. propose a deterministic routing framework where a space–time graph is built from predicted contact information. It starts with a time-varying link function defined as "1" when the link between two nodes is available and "0" otherwise. This function is defined as a function of time, where the time is discretized.

In Ref. 11, the space–time graph is built in multiple layers where the network nodes are replicated at each layer for each time unit t. Each layer has a copy of each network node. A column of these vertices maps to a single network node. A temporal link in the space–time graph connects graph vertices from the same column at successive time intervals. When it is traversed, it indicates that the message is buffered. A spatial link connects two vertices from different columns, representing message forwarding. Forwarding delay is modeled by the number of layers traversed by a spatial link.

The objective of the least-cost routing in this DTN is to find the lowest cost (shortest) path from the source space–time node (column:layer) associated with the message arrival time to a vertex from the column corresponding to the destination DTN node. The end-to-end latency for a message becomes equal to the length of the path traversed in the space–time graph. The routing problem is solved using the Floyd–Wars Hall all-pairs shortest paths algorithm, modified to account for the particular characteristics of the space–time graph. Multiple message sizes are supported by a path-coloring scheme.

One issue with this approach is that time discretization increases the algorithm complexity by a factor of the size of the time horizon T. This space–time routing approach is similar to the Earliest Delivery partial knowledge algorithm from Jain et al.[10] in the way it handles queuing delays with route computation at each hop. Cycles are avoided by verifying the path vector from the message header when computing the next hop.

10.4.3 Delay-Tolerant Routing with Link State

In Ref. 12, Gnawali et al. propose ASCoT, a dynamic routing mechanism for space networks and the Positional Link State routing protocol (PLS) to implement position-based routing that enables the prediction of the trajectories of satellites and other space assets. Link state updates with predicted contacts and their link performances are disseminated in advance in the network through reliable flooding. Nodes execute a modified Dijkstra algorithm to recompute routing tables when link state updates are received.

In Ref. 13, the authors propose a data-centric approach similar to directed diffusion to support proximity routing for space assets in close formation. Note that in deterministic routing techniques using shortest-path algorithms, routing tables and forwarding schedules are recomputed whenever the contact graph state has changed, and selection of the next contact is done for a message at each hop along the path, as opposed to source routing. Thus, loops become possible because nodes may use outdated topology information. Cycles are avoided with path vectors.

For a limited range of applications, deterministic DTN routing protocols are effective where the contact schedule can be accurately modeled and predicted. Otherwise, it is necessary to frequently disseminate nodes' states throughout the network. In networks with constrained capacity or limited connectivity, this becomes very expensive and difficult to implement without an out-of-band broadcast channel. When contacts cannot be accurately predicted, routing must consider stochastic mechanisms that can only hint to predilection for future contacts based on historic information.

10.5 Stochastic Delay-Tolerant Routing

Depending on whether node mission is changed in order to support message relay, stochastic routing techniques can be passive or active. Passive routing techniques do not interfere with node missions and do not change node trajectory to adapt to traffic demands.

Passive routing techniques generally rely on flooding multiple copies of the same message with the objective of eventual delivery. In contrast, active routing techniques coordinate the mission (trajectory) of some nodes to improve capacity with their store-and-carry capability. In general, passive routing techniques trade off delivery performance against resource utilization.

By sending multiple copies of the same message on multiple contact paths, the delivery probability increases and the delay drops at the cost of additional buffer occupancy during message ferrying and higher link capacity usage during contacts. This approach is appropriate when very little or nothing is known about mobility patterns.

10.5.1 Passive Stochastic Routing

First, we present two passive stochastic routing protocols, Epidemic Routing and Spray and Wait, which do not need any information about the network state. For other routing protocols, nodes can memorize contact history and use it to make more informed forwarding decisions. The section then continues with several passive routing protocols that operate with contact estimation.

10.5.1.1 Epidemic Routing

In Ref. 14 Vahdat and Becker propose the Epidemic Routing protocol for message delivery in a mostly disconnected network with mobile nodes. Epidemic routing implements flooding in a DTN, named after a technique for message forwarding that emulates how a disease spreads through direct contact in a population during an epidemic. Even when just one individual of an entire population is initially infected, if the disease is highly contagious and contacts are frequent, over time it will spread exponentially and reach the entire population with a high probability.

In epidemic routing, the "disease" that spreads is a message that must reach one or more destinations. Each node maintains a summary vector with IDs of messages it has already received. When two nodes initiate a contact, they first exchange their summary vectors in the anti-entropy session. Comparing message IDs, each node decides what messages it has not already received that it needs to pull from other nodes.

The second phase of a contact consists of nodes exchanging messages. Messages have a time-to-live (TTL) field that limits the number of hops (contacts) they can pass through. Messages with TTL = 1 are forwarded only to the destination. The main issue with epidemic routing is that messages are flooded in the whole network to reach just one destination. This creates contentions for buffer space and transmission time.

Reserving a fraction of their storage for locally originated messages is an approach to mitigate buffer space contention for nodes. Even so, older messages in buffers will be dropped when new messages are received, reducing the delivery probability for destination nodes that have a low contact rate. An attempt to reduce resource waste is proposed that uses delivery confirmation (ACK) messages that are flooded starting from the destination and piggybacked with regular messages. Whenever a node receives an ACK, it purges the acknowledged message from its buffer, if it is still present.

Node movement is used in epidemic routing to spread messages during contacts. With large buffers, long contacts, or a low network load, epidemic routing is very effective and provides minimal delays and high success rates, as messages reach the destination on multiple paths. End-to-end delay depends heavily on nodes' contact rate (infection rate), which is in turn affected by the communication range and node speed.

To trade off message latency and delivery ratio, different implementations of epidemic routing tune message TTL and buffer allocation. In scenarios with a high message load, the increased contention from forwarding mostly redundant messages reduces the protocol performance.

Epidemic routing is relatively simple to implement and is used in the DTN research literature as a benchmark for performance evaluation.

10.5.1.2 Spray and Wait

In Ref. 15, Spyropoulos et al. present Spray and Wait, a zero-knowledge routing protocol introduced to reduce the wasteful flooding of redundant messages in a DTN. Similar to epidemic routing, this protocol forwards message copies to nodes met randomly during contact in a mobile network. The main difference from epidemic routing is that Spray and Wait limits the total number of disseminated copies of the same message to a constant number L. In the spray phase, for every message originated by a source, L copies are forwarded by the source and other nodes receiving the message upto a total of L distinct relays. In the wait phase, all L nodes storing a copy of the message perform direct transmission.

Direct transmission[16] is a single-copy routing technique in DTNs where the message is forwarded by the current node only, directly to the destination node. Direct transmission has been used for wildlife tracking applications and has minimal overhead, but suffers from unbounded delay as there is no guarantee that the source will ever have contact with the destination node.

Initially, Spray and Wait spreads L copies of a message in an epidemic fashion in order to increase the probability that at least one relay node would have direct contact with the destination node. With a simple Source Spray and Wait heuristic, the source node forwards all L copies to the first L nodes encountered.

Binary Spray and Wait is the optimal forwarding policy in which nodes move randomly with identical and independent probability distribution (i.i.d.). A message will be physically stored and transmitted just once even when a transfer may virtually involve multiple copies. Each message has a header field indicating the number of copies. The paths followed by copies of a message can be represented by a binary tree rooted in the source node.

The transfer contacts are formed by edges in the tree. The more nodes that have multiple copies to distribute, the less the expected end-to-end delay will be. The binary heuristic has the least expected delivery latency in networks with random i.i.d. random mobility. An interesting property of this routing protocol is that, as the network node count M increases, the minimum fraction L/M necessary to achieve the same performance relative to the optimal path decreases. This property makes the Spray and Wait approach very scalable.

10.5.1.3 PROPHET

Spray and Wait performs much better than epidemic routing at higher loads because the limit L of maximum transmissions reduces contention on queue space and transmission time. Some passive DTN routing protocols use delivery estimation to determine a metric for contacts relative to successful delivery, such as delivery probability or delay. Some of these protocols can forgo flooding and deliver single-copy messages by being selective with contact scheduling. The advantage is that considerably less memory, bandwidth, and energy are wasted on end-to-end message delivery.

One of the drawbacks of Spray and Wait is that nodes must keep track of other nodes' movements and contacts, and that network-wide dissemination of this information imposes additional overhead in a network that is already constrained. A representative routing protocol for DTNs that uses delivery estimation is PROPHET, a Probabilistic ROuting Protocol using History of Encounters and Transitivity, proposed by Lindgren et al. in Ref. 17. PROPHET works on the realistic premise that node mobility is not truly random. Instead, it is assumed that nodes in a DTN tend to visit some locations more often than others, and that node pairs that have had repeated contacts in the past are more likely to have contacts in the future.

A probabilistic metric called *delivery predictability* estimates the probability that node A will be able to deliver a message to node B. The delivery predictability vectors are maintained at each node A for every possible destination B.

Two nodes (A and B) exchange the summary vectors (as in epidemic routing) and also the delivery predictability vectors at the beginning of a contact. Node A then updates its own delivery predictability vector using the new information from B, after which it selects and transfers messages from B for which it has a higher delivery probability than B. The delivery probability is updated during a contact so that node pairs that meet more often have a higher value.

Additionally, the delivery predictability has a transitive property that encodes the assumption that if nodes A and B have frequent contacts and nodes B and C have frequent contacts, then node A has a good chance of forwarding messages intended for node C. After exchanging delivery predictability vectors at the beginning of a contact, nodes A and B update their values for each other node C.

As node A begins a contact with node B, it decides to forward a message to B with destination C if $P(B, C) > P(A, C)$. Node A will also keep a copy in its buffer. The buffer has a first-in first-out (FIFO) policy for dropping old messages when new messages are received. Transitive reinforcement of delivery probabilities based on prior contacts make this protocol perform better in simulations than epidemic routing because it reduces the contention for buffer space and transmission time.

Related techniques for delivery probability estimation based on prior contact history are used in MV routing[18] and Zebranet.[19] A novel approach for delivery estimation is the use of a virtual Euclidean mobility pattern space, called MobySpace, proposed by Leguay et al.[20] The idea is that messages in a DTN should be forwarded to another node if this next hop has a mobility pattern similar to the destination node. This concept was adapted from the Content Addressable Network peer-to-peer overlay architecture.[21]

10.5.1.4 MobySpace

In existing works on user mobility in various scenarios where users tend to follow similar trajectories, the authors suggest a model where the node movement follows a power law. This means that the probability that a node is at a location i from a set of N locations is $P(i) = K(1/d)^{n_i}$, where n_i is the preference index for location i, $d > 1$ is the exponent of the power law, and K is a normalization constant. When d is high, nodes tend to visit far fewer locations far more often. When $d \to 1$, nodes have similar preference for all locations. The mobility pattern space has a dimension for each possible location, and the coordinate value of a node's point in this space (MobyPoint) in dimension i is equal to the probability $P(i)$. This model assumes that dwell time at each location is uniformly distributed in a narrow interval.

In MobySpace, two nodes that have a small distance between them are more likely to have a contact than two nodes that are situated further apart. With this insight, the forwarding algorithm simply decides to forward a message during a contact to a node that has a shorter distance to the message destination. Messages take paths through the MobySpace to bring them closer and closer to the destination. Several distance functions have been proposed to measure similarity in nodes' mobility patterns. The Euclidean and the cosine separation distance provide lower delays in simulations.

The MobySpace approach is only effective if nodes exhibit stable mobility patterns. It also fails if a message reaches a local maximum where the current node has a similar mobility pattern with the destination, but a direct contact with the destination is rare due to trajectory synchronization. Such a case is possible in a DTN where nodes are public transportation buses. Although the buses on a line follow the same path and visit the same stations, two buses may get within radio range only at night when they park in the garage.

Two nodes having similar mobility patterns does not mean that they are frequent contacts. A possible solution to this problem is to use the probability (or frequency) of direct contacts with the other nodes as dimensions in the MobySpace. Another approach to deal with the temporal variability of mobility patterns is to supplement MobySpace with conversion of the spatial visit patterns to the frequency domain, representing the dominant visitation frequency and the phase. Other issues with MobySpace include effective dissemination of location probabilities for all nodes in a constrained DTN and high convergence time.

10.5.2 Active Stochastic Routing

In active routing protocols, the trajectory of some nodes are controlled to improve delivery performance with store-and-carry. Mobile nodes pick up messages and ferry them for a distance before another contact brings them closer to the destination. Active routing techniques provide improved flexibility and lower delays with the additional cost of increased protocol and system complexity. Active DTN routing techniques are frequently implemented as optimization problems.

The general objective of an active routing protocol is to maximize network capacity, reduce message latency, and reduce message loss while facing resource constraints. Applications where mobile nodes are controlled to ferry messages can be used in multiple domains. In disaster recovery, mobile nodes (helicopters, UAVs, or personnel) equipped with communication devices capable of storing a large number of messages can be commanded to follow a trajectory that interconnects disconnected user partitions.

10.5.2.1 Meet and Visit (MV) Routing

In wireless sensor networks, mobile nodes can also traverse the sensing area and pick up/deliver measurements, queries, and event messages. In the remainder of this chapter, we review two DTN routing mechanisms that employ active node trajectory control. In Ref. 18, Burns et al. introduce the Meet and Visit (MV) routing scheme, where node trajectory is adjusted according to traffic demands by autonomous agents. MV aims to improve four performance metrics with a multi-objective control approach.

On each controlled mobile node, separate controllers for total bandwidth, unique bandwidth, delivery latency, and peer latency, respectively, are combined through multi-objective control techniques such as null-space or subsumption. Each controller adjusts the node trajectory such that its own objective is maximized.

The Total Bandwidth Controller selects the DTN that has the greatest number of unseen messages amortized by the trip time. This prevents making long trips without a matching load of new messages. The Unique Bandwidth Controller selects a node that has the largest number of new messages not yet forwarded to any other nodes. The Delivery Latency Controller picks the node with the highest average delivery time. The Peer Latency Controller selects the node that is least-visited by an agent such that the traveling time to visit this node does not increase the overall peer latency metric.

The four controllers can be composed to optimize agent missions across performance metrics. To do that, controllers are first ordered according to their importance. With the null-space approach, an agent's subordinate controller actions can be optimized without affecting the performance of the dominant controller's actions. To increase the optimal solution space of the dominant controller, a minimum performance threshold method is used. The actions controlled by the subordinate controller are acceptable as long as the dominant controller's performance is above this threshold.

A different controller composition approach uses a subsumption approach. A controller with a higher priority computes the action space for achieving a specified performance level for its metric. Within this space, the immediate lower priority controller finds its own optimum without changing the performance of any higher priority controllers. MV implements an epidemic dissemination protocol for the network state necessary for the four controllers. Node information is tagged with a time stamp and flooded during contact.

Simulation results have shown that this approach is sufficient for low-bandwidth and latency-estimation errors, but not enough to correctly estimated "last visit" times and location information. MV routing could be further improved with additional offline or out-of-band network states. Another limitation of this approach is the key assumption that contact bandwidth is unlimited.

10.5.2.2 Message Ferrying

In Ref. 22, Zhao et al. describe a proactive Message Ferrying routing method (MF) with 2-hop forwarding and a single ferry. A message ferry is a special mobile node tasked with improving the transmission capacity in a mobile DTN. The authors present two methods for message ferrying in sparse DTNs. In the Node-Initiated Message Ferrying (NIMF) scheme the ferry follows a specific trajectory. Nodes that need to send messages adjust their trajectory periodically to meet the ferry for message up-/download.

The objective of the NIMF node trajectory control mechanism is to minimize message loss due to TTL expiration and buffer limits, while reducing the negative impact of trajectory changes on node mission goals. The first objective can be expressed by knowing message generation/drop rates and by estimating contact times. The second objective can be modeled as the Work Time Percentage (WTP). The WTP represents the fraction of time a node performs its main task. It is assumed that during a detour to meet a ferry, a node does not contribute to its main task. The NIMF controller allows node trajectory changes only when the WTP is above a minimum threshold.

In the Ferry-Initiated Message Ferrying (FIMF) scheme, the ferry responds to requests for contacts broadcast by nodes on a long-range radio channel. The authors show that the ferry trajectory control problem is NP-hard and propose a greedy nearest-neighbor heuristic and a traffic-aware heuristic that optimizes, locally, both location and message drop rates. In Ref. 23, the same authors extend their ferry-based DTN routing method for coordinating multiple message ferries such that traffic demands are met and delay is minimized. Approximations are provided for single-route and multi-route trajectory control. Ferry replacement algorithms for fault-tolerant delivery are further explored in Ref. 24.

10.6 Vehicle Traffic Model

In this section, we discuss vehicle traffic models. Vehicle traffic models are important for DTN routing in vehicle networks because the performance of DTN routing protocols are closely related to the mobility model of the network. The car-following model is used in civil engineering to describe traffic behavior on a single lane under both free-flow and congested traffic conditions.[25] This model assumes that each driver in the following vehicle maintains a safe distance from the leading vehicle and the deceleration factor is also taken into account for braking performance and drivers' behavior. The complete mathematical model is given by

$$S' = L + \beta'V + \gamma V^2$$

where S' is the headway spacing from rear bumper to rear bumper, L is the effective vehicle length in meters, and V is the vehicle speed in meters/second. β' is driver reaction time in seconds, and the γ coefficient is the reciprocal of twice the maximum average deceleration of a following vehicle. Both the β' parameter and the γ coefficient are introduced to ensure that the following vehicle can come to a complete stop if the leading vehicle suddenly brakes. As in many other civil engineering studies, we use a so-called "good driving" rule, which assumes that each vehicle has similar braking performance. In this case, the car following model can be simplified as

$$S' = L + \beta'V.$$

The car-following model has some limitations in modeling freeway traffic behavior for the purpose of wireless networking research, but is one of the most popular models in civil engineering. These limitations can be summarized as follows:

1. The car-following model is limited to the situation where driver reaction time is believed to be a dominant factor. Therefore, it is only an appropriate model under

free-flow traffic or heavy traffic scenarios. Empirical studies[26] confirm that during rush hour β' is typically a small number that represents the reaction time of a driver, following a log-normal distribution.[27] However, in light to moderate traffic, β' can be as large as 50 to 100 sec and cannot be interpreted as driver reaction time.[27] Instead, interarrival time between vehicles should be used to describe this spacing.

2. This is the focus of vehicular safety research in civil engineering. Therefore, the car-following model describes headway spacing between two adjacent vehicles of the same lane (i.e., lane-level spacing). From the network connectivity standpoint, however, we observe that the most relevant metric is spacing from the leading vehicle to the nearest following vehicle on a multilane road (i.e., road-level spacing), regardless of whether the following vehicle is on the same lane or on a different lane from the leading vehicle.

To address both of the aforementioned limitations, the car-following model is extended to the road level by replacing the lane-level reaction time β' with a road-level interarrival time β (the interarrival time of vehicles on any lane on the same road as observed from a fixed observation point). The lane-level car-following model can be generalized as

$$S = L_{min} + \beta V$$

where L_{min} is the minimum spacing between any two adjacent vehicles, which is assumed to be zero in this study. By focusing on road-level intervehicle spacing S, the proposed model not only models rush-hour heavy traffic but also captures the sparse or intermediate traffic during nonrush hour times.

10.7 Vehicle–Roadside Data Access

Although a lot of research has been carried out on intervehicle communication, vehicle–roadside data access is also an important issue in vehicle DTN network. Medium access control (MAC) issues have been addressed in Refs. 2, 28, and 29, where slot-reservation MAC protocols[28,29] and congestion control policies for emergency warning[2] are studied.

In a recent paper on vehicle–roadside data access,[30] the roadside unit (RSU) can act as a router in a delay-tolerant network or as an access point for vehicles to access the Internet. Although this can bring many benefits to drivers, the deployment cost and maintenance cost are very high. As another option, RSU can also be used as a buffer point (or data island) between vehicles. This section focuses on the latter paradigm due to its low cost and easy deployment.

All data on the RSUs are uploaded or downloaded by vehicles in this paradigm. For example, some data, especially those with spacial/temporal constraints, only need to be stored and used locally. Applications that also belong to this case where the data is buffered at the RSUs and will not be sent to the Internet include the following:

1. *Real-time traffic.* Vehicles can observe real-time traffic observations and report them to nearby RSUs. The traffic data are stored at RSUs, providing real-time

query and notication services to other vehicles. The data can be used to provide traffic conditions and alerts such as road congestion and accidents.

2. *Value-added advertisement.* To provide efficient advertisements, stores may want to advertise their sale or activity information in nearby area. Without Internet connection,[4] they can ask the running vehicles to carry and upload the advertisement information to nearby RSUs. At the same time, other vehicles driving around can download these advertisements and visit the stores.

3. *Digital map downloading.* It is impossible for vehicles to install all the most up-to-date digital maps before traveling. This would help to solve the storage limitations of memory cards and changes resulting from frequent road construction. Hence, vehicles driving to a new area may update map data locally for travel guidance.

Vehicles are moving and they only stay in the RSU area for a short period of time. This makes vehicle networks different from traditional data access systems in which users can always wait for the service from the data server. As a result, there is always a time constraint associated with each request. Meanwhile, to make the best use of the RSU and to share the information with as many vehicles as possible, RSUs are often set at roadway intersections or areas with high traffic. In these areas, download (query) requests retrieve data from the RSU, and upload (update) requests upload data to the RSU. Both download and upload requests compete for the same limited bandwidth. As the number of users increases, deciding which request to serve at which time will be critical to system performance. Hence, it is important to design an efficient scheduling algorithm for vehicle–roadside data access.

10.7.1 A Model for Vehicle–Roadside Data Access

An architecture of vehicle–roadside service scheduling is shown in Figure 10.3, where a large number of vehicles retrieve (or upload) their data from (or to) the RSU when they are in communication range. The RSU (server) maintains a service cycle, which is non-preemptive; that is, a service cannot be interrupted until it finishes. When one vehicle enters the RSU area, it listens to the wireless channel.

All vehicles can send requests to the RSU if they want to access the data. Each request is characterized by a 4-tuple: <*v-id*, *d-id*, *op*, *deadline*>, where *v-id* is the identifier of the vehicle, *d-id* is the identifier of the requested data item, *op* is the operation that the vehicle wants to do (upload or download), and *deadline* is the critical time constraint of the request, beyond which the service becomes useless.

All requests are queued at the RSU server upon arrival. Based on the scheduling algorithm, the server serves one request and removes it from the request queue. Unlike traditional scheduling services, data access in vehicular networks has two unique features:

1. The arrival request is only active for a short period of time due to vehicle movement and coverage limitations of RSUs. When vehicles move out of the RSU area, the unserved requests have to be dropped.

2. Data items can be downloaded and uploaded from the RSU server. The download and update requests compete for the service bandwidth.

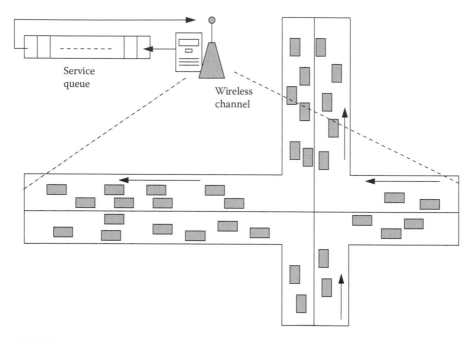

FIGURE 10.3   The architecture of vehicle–roadside service scheduling.

It is assumed that each vehicle knows the service deadline of its request. This is reasonable because when a vehicle with a GPS device enters the coverage area of a RSU, it can estimate its departure time based on the knowledge of its driving velocity and its geographic position. After a vehicle establishes connectivity with one RSU, it can get the geographic information and radio range of the RSU through beacon messages. With its own driving velocity and position information, the vehicle can estimate its departure time, which is its service deadline.

10.7.2 Performance Metrics

The metrics for scheduling algorithms are responsiveness (e.g., average/worst-case waiting time[31–33]) or fairness (e.g., stretch[34,35]) and are commonly used in previous works. In most of these works, requests do not have time constraints, and the data on the server is either not updated, or updated only by the server. However, in the vehicle–roadside data access scenario, requests that are not served within a set time limit will be dropped as the vehicles move out of the RSU area. As update requests compete for bandwidth with other download requests, some data may become stale after an update is missed, degrading service quality. Therefore, we use the following metrics for scheduling vehicle–roadside data access compared with responsiveness and fairness, providing fresh data to more vehicles.

1. *Data quality.* Good data quality means data is not stale. Data become stale if a vehicle has the new version of the data but fails to upload it before the vehicle

moves out of the RSU range. The staleness of the data will degrade the data quality for the download service. In this chapter, we use the percentage of fresh data access to represent the data quality of the system. Therefore, a good scheduling scheme should update data in time and try to avoid data staleness.

2. *Service ratio.* A good scheduling scheme should serve as many requests as possible. The ratio of the number of requests served before the service deadline to the total number of arriving requests is the service ratio.

10.7.3 Roadside Unit Scheduling Schemes

Giving more bandwidth to download requests can provide a higher download service ratio, but a higher update drop ratio and hence low data quality. Therefore, achieving both high service ratio and good data quality is very difficult. If update requests get more bandwidth, the service ratio decreases.

There is always a trade-off between high service ratio and good data quality. Our focus now switches to improving the service ratio. The primary goal of a scheduling scheme is to serve as many requests as possible. We identify two parameters that can be used for scheduling vehicle–roadside data access:

1. *Deadline.* The request is not useful and should be dropped if a request cannot be served before its deadline. The request with an earlier deadline is more urgent than the request with a later deadline.
2. *DataSize.* Usually, vehicles can communicate with the RSU at the same data transmission rate. The data size decides how long the service will last.

Three naive schemes for roadside unit scheduling are as follows:

1. *First Deadline First (FDF).* In this scheme, the request with the most urgency will be served first.
2. *Smallest DataSize First (SDF).* In this scheme, the data with a small size will be served first.
3. *First Come First Serve (FCFS).* In this scheme, the request with the earliest arrival time will be served first.

The service ratios under these three naive scheduling schemes are compared in Figure 10.4. The interarrival time of the requests is determined by the percentage of vehicles that will issue service requests, which is varied along the *x*-axis. As shown in the figure, when the request arrival rate is low, FDF outperforms FCFS and SDF. This is because, when the workload is low, the deadline factor has more impact on the performance.

After the urgent requests are served, other pending requests can still have the opportunity to get services. However, when the request arrival rate increases, the service ratio of FDF drops quickly while SDF performs relatively better. Because the system can always find short requests for service, SDF can still keep a higher service ratio. FCFS does not take any deadline or data size factors into account when making scheduling decisions. It has the worst performance.

Data size and request deadlines are not considered in FCFS. FDF gives the highest priority to the most urgent requests while neglecting the service time spent on those

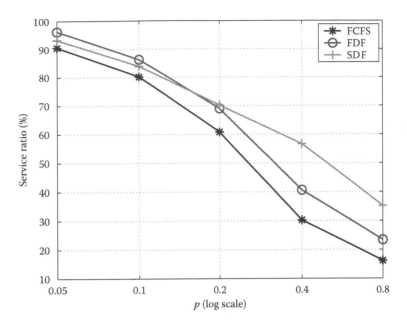

FIGURE 10.4   Service ratio for FCFS, FDF, and SDF schemes.

data items. SDF takes the data size into account but ignores the request urgency. It is clearly shown in the figure that FDF and SDF can only achieve good performance for certain workloads.

This motivates the integration of the deadline and data size to improve the performance of scheduling. None of them can provide a good scheduling as a result. $\mathcal{D} * \mathcal{S}^{30}$ considers both data size and deadlines when scheduling vehicle–roadside data access. From the above observations, there are two principles are:

1. Given two requests with the same deadline, the one asking for a small size of data should be served first.
2. Given two requests asking for data with same size, the one with the earlier deadline should be served first.

Each request is given a service value based on its deadline and data size, called *DS_value*, as its service priority weight:

$$DS\_value = (Deadline - CurrentClock) * DataSize$$

In this equation, the deadline and data size factors are multiplied because these two factors have different measurement scales and/or units. With product, different metrologies will not impose any negative effect on the comparison of two *DS_values*. At each scheduling time, the $\mathcal{D} * \mathcal{S}$ scheme always serves the requests with the minimum *DS_value*.

10.8 Delay-Tolerant Routing in VANETs

Although most of the existing work on vehicle networks is limited to 1-hop or short-range multihop communication, vehicular delay-tolerant networks are useful to other scenarios. For example, without Internet connection, a moving vehicle may want to query a data center ten miles away through a VANET. The widely deployed wireless LANs or infostations[36,37] can also be considered.

Vehicle delay-tolerant networks have many applications, such as delivering advertisements and announcements regarding sale information or remaining stocks at a department store. Information such as the available parking spaces in a parking lot, the meeting schedule at a conference room, and the estimated bus arrival time at a bus stop can also be delivered by vehicle delay-tolerant networks.

For the limited transmission range, only clients around the access point can directly receive the data. However, this data may be beneficial to people in moving vehicles far away, as people driving may want to query several department stores to decide where to go. A driver may query the traffic cameras or parking lot information to make a better travel plan. A passenger on a bus may query several bus stops to choose the best stop for bus transfer. All these queries may be issued miles away from the broadcast site. With a vehicular delay-tolerant network, the requester can send the query to the broadcast site and get a reply from it. In these applications, the users can tolerate up to a minute of delay as long as the reply eventually returns.

The problem of efficient data delivery in vehicular delay-tolerant networks is studied in this section. Specifically, when a vehicle issues a delay-tolerant data query to some fixed site, we must know how to efficiently route the packet to that site and receive the reply with a reasonable delay. We will present a vehicle-assisted data delivery (VADD)[4] based on the idea of carry and forward.[38]

Some of the carry-and-forwarding approaches either pose too much control or no control at all on mobility, and hence are not suitable for vehicular networks. They include the ones proposed for delay-tolerant network.[14,22,38,39] In contrast, VADD makes use of predictable vehicle mobility, which is limited by the traffic pattern and road layout. For example, the driving speed is regulated by the speed limit and the traffic density of the road, the driving direction is predictable based on the road pattern, and the acceleration is bounded by the engine speed. VADD exploits the vehicle mobility pattern to better assist data delivery.

10.8.1 The VADD Protocol

In the model assumed by the VADD protocol, vehicles communicate with each other through a short-range wireless channel, and vehicles can find their neighbors through beacon messages. The packet delivery information such as source ID, source location, packet generation time, destination location, expiration time, and so on, are specified by the data source and placed in the packet header. A vehicle knows its location by triangulation or through a GPS device, which is already popular in new cars and will be common in the future.

Geographical information is also assumed to be available in the vehicles. Vehicles are equipped with preloaded digital maps, which provide street-level maps and traffic statistics such as traffic density and vehicle speed on roads at different times of the day. Such digital maps have already been commercialized. The latest one is developed by Map Mechanics,[40] and includes road speed data and an indication of the relative density of vehicles on each road. Yahoo! is also working on integrating traffic statistics in its new product called SmartView,[41] where real traffic reports of major U.S. cities are available.

It is expected that more detailed traffic statistics will be integrated into digital maps in the near future. The cost of setting up such a vehicular network can be justified by its application to many road safety and commercial applications,[1-3] which are not limited to the proposed delay-tolerant data-delivery applications.

The most important issue is to select a forwarding path with the smallest packet delivery delay. VADD is based on the idea of carry and forward. Although geographical forwarding approaches such as GPSR,[42] which always chooses the next hop closer to the destination, are very efficient for data delivery in ad hoc networks, they may not be suitable for sparsely connected vehicular networks.

Suppose a driver approaches intersection I_a and he wants to send a request to the coffee shop (to reserve a sandwich) at the corner of intersection I_b, as shown in Figure 10.5. To forward the request through $I_a \rightarrow I_c, I_c \rightarrow I_d, I_d \rightarrow I_b$ would be faster than forwarding through $I_a \rightarrow I_b$, even though the latter provides a geographically shortest-possible path. The reason is that, in the case of disconnection, the packet has to be carried by the vehicle, whose moving speed is significantly slower than the wireless communication. In sparsely connected networks, vehicles should try to make use of the wireless communication channel, and resort to vehicles with faster speed. Thus, VADD follows the following basic principles:

1. If the packet has to be carried through certain roads, the road with higher speed should be chosen.

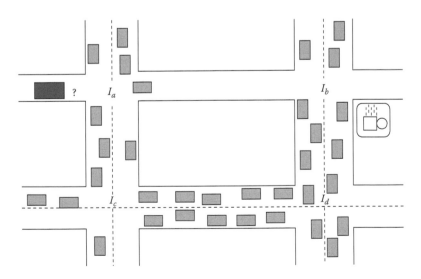

FIGURE 10.5   Find a path to the coffee shop.

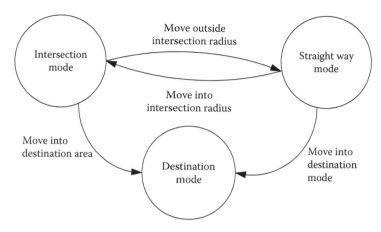

FIGURE 10.6 The transition mode in VADD.

2. Transmit through wireless channels as much as possible.
3. Owing to the unpredictable nature of VANETs, the packet cannot be expected to be successfully routed along the precomputed optimal path, so dynamic path selection should continuously be executed throughout the packet-forwarding process.

VADD has three packet modes (Figure 10.6): Intersection, Straight Way, and Destination, based on the location of the packet carrier (i.e., the vehicle that carries the packet.) By switching between these packet modes, the packet carrier takes the best packet-forwarding path. Among the three modes, the Intersection mode is the most critical and complicated one, because vehicles have more choices at the intersection.

10.9 Data Dissemination in VANETs

Data dissemination protocols[43,44] have been proposed to disseminate information about traffic, obstacles, and hazards on the roads. Similar applications such as real-time video streaming between vehicles have been studied.[45] A conventional way to report accidents or traffic conditions is to use certain infrastructures such as roadside traffic sensors reporting data to a central database, or cellular wireless communication between vehicles and a monitoring center. The problem with this design is the expensive deployment. In addition, these infrastructure-based networks are not scalable due to their centralized nature. VANETs, as an alternative to infrastructure-based vehicle networks, are constructed on-the-fly and do not require any investment besides the wireless network interfaces that will be a standard feature in the next generation of vehicles.

How to exchange traffic information among vehicles in a scalable fashion in VANETs is an interesting but challenging problem that has to be solved. Solutions to this problem can be categorized into two main mechanisms: a flooding-based approach and a dissemination-based approach. In the flooding mechanism, each individual vehicle periodically broadcasts information about itself. Every time a vehicle receives a broadcast message, it stores it and immediately forwards it by rebroadcasting the message. This mechanism is clearly not scalable due to the large volume of messages flooded over the network, especially in high-traffic-density scenarios. The flooding-based mechanism can be further divided into three categories: priority-based approaches, distance-based approaches, and geocast approaches. On the other hand, in the dissemination mechanism, each vehicle broadcasts information about itself and the other vehicles it knows about. Each time a vehicle receives information broadcasted by another vehicle, it updates its stored information to the next broadcast period, at which time it broadcasts its updated information. The dissemination mechanism is scalable, because the number of broadcast messages is limited, and they do not flood the network. The dissemination-based mechanism can be further divided into two categories: approaches utilizing the bidirectional mobility of vehicles and forwarding-based approaches. We can see the classification of mechanisms of multicast/broadcast in VANETs in Figure 10.7.

10.9.1 Flooding-Based Mechanisms

A number of safety applications require communications to a group of vehicles, not just pairwise communications supported by unicast protocols. Safety applications require propagation of information to a large number of nodes quickly and reliably. Flooding is the most common approach for broadcasting without explicit neighbor information. However, flooding is known to be inefficient due to the so-called broadcast storm problem, especially in networks with high node density. Most existing flooding-based information dissemination approaches in VANETs aim to achieve a high message delivery ratio by avoiding contention and collision caused by the broadcast storm phenomena.[43,46,47]

10.9.1.1 Priority-Based Approach

In Ref. 48, the authors study how broadcast performance scales in VANETs and propose a priority-based broadcast scheme that gives higher priority to nodes that need to transmit time-critical messages. The proposed algorithm categorizes nodes in the network

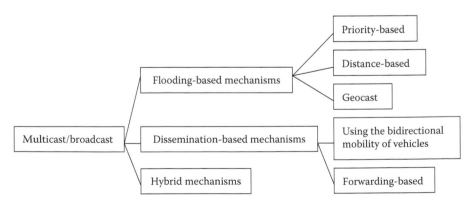

FIGURE 10.7   The classification of multicast/broadcast mechanisms.

into multiple classes with different priorities and schedules the packet transmission accordingly. Although this technique is not designed to solve the broadcast storm problem, it can indirectly mitigate the severity of the storm by allowing nodes with higher priority to access the channel as quickly as possible.

10.9.1.2 Distance-Based Approach

Reference 43 proposed an efficient 802.11-based urban multihop broadcast protocol (UMB) where only the furthest vehicle from the transmitter rebroadcasts the packet such that broadcast redundancy is suppressed.

In Ref. 46, three distance-based mechanisms were proposed:

1. Weighted p-persistence broadcasting
2. Slotted 1-persistence broadcasting
3. Slotted p-persistence broadcasting

The basic broadcast techniques follow either a p-persistence rule or a 1-persistence ($p = 1$) rule. Although the overhead is excessive, most routing protocols designed for multihop ad hoc wireless networks follow the brute-force 1-persistence flooding rule where all nodes in the network rebroadcast the packet with probability 1. On the other hand, the gossip-based approach follows the p-persistence rule where each node rebroadcasts with a predetermined probability p. This method is also referred to as probabilistic flooding.[49]

In weighted p-persistence broadcasting, upon receiving a packet from node i, node j checks the packet ID and rebroadcasts with probability p_{ij} if it is the first time that node j receives the packet; otherwise, the packet is discarded by node j. The forwarding probability, p_{ij}, can be calculated on a per-packet basis using the following expression, $p_{ij} = D_{ij}/R$ where D_{ij} is defined as the relative distance between node i and j, and R is the average transmission range. Unlike the p-persistence scheme, the weighted p-persistence broadcasting assigns nodes that are farther away from the broadcaster higher probability given that the GPS information is available and accessible from the header of a packet. The weighted p-persistence approach is illustrated in Figure 10.8a.

In slotted 1-persistence broadcasting, upon receiving a packet, a node checks the packet ID and rebroadcasts with probability 1 at the assigned timeslot, if it is the first time it receives the packet and it has not received any duplicate packets before its assigned timeslot $T_{S_{ij}}$; otherwise, the packet is discarded. Denoting D_{ij} as the relative distance between node i and j, R as the average transmission range, and N_s the predetermined number of slots, $T_{S_{ij}}$ can be calculated as $T_{S_{ij}} = S_{ij} \times \tau$ where τ is the estimated 1-hop delay, which includes the propagation delay and the medium access delay, and S_{ij} is the assigned slot number which is defined as $S_{ij} = N_s \times \left(1 - \lceil \min(D_{ij}, R)/R \rceil\right)$. The timeslot method follows the same logic as the weighted p-persistence scheme. However, each node uses the GPS information to calculate the waiting time to retransmit instead of calculating the reforwarding probability. For example, in Figure 10.8b, the broadcast coverage is spatially divided into four regions and the nodes located in the farthest region will be assigned a shorter waiting time. Therefore, a node takes on the smallest D_{ij} value if it receives duplicate packets from more than one sender. Similar to the p-persistence scheme, this method needs the transmission range information so as to agree on a certain value of slot size or the number of slots. Note that N_s is a design parameter that needs to be carefully selected.

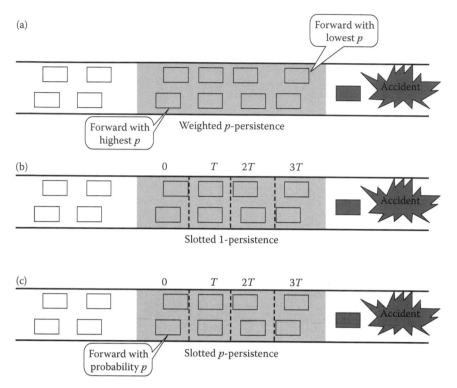

FIGURE 10.8   An example of the broadcast suppression technique.

In slotted p-persistence broadcasting, upon receiving a packet, a node checks the packet ID and rebroadcasts with the predetermined probability p at the assigned timeslot $T_{S_{ij}}$, if it is the first time that it receives the packet and it has not received any duplicate packets before its assigned timeslot; otherwise, the packet is discarded by it. Figure 10.8c illustrates the concept of the slotted p-persistence with four slots.

Reference 46 asserted that the slotted p-persistence scheme can substantially reduce the packet loss ratio at the expense of a slight increase in total delay and reduced penetration rate.

10.9.1.3 Flooding Geographically Defined Information

In many safety applications, vehicle safety alarms are required to be sent to all vehicles within a specific area where protocols for flooding geographically defined information are needed. Geocast is a variation of conventional multicast, which specifies the destination as a geographic position rather than a specific node or multicast addresses. The multicast group (or geocast group) is implicitly defined as the set of nodes within a specified area, which is different from the conventional multicast schemes. That is, a node automatically becomes a member of the corresponding geocast group at a given time if it is within the geocast region at that time.

The IVG protocol proposed in Ref. 50 addresses how to broadcast alarm messages only to vehicles approaching areas of a given accident. An alarm message received is discarded if it is not relevant. Otherwise, it is rebroadcasted if this message is still relevant after a deferred period of time.

10.9.2 Dissemination-Based Mechanisms

Compared to the flooding-based approaches, dissemination-based mechanisms are more scalable because the number of broadcast messages is limited, and they do not flood the network. The dissemination mechanism can either broadcast information to vehicles in all directions, or perform a directed broadcast restricting information about a vehicle to vehicles behind it.

10.9.2.1 Data Dissemination Considering the Bidirectional Mobility of Vehicles

Reference 51 presents a formal model of data dissemination in VANETs and studies how the performance of data dissemination is affected by VANET characteristics, especially the bidirectional mobility on well-defined paths. The analysis as well as simulation results show that dissemination using only vehicles in the opposite direction significantly increases the data dissemination performance.

Without loss of generality, vehicles are assumed to move on bidirectional straight roads with multiple lanes in each direction, as shown in Figure 10.9. It is assumed that a vehicle on the road moves either to the *East* as shown in the lower part of the road

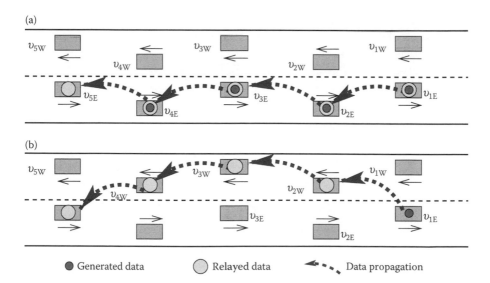

FIGURE 10.9 Dissemination models: (a) the same-dir dissemination model, and (b) the opp-dir dissemination model.

in Figure 10.9 (e.g., v_{1E} and v_{2E}), or to the *West* as shown in the upper part of the road (e.g., v_{1W} and v_{2W} in Figure 10.9). S_E and S_W are the average speeds for *East* and *West* directions, respectively. All transmissions are assumed to be omnidirectional with communication range R. Each vehicle in the model is assumed to be concerned about the road information ahead of it. Information should propagate backwards with respect to the vehicle's direction (i.e., propagates in the opposite direction) in order to accomplish this assumption. Assume that vehicles broadcast data packets periodically every B seconds. For the sake of simplicity, consider the propagation of information about vehicles moving *East* where the direction of propagation is from the east to the west.

There are two types of broadcasted data: generated data and relayed data. Generated data, denoted by a small circle in Figure 10.9, is the vehicle's own data (e.g., ID, speed, and location) and it is updated every new broadcast period. Relayed data, denoted by a large circle, is the stored data about the other vehicles ahead, and is propagated backward within every broadcast period. Three dissemination/propagation models are considered: *same-dir*, *opp-dir*, and *bi-dir*. In the *same-dir* model, every vehicle periodically broadcasts both the store-relayed data and its generated data in the same data packet. When a vehicle broadcasts a data packet, only vehicles moving in the same direction are involved in the propagation of this packet. More specifically, when v_1 broadcasts a data packet, v_2 will propagate it later if and only if the following all apply:

1. v_2 is within the transmission range of v_1
2. v_1 and v_2 are moving in the same direction (i.e., *East*)
3. v_1 is in front of v_2 with respect to their directions (i.e., v_1 is located east to v_2)

Figure 10.9a is an example of how information is propagated from vehicle v_{1E} to vehicle v_{5E}, both moving in the *East* direction in the *same-dir* model. Note that no vehicle from the opposite direction is involved in the dissemination in this model.

On the other hand, in the *opp-dir* model, relayed data and generated data are not broadcast together. Instead, vehicles in the same direction (i.e., *East*) only broadcast their generated data. These generated data are aggregated and propagated backwards by the vehicles in the opposite direction (i.e., *West*). When v_1 broadcasts a packet (i.e., relayed data in the case moving *West*, or generated data in the case moving *East*), v_2 will operate according to the following rules, given that it is within the the transmission range of v_1:

1. If v_1 and v_2 are moving *East*, v_2 will accept the packet if v_1 is located east of v_2. This is the case when v_1 broadcasts its generated data.
2. If v_1 and v_2 are moving *West*, v_2 will accept the packet if v_2 is located west of v_1. This is the case when v_1 relays a packet.
3. If v_1 is moving *East* (or *West*) and v_2 is moving *West* (or *East*), v_2 will accept the packet regardless of the relative position of the vehicles.

The first rule guarantees a fast delivery of the newly generated data to all the vehicles within one hop of the source vehicle. Figure 10.9b is an example of how information is propagated from v_{1E} to v_{5E} in the *opp-dir* model.

The *bi-dir* model combines both the *same-dir* and the *opp-dir* models. In this model, vehicles in the same direction(i.e., *East*) are involved in the propagation of generated and relayed data while vehicles in the opposite direction (i.e., *West*) only propagate relayed data. Information in this model is propagated by vehicles moving in both the same and the opposite directions, which is different from the other mechanisms.

Analysis and simulation show that the performance of the data dissemination model relies on the traffic densities in both directions of the road. When traffic in the opposite direction (e.g., *West*) is not sparse, the *opp-dir* model is more efficient than both the *bi-dir* and the *same-dir* models in terms of latency, network utilization, and average error. Although the *bi-dir* model has better knowledge than the *opp-dir* model in this network configuration, this better knowledge comes with the cost of lower utilization rates, higher latency, and lower accuracy. This indicates that the *opp-dir* model is the most efficient data-dissemination model in terms of scalability, accuracy, and efficiency. However, the *bi-dir* model outperforms both the *opp-dir* and the *same-dir* models when traffic in the opposite direction is sparse.

10.9.2.2 Forwarding-Based Data Dissemination Protocols

Several forwarding-based protocols for data dissemination have been proposed recently. An opportunistic forwarding approach is proposed in Ref. 52. It is asserted that the motion of vehicles on a highway can contribute to successful message delivery, provided that messages can be disseminated in a store-carry-forward fashion. Reference 53 proposes a trajectory-based forwarding scheme. Reference 54 proposes MDDV, a combination of opportunistic forwarding and trajectory-based forwarding, which specifically addresses vehicle mobility. MDDV, a mobility-centric approach for data dissemination in vehicular networks, is designed to operate reliably and efficiently in spite of the highly dynamic, partitioned nature of these networks. MDDV is designed to exploit vehicle mobility for data dissemination, and combines the idea of geographical forwarding, opportunistic forwarding, and trajectory-based forwarding.

A forwarding trajectory is specified as a path from the source to the destination region. The road network can be abstracted as a directed graph where nodes represent intersections and edges represent road segments. One of the MDDV objectives is to deliver messages to their destination regions with low delay. Taking the path with the shortest distance from the source to the destination region would be a naive approach in that information propagation along a road depends largely on the vehicle traffic on it, for example, using vehicle density in addition to the distance between the source–destination pair. A short road distance does not necessarily result in short information propagation delay. High vehicle density often guarantees fast information propagation. Therefore both the traffic condition and the road distance must be taken into consideration. However, vehicle traffic conditions vary from one road segment to another and change over time. The number of lanes gives some indication of the expected vehicle traffic. $d(A, B)$ is defined as the "dissemination length" of a road segment from road node A to B, which takes into consideration the static road information. Denoting $r(A, B)$ as the road length between A and B, i/j as the number of lanes from A/B to B/A, the following heuristic formula is used: $d(A, B) = r(A, B)(m - (m - 1)(i^p + cj^p))$ where $0 < c < 1$. The dissemination length of a road segment is used as the weight for the corresponding link in the abstracted road graph. MDDV uses a forwarding trajectory that is specified as the directed path with the smallest sum of weights from the source to the destination region in the weighted road graph.

The dissemination process consists of two phases: the forwarding phase and propagation phase. In the forwarding phase, the message is forwarded along the forwarding trajectory to the destination region. The propagation phase begins and the message is propagated to each vehicle in an area centered on the destination region before the message time expires, once the message reaches the destination region. In order to deliver the message to the intended receivers before they enter the destination region in order to reduce delay, this area covers the destination region and is usually larger.

At first, Ref. 54 assumes that each vehicle has perfect knowledge concerning the global status of the data dissemination. During the forwarding phase, the message holder closest to the destination region along the forwarding trajectory is called the "message head." The vehicle taking the role of the message head may change over time as the message propagates or vehicles move. With perfect knowledge, every vehicle knows the message head vehicle in real time. Only the message head tries to pass the message to other vehicles that may be closer to the destination region. During the propagation phase, the message is propagated to vehicles without the message in the specified area.

Owing to the lack of perfect knowledge for participating vehicles, the above ideal scenario cannot be implemented. In practice, individual vehicles have no idea about which vehicle is the message head in real time. For instance, as illustrated in Figure 10.10, on a two-way traffic road, the current message head is vehicle 1. In Figure 10.10a, vehicle 1 may run out of the trajectory or may become inoperative; vehicle 2, the immediate follower, may not be aware of this because the network is partitioned. In Figure 10.10b, vehicle 1 is moving away from the destination region (note that the road is bidirectional). Once vehicle 1 passes vehicle 2, vehicle 2 should become the new message head. However, vehicle 2 does not know this unless it receives an explicit notification from vehicle 1. With the assumption that vehicles do not know the location of others, this is difficult to do. In both cases, the message is lost. To address this problem, a group

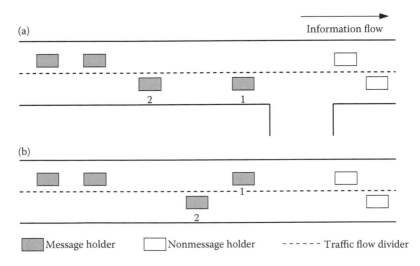

FIGURE 10.10   An example of lack of perfect knowledge.

of vehicles near the real message head can actively forward the message instead of the message head vehicle only. The group membership changes as the actual message head moves toward the destination region. There is a trade-off between delivery reliability and message overhead: larger groups mean higher delivery reliability but higher message overhead, too. Vehicles have to locally determine their own actions based on their approximate knowledge of the global message dissemination status.

Vehicles need to have some information regarding the message dissemination status in order to realize the approximation. Specifically, information concerning the message head is required. However, the message dissemination status changes over time. Vehicles can only expect approximate knowledge, at best. Also a vehicle's knowledge must be updated constantly. A convenient way to exchange such information is to place it in the message. As the message is propagated among vehicles, so does the message dissemination status information. Too much information in the message is cumbersome and expensive, however. To this end, a small amount of data, the message head location and its generation time, called the *message head pair*, is inserted into the message. Every holder of a message maintains a message record containing the message head pair along with other information concerning this message. The message head pair provides the best knowledge of a message holder regarding the message head location.

The actual message head can move either away from or towards the destination region along the forwarding trajectory within a short period of time. But it should move towards the destination region in the long run (because the message head vehicle may change). For simplicity, it is required that the message head location installed by a message holder never moves backward, which means that a message holder can only install a new message head location closer to the destination region than the one currently installed.

To reduce the publication and dissemination of false information, only some vehicles are allowed to generate the message head pair. A message holder is allowed to publish its current location as the message head location if it believes it may be the real message head with some probability. In this sense, a message holder may assume either one of two roles: the message head candidate and nonmessage head candidate. Only a message head candidate can actively publish its current location as the message head location and a nonmessage head candidate can only learn from received messages.

There are rules for a message holder to transit between a message head candidate and nonmessage head candidate. Suppose the current time is t_c, a vehicle's current location is l_c, and a vehicle's installed message head pair is $\langle l, t \rangle$, where l is the message head location and t is the generation time:

1. *Nonmessage head candidate → message head candidate.* During the forwarding phase, one important observation is that a vehicle passing its installed message location a shorter period after the generation time is more likely to be the message head, because after a long period the message may have already been forwarded far away toward the destination region along the trajectory. Thus a nonmessage head candidate becomes a message head candidate if it passes its installed message head location toward the destination region before $t + T_1$, where T_1 is a system parameter. During the propagation phase, message holders moving into the destination region assume the role of the message head candidate.

2. *Message head candidate* → *nonmessage head candidate.* During the forwarding phase, there are two transition rules: (1) if the message head candidate leaves the trajectory or moves away from the destination region along the trajectory, it becomes a nonmessage head candidate; (2) if a message head candidate moves toward the destination region along the trajectory, it stays as a message head candidate until it receives the same message with another message head pair $\langle l_n, t_n \rangle$ where l_n is closer to the destination region than l_c. During the propagation phase, a message head candidate becomes a nonmessage head candidate once it moves out of the destination region.

A message holder updates its installed message head pair with the information from received messages. Two messages differing only in the message head pair are two versions of the same message. One message version with message head pair $\langle l_i, t_i \rangle$ is said to be newer than another message version with message head pair $\langle l_j, t_j \rangle$ if l_i is closer to the destination region than l_j; or $l_i = l_j$ but $t_i > t_j$. A vehicle always updates its installed message head pair with the newer received information. Therefore obsolete/false installations can be eliminated through data exchange.

The data exchange algorithm is defined as the following:

1. *Forwarding phase.* A message holder can be in either one of two dissemination states: the active state and passive state, or not eligible to transmit at all. A message holder in the active state runs the full protocol to actively propagate the message while a message holder in the passive state only transmits the message if it hears some older message version. The active propagation can help populate the message, move the message closer to the destination region, or update dissemination status. The passive updating serves to eliminate false/obsolete information only. Given a message holder's installed message head pair $\langle l, t \rangle$, its current location l_c and the current time t_c, it is in the active state if $t_c < t + T_2$ and l_c is within the distance L_2 from l, and otherwise it is in the passive state if $t_c < t + T_3$ and l_c is within the distance L_3 from l, while $T_2 < T_3$ and $L_2 < L_3$. Otherwise, the message will not be transmitted under any circumstance. T_2, T_3, L_2, and L_3 are system parameters. In this way, the active data propagation is initiated by the fresh generation of a message head pair and is constrained near the message head location (through both geographical and temporal constraints). Data propagation caused by obsolete/false information will eventually stop when the time expires or it is suppressed by updates.

2. *Propagation phase.* A message holder can either be in the active state or not eligible to transmit. A message holder in the active state runs the full protocol. The active propagation serves to deliver the message to intended receivers. Using the same notations as before, a message holder is in the active state if $t_c < t + T_2$ and l_c is within the distance L_2 from l. Every vehicle inside the destination region publishes its own location as the message head location. Therefore this data exchange mechanism limits the active propagation in a region centered on the destination region.

It is important for an opportunistic forwarding mechanism to determine when to store/drop a message. The design decision can influence memory usage, message overhead, and delivery reliability. The decision to store/drop messages can be based on a vehicle's knowledge of its future movement trajectory. For example, a message holder may decide to drop a message if it knows that continually holding the message can no longer contribute to suppress unnecessary message transmissions based on its future movement trajectory, given that vehicles are aware of their own near-future movement trajectory. In MDDV, memory buffers are assumed to be free from limit such that each vehicle stores whatever it overhears. A message is dropped by a vehicle when the vehicle leaves the active state during the propagation phase, leaves the passive state during the forwarding phase, or the message expiration time elapses.

10.9.3 Hybrid Mechanisms

Flooding-based data dissemination mechanisms are unscalable due to the large amount of contention and collision, especially in dense networks. On the other hand, dissemination-based mechanisms are not suited for delay-sensitive safety message dissemination, albeit scalability is achieved. Hence, hybrid mechanisms that combine the strengths of each are proposed. Reference 55 proposes an approach (called Directional Propagation Protocol, or simply, DPP) using clusters of connected vehicles where flooding-based data dissemination mechanisms are used in a cluster and dissemination-based mechanisms are used among clusters.

DPP uses the directionality of data and vehicles for information propagation. DPP comprises three components: a Custody Transfer Protocol (CTP), an Inter-Cluster Routing Protocol, and an Intra-Cluster Routing Protocol. In order to overcome the lack of an end-to-end path between source and destination, the Custody Transfer Protocol is introduced which is derived from delay-tolerant networking concepts. On the one hand, the Inter-Cluster Routing Protocol controls the message exchange between nodes within a cluster. On the other hand, the communication between clusters is governed by the Intra-Cluster Routing Protocol. As illustrated in Figure 10.11, interconnected blocks of vehicles can be formed by vehicles traveling towards the same direction. Gaps are allowed between consecutive blocks. The traffic density has a significant impact on the cardinality of each block. For example, a long continuous block can be formed under dense conditions, while under sparse conditions, the cardinality of each block could be one.

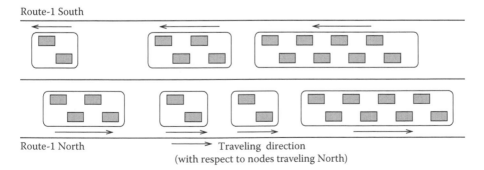

Route-1 South

Route-1 North   ⟶   Traveling direction
(with respect to nodes traveling North)

FIGURE 10.11 An example of blocks of vehicles.

Additionally, vehicles that are within range R and maintain connectivity for a minimum time t are said to be part of a cluster. Thus, a block may comprise several clusters.

Under sparse traffic conditions, gaps between blocks are frequent and network partitions are common, which prevents an end-to-end path between source and destination. Accordingly, the speed of the vehicle that carries the message may influence the data dissemination performance. Under dense traffic conditions, an end-to-end path between source and destination exists with high probability where the data dissemination performance is mainly determined by contentions and collisions.

The effects of speed differentials within the cluster are not considered as the faster vehicles will leave one cluster and join another as they progress on the road. Also, there are intersections on a highway where vehicles may join or leave the clusters. Once a cluster becomes very large, the cluster is split to better manage intracluster traffic.

Each cluster has a header and a trailer, located at the front and rear of each cluster, entrusted with the task of communicating with other clusters. A node at the head or tail of the cluster will elect itself as the header or trailer for our protocol. (Node election is not covered here.) This limits congestion caused by the large number of participating nodes. The remaining nodes in the cluster, nodes that are not header or trailer, are described as intermediate nodes. Within a cluster, communicated messages are shared with all nodes to both facilitate header/trailer replacement and general awareness of disseminated messages.

The intermediate nodes retain a passive role of receiving messages and acknowledgments from opposing blocks and forwarding them to the header or trailer sharing the information within the cluster. Similarly, messages originating from intermediate nodes are immediately routed to header or trailer depending upon the direction in which information needs to propagate. Any duplicate messages received at any of the nodes are dropped. End-to-end path formation can be assumed to be taking place within a cluster.

In most message-passing schemes, a message is buffered until an acknowledgment from the destination is received. However, due to network fragmentation in a VANET and the resultant lack of continuous end-to-end connectivity at any given instant, the message can require buffering for an indeterminate amount of time. The result translates to the requirement for large buffer sizes or dropped messages and difficulty in exchanging acknowledgments. For applications that do not require continuous end-to-end connectivity, a store-and-forward approach can be used.

With the custody transfer mechanism, a message is buffered for retransmission from the originating cluster until it receives an acknowledgment from the next-hop cluster. The custody is implicitly transferred to another cluster that is in front along the direction of propagation and is logically the next hop in terms of the message path. The traffic in the opposing direction acts as a bridge but is never given custody of the message. The custody is not released until an acknowledgment is received from the cluster in front. Once the message reaches the next-hop cluster, it has custody of the message and the responsibility for further relaying the message is vested with this cluster. The custody of the message may be accepted or denied by a cluster by virtue of it being unable to satisfy the requirements of the message.

The propagation is called *reverse propagation* if the data are headed in a direction opposite to the direction of motion of the vehicles and *forward propagation* if data are headed along the direction of motion of the vehicles. In forward propagation, as

Route-1 South

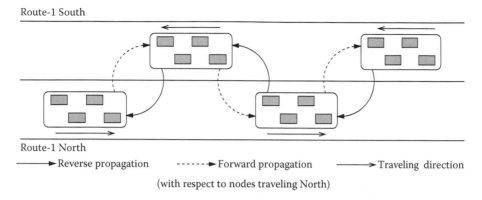

Route-1 North

——►Reverse propagation   ----→Forward propagation   ——→Traveling direction

(with respect to nodes traveling North)

FIGURE 10.12   An example of forward propagation and reverse propagation.

illustrated in Figure 10.12, the vehicle is assumed to be traveling along the N direction and the data are also to be propagated in the N direction. The data can travel at a minimum rate of the speed of the vehicle because the data are traveling along with the vehicle. The data are propagated to the header of the cluster. The header now tries to propagate the data further along the N direction, trying to communicate with other clusters located ahead of this cluster. If the clusters are partitioned, the header attempts to use the clusters along the S direction, which may overlap with other clusters along the N direction to bridge this partition. Thus, the data are propagated to nodes traveling along the N direction that are otherwise partitioned from each other, by using clusters along the S direction. This temporary path occurs due to opportunistic contact with nodes in the overlapping clusters. Once the data are forwarded to the next hop and an acknowledgment is received, the custody is transferred to that cluster. The entire process is repeated until the data reaches its required destination. The reverse propagation scheme can be modeled as an extension of the forward propagation scheme.

10.10 Conclusion

In this chapter, we studied the problem of efficient data delivery (unicast) and dissemination (multicast) in delay-tolerant vehicle networks.

First, we described the concepts of delay-tolerant networks (DTNs) including the characteristics of these networks, store-carry-forward routing protocols. Because VANETs are a special type of DTN, which result from its unique mobility pattern, we further studied the vehicle traffic model. Delay-tolerant vehicle networks can be either infrastructure-based or infrastructure-free. We studied the role of roadside units in DTN routing in infrastructure-based vehicle networks. We also studied routing protocols such as VADD in infrastructure-free vehicle networks. As an emphasis, we studied the problem of data dissemination in VANETs where flood-based mechanisms, dissemination-based mechanisms, and hybrid mechanisms were visited.

We believe that routing and data dissemination are still challenging problems due to the unique characteristics of vehicle networks, such as high node mobility, high probability of network disconnection, and partition and lane-based node move pattern. This chapter summarizes some, but not all of the important findings in the vehicle networks community regarding routing and data dissemination. We hope our work lays the groundwork for future study in this area.

References

1. Xu, Q., Mark, T., Ko, J., and Sengupta, R., Vehicle-to-Vehicle Safety Messaging in DSRC, in Proc. of ACM VANET, Philadelphia, PA, USA, 2004.
2. Yang, X., Liu, J., Zhao, F., and Vaidya, N., A Vehicle-to-Vehicle Communication Protocol for Cooperative Collision Warning, in Proc. of ACM MOBIQUITOUS, Boston, MA, USA, August 2004.
3. Yin, J. et al., Performance Evaluation of Safety Applications Over DSRC Vehicular Ad Hoc Networks, in Proc. of ACM VANET, Philadelphia, PA, USA, 2004.
4. Zhao, J., and Cao, G., VADD: Vehicle-Assisted Data Delivery in Vehicular Ad Hoc Networks, in Proc. of IEEE INFOCOM, Barcelona, Spain, 2006.
5. Namboodiri, V., Agarwal, M., and Gao, L., A Study on the Feasibility of Mobile Gateways for Vehicular Ad-Hoc Networks, in Proc. of ACM VANET, Philadelphia, PA, USA, 2004.
6. Clausen, T. and Jacquet, P., Optimized Link State Routing Protocol OLSR, in RFC 3626, IETF Network Working Group, 2003.
7. Fall, K., A Delay-tolerant Network Architecture for Challenged Internets, in Proc. of ACM SIGCOMM, Karlsruhe, Germany, 2003.
8. Cerf, V. et al., Delay Tolerant Network Architecture, available at http://www.ietf.org/rfc/rfc4838.txt, 2007.
9. Cardei, I., Liu, C., and Wu, J., Routing in Wireless Networks with Intermittent Connectivity, in Encyclopedia of Wireless and Mobile Communications, CRC Press, Taylor & Francis Group, Boca Raton, FL, USA, 2007.
10. Jain, S., Fall, K., and Patra, R., Routing in a Delay Tolerant Network, in Proc. of ACM SIGCOMM, Portland, OR, USA, 2004.
11. Merugu, S., Ammar, M., and Zegura, E., Routing in Space and Time in Networks with Predictable Mobility, in GIT-CC-04-7, 2004.
12. Gnawali, O., Polyakov, M., Bose, P., and Govindan, R., Data Centric, in Proc. of IEEE Aerospace Conference, Big Sky, MT, USA, 2005.
13. Intanagonwiwat, C., Govindan, R., and Estrin, D., Directed Diffusion: A Scalable and Robust Communication Paradigm for Sensor Networks, in Proc. of ACM MOBICOM, Boston, MA, USA, 2000.
14. Vahdat, A. and Becker, D., Epidemic Routing for Partially Connected Ad Hoc Networks, in Technical Report, April 2000.
15. Spyropoulos, T., Psounis, K., and Raghavendra, C.S., Spray and Wait: An Efficient Routing Scheme for Intermittently Connected Mobile Networks, in Proc. ACM SIGCOMM Workshop on Delay-Tolerant Networking, Philadelphia, PA, USA, 2005.
16. Shah, R.C., Roy, S., Jain, S., and Brunette, W., Data Mules: Modeling a Three-tier Architecture for Sparse Sensor Networks, in Proc. IEEE International Workshop on Sensor Network Protocols and Applications, Anchorage, AK, USA, 2003.

17. Lindgren, A., Doria, A., and SchelSn, O., Probabilistic Routing in Intermittently Connected Networks, in Poster of ACM MOBIHOC, Annapolis, MD, USA, 2003.

18. Burns, B., Brock, O., and Levine, B.N., Mv Routing and Capacity Building in Disruption Tolerant Networks, in Proc. IEEE INFOCOM, Miami, FL, USA, 2005.

19. Juang, P. et al., Energy-efficient Computing for Wildlife Tracking: Design Tradeoffs and Early Experiences with Zebranet, in Proc. of ASPLOS, San Jose, CA, USA, 2002.

20. Leguay, J., Friedman, T., and Conan, V., Evaluating Mobility Pattern Space Routing for DTNs, in Proc. of IEEE INFOCOM, Barcelona, Spain, 2006.

21. Ratnasamy, S., Francis, P., Handley, M., Karp, R., and Shenker, S., A Scalable Content Addressable Network, in Proc. of ACM SIGCOMM, San Diego, CA, USA, 2001.

22. Zhao, W., Ammar, M., and Zegura, E., New directions: A Message Ferrying Approach for Data Delivery in Sparse Mobile Ad Hoc Networks, in Proc. of ACM MOBIHOC, Roppongi Hills, Tokyo, Japan, 2004.

23. Zhao, W., Ammar, M., and Zegura, E., Controlling the Mobility of Multiple Data Transport Ferries in a Delay-Tolerant Network, in Proc. of IEEE INFOCOM, Miami, FL, USA, 2005.

24. Yang, J., Chen, Y., Ammar, M., and Lee, C., Ferry Replacement Protocols in Sparse MANET Message Ferrying Systems, in Proc. of IEEE WCNC, New Orleans, LA, USA, 2005.

25. Wisitpongphan, N., Bai, F., Mudalige, P., Sadekar, V., and Tonguz, O.K., On the Routing Problem in Disconnected Vehicular Networks, in Proc. of IEEE INFOCOM Minisymposia, Anchorage, AK, USA, 2007.

26. *Highway Capacity Manual 2000*, Transportation Research Board, May 2000.

27. Sahoo, P.K., Rao, S.K., and Kumar, V.M., A Study of Traffic Characteristics on Two Stretches of National Highway No 5, Indian Highways, Indian Road Congress, 1996.

28. Verdone, R., Multi-Hop R-Aloha for Inter-Vehicle Communication at Millimeter Waves, *IEEE Transactions on Vehicular Technology*, 46(4), 992–1005, November 1997.

29. Lott, M., Halmann, R., Schulz, E., and Radimirsch, M., Medium Access and Radio Resource Management for Ad Hoc Networks Based on UTRA TDD, in Poster of ACM MOBIHOC, Long Beach, CA, USA, 2001.

30. Zhang, Y., Zhao, J., and Cao, G., On Scheduling Vehicle-Roadside Data Access, in Proc. of ACM VANET, Montreal, QC, Canada, 2007.

31. Su, C. and Tassiulas, L., Broadcast Scheduling for Information Distribution, in Proc. of IEEE INFOCOM, Kobe, Japan, 1997.

32. Aksoy, D. and Franklin, M., R × W: A Scheduling Approach for Large-scale On-demand Data Broadcast. *IEEE/ACM Transactions on Networking*, 7(6), 846–880, 1999.

33. Gandhi, R., Khuller, S., Kim, Y., and Wan, Y., Algorithms for Minimizing Response Time in Broadcast Scheduling, Algorithmica, 2004.

34. Acharya, S. and Muthukrishnan, S., Scheduling On-demand Broadcasts: New Metrics and Algorithms, in Proc. of ACM MOBICOM, Dallas, TX, USA, 1998.

35. Wu, Y. and Cao, G., Stretch-Optimal Scheduling for On-demand Data Broadcasts, in Proc. of IEEE ICCCN, Scottsdale, AZ, USA, 2001.

36. Frenkiel, R.H., Badrinath, B.R., Borras, J., and Yates, R., The Infostations Challenge: Balancing Cost and Uiquity in Delivering Wireless Data, *IEEE Personal Communications Magazine*, 7(2), 66–71, April 2000.

37. Goodman, D., Borras, J., Mandayam, N., and Yates, R., INFOSTATIONS: A New System Model for Data and Messaging Services, in Proc. of IEEE VTC, Braunschweig, Germany, May 1997.
38. Davis, J., Fagg, A., and Levine, B., Wearable Computers as Packet Transport Mechanisms in Highly-Partitioned Ad-hoc Networks, in Proc. of Fifth International Symposium on Wearable Computers, Zurich, Switzerland, October 2001.
39. Li, Q. and Rus, D., Sending Messages to Mobile Users in Disconnected Ad-hoc Wireless Networks, in Proc. of ACM MOBICOM, Boston, MA, USA, 2000.
40. U.S. Census Bureau, Tiger, Tiger/line and tiger-related products, http://www.census.gov/geo/www/tiger/.
41. SmartView, available at http://maps.yahoo.com/smartview.php.
42. Karp, B. and Kung, H.T., GPSR: Greedy Perimeter Stateless Routing for Wireless Networks, in Proc. of ACM MOBICOM, Boston, MA, USA, 2000.
43. Korkmaz, G., Ekici, E., Ozguner, F., and Ozguner, U., Urban Multi-Hop Broadcast Protocol for Inter-Vehicle Communication Systems, in Proc. of ACM VANET, Philadelphia, PA, USA, October 2004.
44. Xu, B., Ouksel, A., and Woflson, O., Opportunistic Resource Exchange in Inter-vehicle Ad Hoc Networks, in Proc. of IEEE MDM, Berkeley, CA, USA, 2004.
45. Ghandeharizadeh, S., Kapadia, S., and Krishnamachari, B., PAVAN: A Policy Framework for Content Availability in Vehicular Ad-hoc Networks, in Proc. of ACM VANET, Philadelphia, PA, USA, October 2004.
46. Wisitpongphan, N. et al., Broadcast Storm Mitigation Techniques in Vehicular Ad Hoc Wireless Networks, *IEEE Wireless Communications*, 14(6), 84–94, December 2007.
47. Korkmaz, G., Ekici, E., and Ozguner, F., An Efficient Fully Ad-hoc Multi-hop Broadcast Protocol for Inter-vehicular Communication Systems, in Proc. of IEEE ICC, Istanbul, Turkey, 2006.
48. Torrent-Moreno, M., Jiang, D., and Hartenstein, H., Broadcast Reception Rates and Effects of Priority Access in 802.11-Based Vehicular Ad-Hoc Networks, in Proc. of ACM VANET, Philadelphia, PA, USA, 2004.
49. Haas, Z., Halpern, J.Y., and Li, L., Gossip-based Ad Hoc Routing, in Proc. of IEEE INFOCOM, New York, NY, USA, 2002.
50. Bachir, A. and Benslimane, A., A Multicast Protocol in Ad Hoc Networks Inter-Vehicle Geocast, in Proc. of IEEE VTC, Orlando, FL, USA, 2003.
51. Nadeem, T., Shankar, P., and Iftode, L., A Comparative Study of Data Dissemination Models for VANETs, in Proc. of ACM MOBIQUITOUS, San Jose, CA, USA, 2006.
52. Chen, Z., Kung, H., and Vlah, D., Ad Hoc Relay Wireless Networks Over Moving Vehicles on Highways, in Proc. of ACM MOBIHOC, Long Beach, CA, USA, 2001.
53. Niculescu, D. and Nath, B., Trajectory Based Forwarding and Its Applications, in Proc. of ACM MOBICOM, San Diego, CA, USA, 2003.
54. Wu, H., Fujimoto, R., Guensler, R., and Hunter, M., MDDV: A Mobility-Centric Data Dissemination Algorithm for Vehicular Networks, in Proc. of ACM VANET, Philadelphia, PA, USA, 2004.
55. Little, T.D.C. and Agarwal, A., An information Propagation Scheme for VANETs, in Proc. of IEEE Intelligent Transportation Systems, Vienna, Austria, 2005.

11

Localization in Vehicular Ad-Hoc Networks

Azzedine Boukerche
School of Information Technology and Engineering University of Ottawa

Horacio A.B.F. Oliveira, Eduardo F. Nakamura, and Antonio A.F. Loureiro
Department of Computer Science Federal University of Minas Gerais

11.1 Introduction

A new kind of ad hoc network is hitting the streets: Vehicular ad hoc networks (VANETs).[1-4] In these networks, vehicles communicate with each other and, possibly, with roadside infrastructure to provide a long list of applications varying from traffic safety to driver assistance and Internet access. In these networks, real-time position knowledge of nodes is an assumption made by most protocols, algorithms, and applications. This is a reasonable assumption, because GPS receivers can be easily installed in vehicles, a number of which already come with this technology. However, as VANETs advance into critical areas and become more dependent on localization systems, some undesirable problems with GPS begin to surface, including not always being available or not being robust enough for some applications. For this reason, a number of other localization techniques such as dead reckoning, cellular localization, and image/video localization, to cite a few, have been used in VANETs to overcome such limitations. A common factor in all these cases is the possibility of using data fusion techniques to compute an accurate vehicle position, creating a new paradigm for localization where different localization techniques are combined into a single solution that is more robust and precise than the individual approaches. In this chapter, we further discuss this subject, describe each localization technique, and show how they can be combined by means of data fusion techniques.

A number of interesting and desired applications for intelligent transportation systems (ITS) have been motivating the development of VANETs. In these networks, vehicles are equipped with a communication device that allows two types of communication: the exchange of messages with others in vehicle-to-vehicle communication (V2V) and the exchange of messages with a roadside network infrastructure in vehicle-to-roadside communication (V2R).

A number of applications are envisaged in these networks, some of which are already possible in some vehicle designs (Figure 11.1):

1. Vehicle collision warning systems
2. Security distance warning
3. Driver assistance
4. Cooperative driving
5. Cooperative cruise control
6. Dissemination of road information
7. Internet access
8. Map location
9. Automatic parking
10. Driverless vehicles

All these applications require, or can take advantage of, some sort of localization technique. For instance, map location is usually done using GPS receivers with a geographic information system, while security distance warning and automatic parking can be implemented by means of distance estimator sensors or image/video processing.

As ITS and VANET technology advances toward more critical applications like vehicle collision warning systems (CWS) and driverless vehicles, it is likely that a robust and

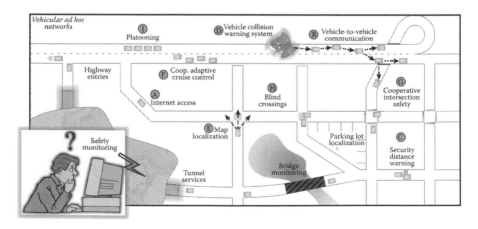

FIGURE 11.1   Several VANET applications.

highly available localization system will be required. Unfortunately, GPS receivers are not the best solution in these cases, because their accuracy ranges from 20 or 30 m and they do not work in indoor or dense urban areas where there is no direct line of visibility to satellites. For these reasons and, of course, for security reasons, GPS information is likely to be combined with other localization techniques such as dead reckoning, cellular localization, and image/video localization. This combination of localization information from different sources can be carried out using such data fusion techniques such as the Kalman filter and the particle filter.

In this chapter, we discuss the localization requirements of a number of VANET applications. We show several proposed localization techniques that can be used to estimate the position of a vehicle, and we highlight their advantages and disadvantages when applied to VANETs. By concluding that none of these techniques can achieve individually the desired localization requirements of a critical VANET application, we show how the location information from multiple sources can be combined into a single position that is more accurate and robust by using data fusion techniques.

The remainder of this chapter is organized as follows. In the next section, we identify the location information requirements of several VANET applications, while in Section 11.3 we show how these positions can be computed through several localization techniques. Finally, Section 11.4 shows how Data Fusion techniques can be used to combine the position information gained from these multiple sources. Section 11.5 provides our conclusions and future directions for localization systems in VANETs.

11.2 Location-Aware VANET Applications

Most VANET applications consider the availability of real-time updated position information. Unlike other networks, this position availability is a very plausible assumption in VANETs, because GPS receivers can be installed easily in vehicles, a number of which are already deployed with this technology. In this section we divide VANET applications into three main groups and show how position information is used by the protocols and algorithms in these applications, as well as the location requirements for each of them.

11.2.1 Vehicle Communication

Vehicle communication is the main goal of VANETs. It includes V2V and V2R communication and provides services like information routing and the data dissemination of incidents, road congestion, and so on.

Most *routing* protocols for VANETs[5] use position information in order to improve their performance and be compliant with VANET requirements, such as dynamic topology changes and frequent network fragmentation. This routing technique has long been used in ad hoc networks[6-9] and most of these can also be applied to VANETs. A classical example is *greedy forwarding*[8] in which, at each step, location information is used to forward a packet to the neighbor nearest to the destination node. But some geographic routing protocols have also been designed specifically for VANETs, taking advantage of more geographical knowledge like maps[10,11] and movement information.[12] Routing techniques are also used to access local infrastructure networks, which can have Internet connections (Figure 11.1, label A). In these cases, position information, as well as future trajectory knowledge, can be used to assist routing.

A number of protocols[13,14] that aim to inform both near and far vehicles about transit conditions—such as road flow, traffic congestion, possibly dangerous situations, and so on—have been proposed for VANETs' *data dissemination*. Most of these protocols use location knowledge, mainly to ensure that locally disseminated information reaches only the vehicles that are interested in that information. Driver direction can also be used, as proposed by the ODAM algorithm.[13] In Figure 11.1, label B, road information about a dangerous situation is disseminated to interested vehicles.

With regard to *location requirements*, these algorithms, based on communication, will accept location errors mostly within a few meters—like normal GPS receivers—because the long transmission ranges of the vehicles will reduce the effects of some meters of localization inaccuracy.

11.2.2 Driver Assistance

In driver assistance applications, VANET resources are used to enhance the driver's perception and knowledge of the road and environment. In these applications, the driver is informed about the surrounding environment but he/she maintains full control of the vehicle (except in emergency and other requested procedures).

Vehicle collision warning systems[1,13] are one of the most interesting applications of VANETs for driver assistance. One component of these systems is the *security distance warning*, in which the driver is warned when a minimum distance to another vehicle is reached (Figure 11.1, label C). It can also implement an emergency break when the distance between two vehicles or between a vehicle and an obstacle decreases too quickly, as in Figure 11.1, label D. Another part of these systems acts when a collision has already occurred and we need to warn nearby vehicles (warn messages), so they can avoid pile-up collisions (Figure 11.1, label D). In these cases, multihop communication can be used to disseminate collision information. With regard to *localization requirements*, because they provide a critical application for safe driving, these applications require robust and reliable local distance estimation, which can be done using sensors and cameras. GPS positions, which are less accurate, can be used to warn distant vehicles about the location of an accident.

A widely known and already in-use driver assistance application is *map localization*, in which the current position of the vehicle is shown on a map. In these applications, a path direction between two points of the city, for instance, can be drawn on a map indicating the current location of the vehicle. This application can assist drivers in situations where they find themselves lost in a unknown part of the city, as depicted in Figure 11.1, label E. With regard to *localization requirements*, GPS localization is a proven working technology for this application, because map knowledge can be used to overcome GPS inaccuracy.

Another driver assistance application is *vision enhancement*, in which drivers are given a clear view of vehicles and obstacles on foggy days, and can learn about the existence of vehicles hidden by obstacles, buildings, and other vehicles. With regard to *localization requirements*, a localization system with some meters of accuracy combined with map knowledge can be used to locate distant and/or hidden vehicles, while sensors can be used to locate obstacles on the road as well as accurately estimate distances to near vehicles.

Automatic parking is an application through which the vehicle can park itself without the need of driver intervention. With regard to *localization requirements*, in order to park automatically, a vehicle needs sensors for distance estimation and/or video/image processing.

11.2.3 Cooperative Driving

In cooperative driving applications, vehicles in a VANET exchange messages to drive and share the available space in the road cooperatively. In these applications, the vehicle assumes partial or total control over driving. The main feature of these applications is that vehicles must cooperate with each other in order to accomplish a common goal.

In *cooperative adaptive cruise control*, the vehicle maintains the same speed whether traveling up or down a hill, without requiring driver intervention. Usually, the driver sets the speed and the system will take over, but in this case, vehicles can cooperate among themselves to set this speed adaptively (Figure 11.1, label F). This application only takes care of speed, while the driver still has to control the direction of the vehicle.

The *localization requirements* for this kind of application involve a certain degree of confidence in distance estimation among vehicles. Because nearby GPS receivers have correlated errors (they have all the same approximated error magnitude and direction), they can be used to exchange positions in order to compute the distance between them. Sensors can also be used to increase the system's confidence.

Another interesting application of VANETs is *cooperative intersection safety*, in which vehicles arriving at a road intersection exchange messages in order to make safe crossroads, as depicted in Figure 11.1, label G. Besides ensuring a safe crossroads, it is also possible to construct a *blind crossroads*, where there is no light control and the vehicles cooperate with each other to make a cooperative crossroads (Figure 11.1, label H).

The *localization requirements* for this include position information as part of the decision-making process. The localization accuracy must allow the application to differentiate between the lanes and street sides.

Vehicle following or *platooning* is a technique used to make one or more vehicles follow a leader vehicle to form a train-like system, as shown in Figure 11.1, label I. This application can be useful in situations where two or more vehicles are going to the same location. The *localization requirements* for this include ensuring a minimum distance between vehicles. Also, vehicles must track the position of the vehicle in front of them. Video/image processing as well as GPS receivers can be used in this application.

11.3 Localization Techniques for VANETs

A number of localization techniques have been proposed for computing the position of mobile nodes. An interesting aspect of VANETs is that most localization techniques can be applied easily to these networks. Figure 11.2 depicts a number of localization techniques that can be used by vehicles to estimate their positions: map matching, dead

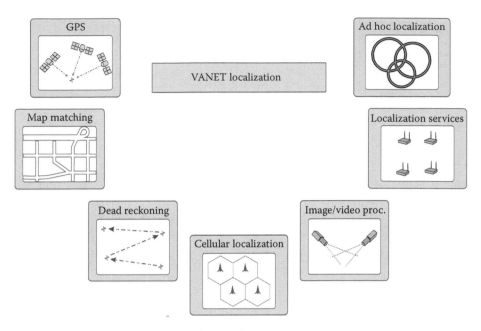

FIGURE 11.2 Several localization techniques for VANETs.

reckoning, cellular localization, image/video processing, localization services, and relative distributed ad hoc localization. All of these techniques have their pros and cons. In this section, we briefly explain each of these techniques and discuss when and how they can be used to localize vehicles in intelligent transport systems.

11.3.1 Global Positioning System—GPS/DGPS

GPS, the global positioning system,[15,16] is composed of 24 satellites that operate in orbit around the Earth. Each satellite circles the Earth at a height of 20,200 km and makes two complete rotations every day. The orbits were defined in such a way that in each region of the Earth one could "see" at least four satellites in the sky.

A GPS receiver is a piece of equipment that is able to receive the information constantly sent by the satellites, to estimate its distance to at least four known satellites using a technique called time of arrival (ToA), and, finally, to compute its position using trilateration. Once these procedures are executed, the receiver is able to know its latitude, longitude, and altitude.

The main solution for VANET localization is to equip each vehicle with a GPS receiver (Figure 11.3, label A). This is a reasonable solution because GPS receivers can be installed easily in vehicles, a number of which already come with this technology. However, as VANETs advance into critical areas and become more dependent on localization systems, GPS receivers display some undesirable problems such as not always being available and not being robust enough for some applications.

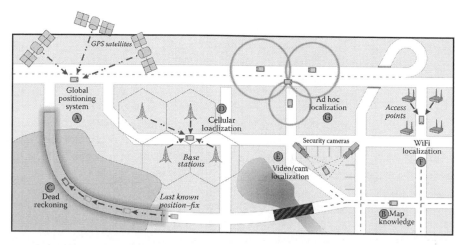

FIGURE 11.3   Examples of localization techniques applied in VANETs.

In order to function properly and compute its position, a GPS receiver needs access to at least three satellite signals for two-dimensional (2D) positioning and at least four satellite signals for a three-dimensional (3D) position computation. At first sight, this is not a major issue, because the number of visible satellites usually varies between four and eleven. However, the problem is that these signals are easily disturbed or blocked by obstacles including buildings, rocks, dense foliage, electronic interference, and so on. The result is position inaccuracy or unavailability in dense urban environments (urban canyons), tunnels, indoor parking lots, forests, and any other indoor, underground, or underwater environment.

Also, GPS receivers have a localization error of ±10 to 30 m.[15,16] While this is a reasonable level of precision for most applications, it is definitely not enough for critical VANET applications like the ones presented in Section 11.2. One positive aspect of these errors is that nearby GPS receivers tend to have the same localization error oriented in the same direction. In other words, nearby GPS receivers have correlated errors. If we put a GPS in an already known location, this GPS receiver can compute its position using the information from the satellites and compare the computed position with its known position. The difference between these two positions can be broadcast and all nearby GPS receivers can correct their computed positions based on the broadcast differential information. This technique is known as differential GPS (DGPS), and fixed ground-based reference stations are used to broadcast this differential information. The use of DGPS can lead to a submeter precision, which is sufficient for most VANET critical applications, but it requires the installation of ground-based reference stations in order to work.

Owing to these limitations, in VANETs, GPS information needs to be combined with different sources of position information and/or geographic knowledge.[17,18] In the next sections, we show several techniques that can be used as sources of position information to improve GPS localization or to completely replace it in locations where GPS is not available.

11.3.2 Map Matching

Current advances in geographic information systems (GIS) have allowed the collection and storage of, as well as access to, very accurate geographic data even to less powerful devices. This technology has been successfully applied to store cities' map information in recently developed map localization systems for vehicle navigation (Figure 11.3, label B).

Aside from the fact that this map knowledge is not a localization technique by itself, it can be used to improve the performance of many positioning systems, such as GPS. First, by limiting the estimated vehicle positions to roads or other places with vehicle access, it is possible to decrease the error of estimated positions. However, the main application of map knowledge in localization is the *map matching* technique.[17,18]

In the map matching technique, several positions obtained over regular periods of time can be used to create an estimated trajectory. This estimated trajectory is then compared to known digital map data to find the most suitable path geometry in the map that matches the trajectory. Using this technique, position information (e.g., from GPS) can be accurately depicted on the map.

11.3.3 Dead Reckoning

By using *dead reckoning*,[12,18] the current position of a vehicle can be computed based on its last known location and using such movement information as direction, speed, acceleration, distance, time, and so on. The last known position, also known as a fix, can be obtained, for instance, by using GPS receivers (which are most common) or by locating a known reference (road crosses, parking lots, home, and so on) on a digital map. Displacement information can be obtained by sensors including odometers, while direction can be estimated easily using other sensors such as digital compasses and gyroscopes.

In practical VANETs, dead reckoning can be used only for short periods of GPS unavailability, or combined with map knowledge. The reason for avoiding the use of this technique over long periods of time is that it can accumulate errors easily. For instance, positioning errors from 10 to 20 m can be reached in only 30 sec after the last position fix when traveling at about 100 km/h.[19,20]

Because dead reckoning accumulates errors rapidly over time and distance, it is considered as a backup system for periods of GPS outage, as shown in Figure 11.3, label C, in which a vehicle enters a tunnel and loses its GPS connection. In this example, the last GPS computed position is used as a position fix. Another viable application of dead reckoning, as noted above, is to combine it with map knowledge. In these cases, the position restrictions can be applied to decrease dead reckoning errors, and traffic patterns can be used to match the estimated path with the known map information (map matching).[18]

11.3.4 Cellular Localization

Cellular localization[21-24] takes advantage of the mobile cellular infrastructure available in most urban environments to estimate the position of an object. The known applications of this technology include locating mobile phones, tracking domestic animals, and vehicle localization.

In order to function properly, mobile cellular systems require the installation of a communication infrastructure composed of a number of cellular base stations distributed along the covered area. Every base station is responsible for providing communication to mobile phones located in its area. While mobile phones move around a city, they keep changing their base station when the signal strength from a new base station becomes greater then the one in use. This procedure is called handoff.

Although only one base station is used in communication, usually several base stations can listen to and communicate with a mobile phone at any time. This fact allows a number of localization techniques to be used to estimate the position of the mobile phone. A well-known technique called *received signal strength indicator* (RSSI) uses the strength of the received signals to derive the distance to the base stations. It is also possible to estimate a distance based on the time it takes for a signal to leave the sender and arrive at the base station (time of arrival, ToA) or the difference between the times it takes for a single signal to arrive at multiple base stations (time difference of arrival, TDoA). Once we have the distances from the mobile phone to at least three base stations, it is possible to compute the position of the mobile phone using such techniques as trilateration and multilateration (Figure 11.3, label D).

Another common approach is possible when directive antennas or antenna arrays are used as base stations. In this case, the angle at which the signal arrives at a base station can be estimated. Based on the angle of arrival (AoA) of a signal to three different base stations, we can compute the position of the signal source.

Fingerprinting is a localization technique based on a pretraining phase in which signal characteristics from base stations are recorded at each location. After this information is recorded, a mobile node can find in the database the position that best matches its current signal characteristics. This is a very interesting solution for small- or medium-sized areas, because it achieves errors smaller than 5 m in indoor environments.[24] For large urban areas like those covered in VANET applications it has questionable applicability, but in some recent studies,[22] an average accuracy of 94 m is achieved after a 60 h calibration drive in a metropolitan area.

Cellular localization is usually less accurate than GPS. The accuracy depends on a number of factors like the current urban environment, the number of base stations detecting the signal, the positioning algorithm used, and so on. However, in most cases the average localization error will be between 90 and 250 m.[22]

11.3.5 Image/Video Processing

Image and video information sources and data processing techniques can also be used for localization purposes, especially in mobile robot guidance systems.[25] However, in some cases, cameras are already available in security systems implemented in parking lots and tunnels, as shown in Figure 11.3, label E. Commonly, these image/video processing techniques are used to feed data fusion algorithms to estimate and predict (track) a vehicle's location.[26] In fact, both image and video information are actual sources from which we can compute the location parameters of a vehicle. For instance, Chausse et al.[27] use the vision algorithm[28] to detect the sides of lanes in video images. It estimates precisely the vehicle geometrical parameters in a local reference system, including lane width, road lateral curvature, distance of the vehicle from the left side of the lane, vehicle's direction angle, and the camera inclination angle. These local data are transformed in order to be expressed in a precise digital map of the environment. Such information is used to feed a data fusion module that estimates the vehicles' locations. The use of data fusion for VANET localization systems will be discussed further in Section 11.4.

11.3.6 Infrastructured Localization Services

There are places where GPS is not available or not precise enough for local applications. In VANETs, as mentioned in Section 11.3.1, these places include tunnels, urban canyons, and parking lots. In these cases, an infrastructure for a communication and positioning service can be implemented to perform the localization of vehicles, as shown in Figure 11.3, label F.

A localization service can be implemented using any known infrastructured localization system like the Cricket Location-Support System,[29] RADAR,[30] ultra-wideband localization,[31] or WiFi localization.[32,33] In Ref. 33, Thangavelu et al. propose a system called VETRAC, a vehicle tracking and location identification system designed for VANETs that uses WiFi access points as a communication infrastructure and also as landmarks when positioning vehicles. The proposed system can be used in tunnels, university campuses, airports, and so on.

In most cases, localization services are likely to take advantage of the communication system in use to compute a vehicle's position based on signal propagation characteristics (e.g., strength, time, or fingerprint). However, other indoor localization systems such as image/video processing (explained in Section 11.3.5) or laserscanners can also be used.

Probably the most challenging and important task in VANET localization is the development of infrastructured localization systems to be used in tunnels, which is one of the most critical VANET environments. Tunnels are normally used to connect important regions separated by natural environments with difficult access and are generally the only path between these regions. Thus, a damaged tunnel can have an enormous impact on a city or a region. Also, owing to the limited access inside a tunnel, emergency rescue operations can become very difficult and even dangerous. In these scenarios, collision avoidance is crucial, and all available information about the state of these tunnels' infrastructure as well as the number and location of all vehicles inside these tunnels are key information for rescue teams in the case of emergency operations.

VANETs can also use wireless sensor networks (WSNs) as the base for a VANET localization infrastructure. The reason for this is that WSNs can also be used to monitor other road variables like movement, temperature, smoke, visibility, and noise. Thus, these networks are ideal for monitoring critical environments as well as for emergency operations, as shown by a number of works.[34,35] Also, the use of sensor networks as a roadside communication infrastructure is a frequently envisioned scenario in many intelligent transportation systems. A number of WSN features can also be used to improve the performance and accuracy of an infrastructured VANET localization system. For instance, movement sensors can be used to send localization packets only when vehicles are present. These sensors can also be used to increase the localization accuracy by making nodes exchange their sensors' movement detection level. Finally, a WSN used as a VANET localization infrastructure will provide a complete safety monitoring system for these critical scenarios, being able not only to monitor important environment and structural variables like movement, temperature, smoke, visibility, noise, pressure, and structural health, but also the location of all vehicle nodes at a given moment.

11.3.7 Relative Distributed Ad Hoc Localization

Local relative position maps can be constructed by a vehicle estimating the distances between its neighbors and exchanging this distance information with nearby nodes in multihop communication. With this dynamic position map, a vehicle can locate itself in relation to nearby vehicles as well as locate the vehicles in its vicinity (Figure 11.3, label G). This type of relative localization has been used mostly in ad hoc and sensor networks, but recently a number of solutions[19,20,36] have been proposed for VANETs.

In Ref. 36, a distributed localization algorithm is proposed to assist GPS-unequipped vehicles in estimating their positions based on nearby GPS-equipped vehicles. To estimate a position, a vehicle not equipped with GPS needs to communicate with at least three GPS-equipped vehicles in its vicinity in order to estimate distances and gather their position information. When the number of nearby GPS-equipped vehicles is less than three, the author shows how to estimate at least the direction of the vehicle and the distance from an event based on the small amount of available information. The proposed algorithm can successfully estimate the position of vehicles not equipped with GPS, but it is hard to identify situations where vehicles have network cards to communicate with other vehicles but have no GPS equipment. Also, the direction of the cars can be easily estimated by exchanging digital compass or gyroscope information.

Parker et al.[20] propose another distributed VANET localization system where distances among vehicles are estimated by using RSSI, and this information is used by an optimization algorithm to improve the initial position estimation of the vehicles (obtained, for instance, by GPS). This technique is primarily intended to improve GPS initial position estimations, but because nearby GPS receivers tend to have correlated errors, estimating distances by RSSI will hardly improve positions. However, this solution can also be used to improve positions computed by the dead reckoning technique during GPS outages.

A number of distributed relative ad hoc localization systems have been proposed recently for ad hoc and sensor networks,[37,38] but only a few of them can be applied to highly mobile and dynamic networks like VANETs. In Ref. 19, Kukshya et al. propose an architecture for relative positioning of a cluster of vehicles that does not require any GPS information and that is suitable for VANETs. This architecture also relies on distance estimation measurements.

Most VANET applications can work with relative positioning, but most of them would also work better using global positioning. In these cases, relative positions can usually be converted into global positions when some vehicles with GPS or accurate global positions are available, as done in Refs. 20 and 36.

11.4 Data Fusion in VANET Localization Systems

One of the most appealing problems to be solved by VANETs is how to provide an any-time, anywhere, fine-grained, and reliable localization system to be used by vehicles in a VANET for critical safety and emergency applications. An *anytime* requirement means that the localization system must be free of delays when computing current positions and synchronizing the clocks of the vehicles (e.g., no startup delay). This requirement is critical, because the high mobility of VANETs means that slightly out-dated position information cannot be used and could even be dangerous. To be avail-able *anywhere* is a challenge in a VANET localization system. It means that the localization system cannot rely only on satellite infrastructure, as it would not work in environments without direct visibility to satellites. Also, it cannot rely only on local infrastructured localization techniques, because it would not be available in places without this infrastructure. Finally, a *fine-grained* localization system ensures a low localization and synchronization error for vehicles, which enables most critical VANET applications with some degree of confidence. A number of localization techniques already exist for these proposals, but most of them have not been evaluated in VANETs and none of them has all the desired features of *anytime, anywhere* availability, and *fine-grained* localization.

Although several interesting solutions have been reported in the literature, they do not satisfy all the requirements of critical applications at the same time. Table 11.1 briefly compares some of these critical requirements of the cited localization techniques. Thus, it is clear that a single technique will not be enough to provide a localization system with all of the features requested by critical VANET applications. As a result, ways to combine different localization techniques and protocols in a single localization system will be required. Data fusion techniques are the natural choice for technique combination aimed at acquiring improved data.[26]

Data fusion can be simply defined as the combination of multiple sources to obtain improved information (cheaper, greater quality, or greater relevance).[26] Data fusion is commonly used for detection and classification tasks in different application domains, such as robotics and military applications.[39] Lately, these mechanisms have been used in previously unpredicted applications such as intrusion detection[40] and denial of service

TABLE 11.1   Localization Techniques: A Comparison

Technique	Position	Synchronization	Availability	Accurate	Monitoring
GPS	×	×	–	–	–
Differential GPS	×	×	–	×	–
Map knowledge	×	–	–	–	–
Dead reckoning	×	–	×	–	–
Cellular localization	×	×	–	–	×
Video/cam localization	×	–	–	×	×
Infrastructured localization	×	–	–	×	–
Ad hoc localization	×	–	–	–	–

(DoS) detection.[41] Within the domain of WSN, simple aggregation techniques (e.g., *maximum*, *minimum*, and *average*) have been used to reduce the overall data traffic to save energy.[42]

Data fusion techniques such as *Kalman filters*, *particle filters*, and *belief theory* have also been used to improve location estimations in many sensor-based systems.[26] For instance, the SAFESPOT[43] approach for the accurate relative positioning of vehicles forsees the use of data fusion to help with accurate position estimation in VANETs. The key idea is to combine information from a cooperative VANET using a data fusion module to allow vehicular safety applications to determine, not only a vehicle's location, but also the lane in which it is traveling. The general idea behind a location system based on data fusion is to combine several information sources to provide an accurate location estimation. Further details about data fusion techniques are reviewed by Nakamura et al.[26]

Fernandez-Madrigal et al.[44] use particle filters to cope with vehicle localization in combined indoor and outdoor scenarios. In such scenarios, the authors assess the performance of ultra wide band (UWB) sensor technology for indoor positioning and GPS for outdoor areas, and evaluate the use of particle filters to fuse observations from these two types of sensors for vehicle localization. Chausse et al.[27] show how to use particle filters to combine GPS localization with data extracted from vision systems to determine a vehicle's location on the road. The combined information is transformed into a global reference using a map of the environment.

In the context of vehicle localization for production and logistic applications, Fuentes Michel et al.[45] apply Kalman filters to track the position of all vehicles when picking up or putting down items by combining a wireless local positioning system with an optical scan match approach.

Aimed at improving security on the roads, Ammoun et al.[46] use a Kalman filter for trajectory prediction and the estimation of a vehicle's location to evaluate and anticipate the risk of collision at a crossroad. The authors show that despite unavoidable latencies and positioning errors, the application performance is still acceptable when a Kalman filter is used for trajectory prediction and estimation.

Najjar and Bonnifait[47] use belief theory and Kalman filters to provide accurate position estimations for a vehicle relative to a digital road map. In this method, the Kalman filter is used to combine the anti-lock braking system (ABS) measurements with a GPS position, which is then used to select the correct roads. The selection strategy fuses distance, direction, and velocity measurements using belief theory. A new observation is then built and the vehicle's approximate location is adjusted by a second Kalman filter.

11.5 Conclusions

In this chapter, localization systems were studied from the viewpoint of VANETs. We showed how GPS receivers, the most common source of localization in VANETs, can become erroneous or unavailable in a number of situations. We discussed how these localization inaccuracies can affect most VANET applications, especially critical ones. A number of other localization systems are available to be used by vehicles to estimate their positions: map matching, dead reckoning, cellular localization, image/video processing, localization services, and relative distributed ad hoc localization. These techniques have their pros and cons. In this chapter, we argued that future localization systems for VANETs are likely to use some kind of data fusion technique to provide position information for vehicles that is accurate and robust enough to be applied in VANET critical applications. Then, we sketched out how data fusion techniques can be used to compute an accurate position based on a number of relatively inaccurate position estimations.

References

1. Biswas, S., Tatchikou, R., and Dion, F., Vehicle-to-vehicle wireless communication protocols for enhancing highway traffic safety, *Communications Magazine*, 44(1), 74–82, 2006.
2. Blum, J.J., Eskandarian, A., and Hoffman, L.J., Challenges of intervehicle ad hoc networks, *IEEE Transactions on Intelligent Transportation Systems*, 5(4), 347–351, 2004.
3. Kiess, W., Rybicki, J., and Mauve, M., On the Nature of Inter-Vehicle Communication, in WMAN 2007: Proceedings of the 4th Workshop on Mobile Ad-Hoc Networks, March 2007, pp. 493–502.
4. Luo, J. and Hubaux, J. P., A Survey of Inter-Vehicle Communication, technical report IC/2004/24, School of Computer and Communication Sciences, EPEL, 2004.
5. Chennikara-Varghese, J., Chen, W., Altintas, O., and Cai, S., Survey of Routing Protocols for Inter-Vehicle Communications, in Mobile and Ubiquitous Systems— Workshops, 2006, 3rd Annual International Conference, 2006, pp. 1–5.
6. Karp, B. and Kung, H.T., GPSR: Greedy Perimeter Stateless Routing for Wireless Networks, in 6th International Conference on Mobile Computing and Networking, Boston, MA, 2000, pp. 243– 254.
7. Ko, Y.B. and Vaidya, N.H., Location-aided routing (LAR) in mobile ad hoc networks, *Mobile Computing and Networking*, 66–75, 1998.
8. Navas, J.C. and Imielinski, T., Geocast—Geographic Addressing and Routing, in MobiCom '97: Proceedings of the 3rd Annual ACM/IEEE International Conference on Mobile Computing and Networking, ACM, New York, NY, 1997, pp. 66–76.
9. Yu, Y., Govindan, R., and Estrin, D., Geographical and Energy Aware Routing: A Recursive Data Dissemination Protocol for Wireless Sensor Networks, technical report CSD-TR-01-0023, UCLA Computer Science Department, 2001.
10. Lochert, C. et al., A Routing Strategy for Vehicular Ad Hoc Networks in City Environments, in IVS '03: IEEE Intelligent Vehicles Symposium, 2003, pp. 156–161.
11. Tian, J., Han, L., and Rothermel, K., Spatially Aware Packet Routing for Mobile Ad Hoc Inter-Vehicle Radio Networks, in ITS '03: IEEE Intelligent Transportation Systems, Vol. 2, 2003, pp. 1546–1551.
12. King, T., Füßler, H., Transier, M., and Effelsberg, W., Dead-Reckoning for Position-Based Forwarding on Highways, in Proceedings of the 3rd International Workshop on Intelligent Transportation (WIT 2006), Hamburg, Germany, March 2006, pp. 199–204.

13. Benslimane, A., Optimized Dissemination of Alarm Messages in Vehicular Ad-Hoc Networks (VANET), in Proceedings of IISNMC, 2004, pp. 655–666.
14. Sun, M.T. et al. GPS-Based Message Broadcast for Adaptive Inter-Vehicle Communications, in Vehicular Technology Conference, 2000, IEEE VTS-Fall VTC 2000, 52nd, Vol. 6, 2000, pp. 2685–2692.
15. Hofmann-Wellenho, B., Lichtenegger, H., and Collins, J., *Global Positioning System: Theory and Practice*, 4th Edn., Springer-Verlag, 1997.
16. Kaplan, E.D., *Understanding GPS: Principles and Applications*, Artech House, 1996.
17. Jagadeesh, G.R., Srikanthan, T., and Zhang, X.D., A map matching method for GPS-based real-time vehicle location, *Journal of Navigation*, 57, 429–440, 2005.
18. Krakiwsky, E.J., Harris, C.B., and Wong, R.V.C., A Kalman Filter for Integrating Dead Reckoning, Map Matching and GPS Positioning, in Position Location and Navigation Symposium, 1988, Record. "Navigation into the 21st Century," IEEE PLANS '88, 1988, pp. 39–46.
19. Kukshya, V., Krishnan, H., and Kellum, C., Design of a System Solution for Relative Positioning of Vehicles Using Vehicle-to-Vehicle Radio Communications During GPS Outages, in Vehicular Technology Conference, 2005, VTC-2005-Fall. 2005 IEEE 62nd, Vol. 2, 2005, pp. 1313–1317.
20. Parker, R. and Valaee, S., Vehicle Localization in Vehicular Networks, in Vehicular Technology Conference, 2006. VTC-2006 Fall. 2006 IEEE 64th, 2006, pp. 1–5.
21. Caffery, J.J. and Stuber, G.L., Overview of radiolocation in CDMA cellular systems, *IEEE Communications Magazine*, 36(4), 38–45, 1998.
22. Chen, M. et al., Practical Metropolitan-Scale Positioning for GSM Phones, in Proceedings of 8th Ubicomp, Orange County, CA, September 2006, pp. 225–242.
23. Song, H.-L., Automatic vehicle location in cellular communications systems, *IEEE Transactions Vehicular Technology*, 43(4), 902–908, 1994.
24. Varshavsky, A., Are GSM Phones THE Solution for Localization? in Proceedings of the 7th IEEE Workshop on Mobile Computing Systems and Applications, 2006, WMCSA '06, 2006, pp. 20–28.
25. Schmitt, T., Hanek, R., Beetz, M., Buck, S., and Radig, B., Cooperative probabilistic state estimation for vision-based autonomous mobile robots, *IEEE Transactions on Robotics and Automation*, 18(5), 670–684, 2002.
26. Nakamura, E.F., Loureiro, A.A.F., and Frery, A.C., Information fusion for wireless sensor networks: Methods, models, and classifications, *ACM Computing Surveys*, 39(3), 9/1–9/55, 2007.
27. Chausse, F., Laneurit, J., and Chapuis, R., Vehicle Localization on a Digital Map Using Particles Filtering, in Proceedings of Intelligent Vehicles Symposium, 2005, June 2005, pp. 243–248.
28. Chapuis, R., Laneurit, J., Aufrere, R., Chausse, F., and Chateau, T., Accurate Vision Based Road Tracker, in Proceedings Intelligent Vehicle Symposium, 2002, Vol. 2, June 2002, pp. 666–671.
29. Priyantha, N.B., Chakraborty, A., and Balakrishnan, H., The Cricket Location support System, in MOBICOM '00: Proceedings of the 6th ACM Conference on Mobile Computing and Networking, Boston, MA, August 2000, pp. 32–43.

30. Bahl, P. and Padmanabhan, V.N., RADAR: An In-Building RF-Based User Location and Tracking System, in INFOCOM 2000, Proceedings of the 9th Annual Joint Conference of the IEEE Computer and Communications Societies, Vol. 2, Tel Aviv, Israel, March 2000, pp. 775–784.
31. Lee, J.-Y. and Scholtz, R.A., Ranging in a dense multipath environment using a UWB radio link, *Selected Areas in Communications*, 20(9), 1677–1683, 2002.
32. Cheng, Y. C., Chawathe, Y., LaMarca, A., and Krumm, J., Accuracy Characterization for Metropolitan-Scale Wi-Fi Localization, in MobiSys '05: Proceedings of the 3rd International Conference on Mobile Systems, Applications, and Services, ACM Press, New York, NY, 2005, pp. 233–245.
33. Thangavelu, A., Bhuvaneswari, K., Kumar, K., SenthilKumar, K., and Sivanandam, S.N. Location Identification and Vehicle Tracking Using VANET (VETRAC), in Signal Processing, Communications, and Networking, 2007, ICSCN '07, 2007, pp. 112–116.
34. Boukerche, A., Pazzi, R.W.N., and Araujo, R.B., A Fast and Reliable Protocol for Wireless Sensor Networks in Critical Conditions Monitoring Applications, in MSWiM '04: Proceedings of the 7th ACM International Symposium on Modeling, Analysis and Simulation of Wireless and Mobile Systems, ACM Press, New York, NY, 2004, pp. 157–164.
35. Boukerche, A., Silva, F.H.S., Araujo, R.B., and Pazzi, R.W.N., A Low Latency and Energy Aware Event Ordering Algorithm for Wireless Actor and Sensor Networks, in MSWiM '05: Proceedings of the 8th ACM International Symposium on Modeling, Analysis and Simulation of Wireless and Mobile Systems, ACM Press, New York, NY, 2005, pp. 111–117.
36. Benslimane, A., Localization in Vehicular Ad Hoc Networks, in Proceedings of Systems Communications, 2005, pp. 19–25.
37. Capkun, S., Hamdi, M., and Hubaux, J.P., GPS-free positioning in mobile ad hoc networks, *Cluster Computing*, 5(2), 157–167, 2002.
38. Savvides, A., Han, C.C., and Strivastava, M.B., Dynamic Fine-Grained Localization in Ad-Hoc Networks of Sensors, in MobiCom '01: Proceedings of the 7th ACM/IEEE International Conference on Mobile Computing and Networking, ACM, Rome, Italy, 2001, pp. 166–179.
39. Brooks, R.R., and Iyengar, S.S., *Multi-Sensor Fusion: Fundamentals and Applications with Software*, Prentice-Hall, Upper Saddle River, NJ, 1998.
40. Bass, T., Intrusion detection systems and multisensor data fusion, *Communications of the ACM*, 43(4), 99–105, 2000.
41. Siaterlis, C. and Maglaris, B., Towards Multisensor Data Fusion for DoS Detection, in Proceedings of the 2004 ACM Symposium on Applied Computing (SAC), Nicosia, Cyprus, 2004, pp. 439–446.
42. Intanagonwiwat, C., Govindan, R., and Estrin, D., Directed Diffusion: A Scalable and Robust Communication Paradigm for Sensor Networks, in MobiCom '00: 6th ACM International Conference on Mobile Computing and Networking, ACM Press, Boston, MA, August 2000, pp. 56–67.

43. Schubert, R., Schlingelhof, M., Heiko Cramer, H., and Wanielik, G., Accurate Positioning for Vehicular Safety Applications—The Safespot Approach, in Vehicular Technology Conference, 2007, VTC2007-Spring, IEEE 65th, Dublin, Ireland, April 2007, pp. 2506–2510.
44. Fernandez-Madrigal, J.A., Cruz-Martin, E., Gonzalez, J., Galindo, C., and Blanco, J.L., Application of UWB and GPS Technologies for Vehicle Localization in Combined Indoor–Outdoor environments, in ISSPA '07: International Symposium on Signal Processing and its Applications, Sharja (U.A.E.), Febuary 2007.
45. Fuentes Michel, J.C., Christmann, M., Fiegert, M., Gulden, P., and Vossiek, M., Multisensor Based Indoor Vehicle Localization System for Production and Logistic, in 2006 IEEE International Conference Multisensor Fusion and Integration for Intelligent Systems, Heidelberg, Germany, 2006, pp. 553–558.
46. Ammoun, S., Nashashibi, F., and Laurgeau, C., Crossroads risk assessment using GPS and inter-vehicle communications, *IET Intelligent Transport Systems*, 1(2), 95–101, 2007.
47. El Najjar, M.E. and Bonnifait, P., A road-matching method for precise vehicle localization using belief theory and kalman filtering, *Auton. Robots*, 19(2), 173–191, 2005.

V

Simulation

12

Vehicular Mobility Models

Marco Fiore
*Department of Electronics
Polytechnic Institute
of Torino*

12.1 Introduction

The most striking aspect of envisioning the realization of metropolitan-scale vehicular networks is the level of mobility that such systems would be required to support, both qualitatively and quantitatively. As a matter of fact, no other family of mobile networks, either existing or conceived, features hundreds, or even thousands of nodes that travel at speeds up to tens of kilometers per hour, that alternate high-velocity intervals with full stop periods, and whose freedom of movement is constrained by complex (car-to-car) interactions, precise (road) topologies, and detailed (driving) rules.

The resulting node's motion is absolutely unique and, from a telecommunication point of view, fascinating and challenging at the same time, as it affects the evolution of network connectivity over space and time in unprecedented ways. Considering that, in turn, connectivity dynamics play a major role in determining the performance of networking protocols, the reason why studies on vehicular networking cannot be separated from mobility analysis is easily understood.

At the current early stages in the design of large-scale intervehicular communication systems, such investigations on the relationships between mobility and performance of networking techniques are necessarily bound to simulated environments. Indeed, logistic difficulties, economic issues, and technology limitations render experimental studies too complex, time-consuming, and economically expensive, considering that most research projects in the field of vehicular networking deal with novel schemes of unproven reliability, requiring the cooperation of tens of cars and, possibly, a roadside telecommunication infrastructure.

Shifting vehicular networking studies to the simulation domain, however, introduces an intermediate element of complexity, represented by the need to reproduce the aforementioned peculiar car motion within network simulation environments. At first quite an overlooked factor, mobility modeling has gradually been imposed as a critical step for simulation-based evaluations of vehicular communication protocols to reach sufficient levels of reliability and consistency with real-world implementations.

The degree of complexity of car movement descriptions found in the vehicular networking literature in recent years has been constantly increasing. The tendency to employ simplistic representations based on stochastic behaviors of mobile entities, a common practice in pioneering works in the field, is today giving way to more complex models, often borrowed from the vast traffic flow theory literature on the topic of analytical representation of vehicular mobility. Lately, dedicated software tools, designed for interaction with network simulators, have been defined to generate movements mimicking those of real-world vehicles. They employ microscopic descriptions of car mobility that account for a number of factors affecting the behavior of drivers in everyday traffic, and whose properties have been validated through comparison against measured traces. This notwithstanding, it is not uncommon to still find studies relying on mobility models of unproven reliability, but largely diffused due to their use in early works on vehicular networking.

In this chapter, we will provide an insight into the issue of vehicular mobility modeling oriented to network simulation, presenting the rationale at the base of different classes of car motion analytical descriptions in Section 12.2, discussing their level of realism in Section 12.3, determining their suitability to vehicular network simulation in Section 12.4, and finally drawing some conclusions in Section 12.5.

12.2 Vehicular Mobility Modeling

Various approaches can be adopted in modeling the movement of vehicles, and they all undergo a common trade-off between complexity and precision. A common classification is then based on the level of detail of the motion representation, following an approach widely employed in other fields of research such as physics and economics, and distinguishing between macroscopic, mesoscopic, and microscopic levels of analysis. Accordingly, mobility models can be separated into the following categories:

1. *Macroscopic models.* Vehicular traffic is regarded as a continuous flow, and gross quantities of interest, such as the density or the mean velocity of cars, are modeled, often using formulations borrowed from fluid dynamics theory.
2. *Mesoscopic models.* Individual mobile entities are modeled at an aggregate level, exploiting gas-kinetic and queuing theory results or macroscopic-scale metrics, such as velocity/density relationships, to determine the motion of vehicles.

3. *Microscopic models.* Each vehicle's movement is represented in great detail, its dynamics being treated independently from those of other cars, except for those near enough to have a direct impact on the driver's behavior. Microscopic models are able to reproduce fine-grained real-world situations, such as front-to-rear car interaction, lane changing, flows merging at ramps, and intersections.

Although macroscopic and mesoscopic descriptions are employed to capture the dynamics of large-scale vehicular systems, such as those occurring over road topologies covering whole regions or countries, microscopic models, due to their high computational cost, are usually applied to reproduction of traffic in smaller areas, such as single highways or urban areas.

However, the traditional branching of models into macroscopic, mesoscopic, and microscopic becomes less meaningful when considering vehicular mobility models employed in network simulation. As a matter of fact, given the reduced spatial scale of short- and middle-range communication techniques envisioned for employment in vehicle-based networked systems, vehicular network simulations often require a high level of detail in terms of car motion representation. The necessity of precision of the order of meters in the definition of vehicles' absolute and relative positions bounds the mobility descriptions to be used for network simulation to the microscopic or, at most, mesoscopic domain. Thus, a different, better fitting classification could be constructed by differentiating on the nature of the diverse analytical representation of car motion encountered in the vehicular networking literature. We propose the following categorization:

1. *Stochastic models.* Vehicle movement is regarded at a microscopic level and is (1) constrained on a graph representing the road topology, and (2) random, in the sense that mobile entities follow casual paths over the graph, traveling at randomly chosen speed. Stochastic models are the most trivial way to mimic car mobility, and were introduced by pioneering works in the field of vehicular networking.
2. *Traffic stream models.* Vehicular mobility is observed from a high level and treated as a continuous phenomenon. Traffic stream models determine cars' speeds, leveraging fundamental hydrodynamic physics relationships between the velocity, density, and outflow of a fluid, and thus fall into the macroscopic or mesoscopic categories defined before.
3. *Car-following models.* The behavior of each driver is computed on the basis of the state (position, speed, and acceleration) of the surrounding vehicles. Car-following models date back to the 1950s, and represent the most common way to analytically describe microscopic-level mobility in vehicular traffic flow theory.
4. *Flows-interaction models.* Built upon car-interaction representations belonging to the stochastic and car-following classes above, and thus falling into the microscopic category as well, flows interaction characterizes the dynamics of vehicular flows merging, for example, at highway ramps or urban intersections.

In the remainder of this chapter, we will use this second classification to separate vehicular mobility models. Next, we will discuss the categories in detail, presenting and explaining representative instances for each class.

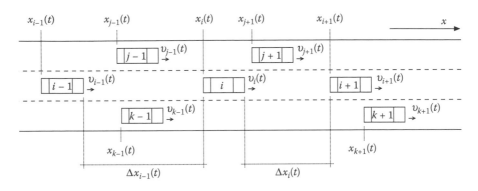

FIGURE 12.1 Notation for models' formal definition.

In particular, the formal definitions of models in the following employ the notation depicted in Figure 12.1. There, the index i refers to the vehicle under investigation, while $i \pm 1$ identifies the front (+) and back (−) vehicles on the current lane. Furthermore, for a generic vehicle i at time t, $x_i(t)$, and $v_i(t)$ represent its position and speed, meaning that its instantaneous acceleration can be expressed as $dv_i(t)/dt$. The front bumper to back bumper distance between i and $i + 1$ is identified as $\Delta x_i(t)$, while the relative speed $v_{i+1}(t) - v_i(t)$ is denoted by $\Delta v_i(t)$. Note that, according to its definition, in the following a positive $\Delta v_i(t)$ will always mean that the distance of car i from its leading vehicle $i + 1$ is growing. The back and front cars on the *left* lane with respect to the one vehicle i is traveling on are denoted by $j - 1$ and $j + 1$, respectively. The back and front cars on the *right* lane with respect to the one vehicle i is traveling on are denoted by $k - 1$ and $k + 1$, respectively. Also, we indicate common model input parameters as summarized in Table 12.1.

TABLE 12.1 Parameter Notation

Parameter	Symbol
Acceleration	a
Deceleration	b
Maximum allowed/desired speed	v_{max}
Minimum allowed/desired speed	v_{min}
Bumper-to-bumper safety distance	Δx_{safe}
Safety time headway	Δt_{safe}
Driver's reaction time	τ
Time step (discrete-time models)	Δt
Space step (discrete-space models)	Δx

12.2.1 Stochastic Models

We classify as stochastic models all those mobility descriptions that constrain random movements of vehicles on a graph. Given the vehicular environment, the graph is intended to represent a road topology, while the movement is random in a sense that vehicles, individually or with group dynamics, follow casual paths over the graph, usually traveling at randomly chosen speed. Because stochastic models regard each vehicle as an independent entity, they fall into the category of microscopic descriptions.

Stochastic models are a product of vehicular networking research, as they were first employed by works introducing the basic concepts of a vehicle-based communication infrastructure. They represent the most intuitive way to describe a vehicular mobility when no particular requirements in terms of realism are to be met.

The habit to avoid realism constraints, together with ease of implementation and low computational costs, is the main cause of the success that stochastic models encountered in the early stages of vehicular networking analysis. In such studies, the performance of communication techniques under stochastic vehicular mobility models were often compared against those obtained under fully random mobility models, that is, models that do not constrain the random cars movement over a graph, such as the *random walk*[1] or the *random waypoint*.[2] These tests showed that mobility characterization has a major impact on the performance of networking schemes, but they could hardly validate the realism of the stochastic models they employed. This notwithstanding, stochastic models' diffusion has been so capillary in seminal works that the most successful among them are still regularly present in today's networking literature.

Obviously, analytical representations belonging to this category either completely ignore or tackle in a simplistic way basic aspects of vehicular mobility such as car-to-car interaction and intersection modeling, with the result that they cannot reproduce even basic phenomena encountered in vehicular traffic.

One of the first examples of stochastic vehicular mobility descriptions is the City Section model, introduced by Davies.[3] It constrains cars movement on a grid-shaped road topology, in which all edges are considered to be bidirectional, single-lane roads. Vehicles randomly select one of the intersections of the grid as their destination and move towards it at constant speed, with (at most) one horizontal and one vertical movement, as depicted in Figure 12.2. The speed depends on the road the vehicle is traveling on. Two road classes, high-speed and low-speed, are allowed, and each vehicle sets its speed to a high or low value accordingly.

The *Constant Speed Motion model*[4,5] is another typical example of a stochastic model, as it describes a random vehicular movement on a graph, representing the road topology. No particular constraint is forced on the graph nature, so that it can embody different levels of realism. Examples of graphs that can be employed with the Constant Speed Motion model are shown in Figure 12.3.

A car's motion is structured in *trips*, that is, movements between vertices of the graph, referred to as *destinations* and randomly selected. At the beginning of each trip, a vehicle *i* chooses its next destination, computes the route to it by running a shortest path algorithm

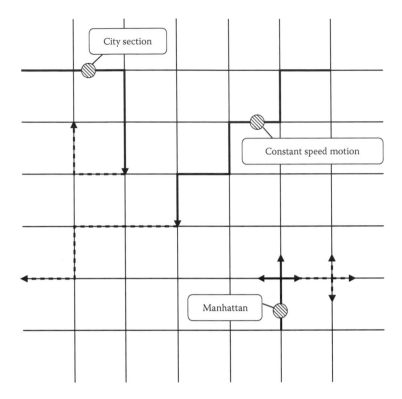

FIGURE 12.2 Different mobility descriptions on a grid topology. For each model, a possible two-phase movement is illustrated, the first phase being identified by the continuous line, the second phase corresponding to the dashed line. City Section selects a destination and then forces a two-step (horizontal plus vertical) motion to reach it. Constant Speed Motion chooses a destination and computes a shortest path to it. Manhattan determines the route of the vehicle via probabilistic choices at each intersection.

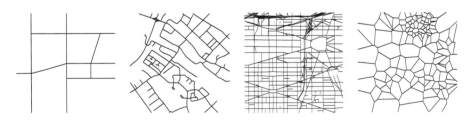

FIGURE 12.3 Examples of graphs mimicking road topologies to be used with the Constant Speed Motion model. From left to right, a user-defined graph, a graph extracted from a Geographic Data File (GDF)[74] map, a graph extracted from a U.S. Census Bureau TIGER database map,[7] a graph obtained from a clustered Voronoi tessellation. [From Fiore, M. et al. Vehicular Mobility Simulation for VANETs, in SCS/IEEE Annual Simulation Symposium (ANSS), Norfolk, VA, March 2007. With permission.]

on the graph, with link costs possibly biased by parameters such as the road length, speed limits, traffic congestion, and so on, as in Figure 12.2. It then sets its speed to

$$v_i = v_{min} + \eta(v_{max} - v_{min}) \tag{12.1}$$

where η is a uniformly distributed random variable in $[0, 1]$. Such a speed value v_i is selected once at the beginning of each *trip* and kept constant until the destination is reached.

The *Saha* model[6] represents vehicular traffic as a random mobility of cars over real road topologies extracted from the maps of the U.S. Census Bureau TIGER database.[7] An example of a road graph obtained in this way is depicted in Figure 12.3. According to Saha's proposal, vehicles select one point over the graph as their destination and compute the shortest path to get there. The sequence of edges is obtained by weighting the cost of traveling on each road on its speed limit (which is recorded by the TIGER format) and on the number of vehicles already moving on it, in a way to reproduce the real-world tendency of drivers to avoid congested paths. The speed of a mobile entity is set to a constant value in the range $[v_{max} - \varepsilon, v_{max} + \varepsilon]$, where v_{max} is the speed limit of the road on which the car is moving. All roads are considered bidirectional and single lane, and no car-to-car interaction is modeled. This model can thus be regarded as a particular case of the Constant Speed Motion model presented above, in which the road topology graph, the cost of traversing vertices in the route computation, and the speed range are constrained.

If the Constant Speed Motion and Saha models focus on improving the level of realism of the road topology, Bai et al.[8] introduce two stochastic vehicular mobility models that enhance the quality of individual cars' motion representation. The *Freeway* model is designed for road topology graphs representing noncommunicating, bidirectional, multilane freeways traversing the entire simulated area. The movement of each vehicle is restricted to the lane it is moving on, and the following speed management rules apply to vehicle i:

1. *Speed update.* The speed is varied by a random acceleration of maximum magnitude a. If we define as η a random variable uniformly distributed in $[-1, 1]$, then this rule can be expressed as

$$v_i(t + \Delta t) = v_i(t) + \eta a \Delta t \tag{12.2}$$

2. *Speed bounding.* At any time, the speed of a vehicle cannot be lower than a minimum value v_{min} and cannot exceed a maximum value v_{max}. This constraint is enforced as

$$v_i(t + \Delta t) = \min\{\max[v_i(t + \Delta t), v_{min}], v_{max}\} \tag{12.3}$$

3. *Speed reduction.* In order to avoid overlapping, that is, a collision situation, with the front vehicle, a minimum safety distance must be maintained. Formally

$$v_i(t + \Delta t) = \begin{cases} v_{i+1}(t) - a/2 & \text{if } \Delta x_i(t) \leq \Delta x_{safe} \\ v_i(t + \Delta t) & \text{otherwise.} \end{cases} \tag{12.4}$$

Each vehicle starts its movement at one end of a lane, with a speed that is at first selected as uniformly distributed in an interval $[v_{min}, v_{max}]$, and ends it once it reaches the other extremity of the same lane. Then a new movement, on a randomly selected lane, is started over. The second description in Ref. 8 is the *Manhattan* mobility model. It employs the same speed management rules as the Freeway model, extending it to an urban scenario. As a matter of fact, similar to the City Section model seen before, the Manhattan model uses a grid road topology, but it adopts a probabilistic approach in the selection of car's movements, originally introduced in Ref. 9. At each intersection, a vehicle chooses to keep moving in the same direction with probability 1/2 and to turn left or right with probability 1/4 in each case. This approach thus abandons the concept of trip, in favor of an intersection-by-intersection decision on the route of a vehicle, as detailed in Figure 12.2.

Also falling in the category of stochastic models, Zhou et al.[10] propose the *Real Track* mobility model, derived from the *Virtual Track* model.[11] The Virtual Track model binds a car's movement over a graph, whose vertices are referred to as switch stations, and whose edges are defined virtual tracks. The edges not only have a length equal to the distance between the switch stations they connect, but also a predefined customizable width. Thus, they can be graphically represented as rectangles rather than lines. Nodes move in groups, according to the *Reference Point Group Mobility* (RPGM) model,[12] which defines a common direction for the group, and then adds some bounded randomness to the movement of the single vehicle within the group with respect to the common direction. Groups of cars are allowed to move following a *Random Waypoint* model[2] from one switch station to another switch station only within the virtual tracks, and only in the direction of the next switch station. To enforce the last rule, the Random Waypoint of each group's common direction is biased so that the next destination must be nearer than the current position to the target switch station. At switch stations, cars may leave their current group and join other groups. The Real Track model applies the Virtual Track model to real road topologies, extracted, as was the case for Saha's model, from the U.S. Census Bureau TIGER database. Intersections are mapped into switch stations and roads into virtual tracks, to which a fixed width is assigned. Figure 12.4 shows a simple example of Real Track movement. The idea at the base of the Real Track model is to reproduce the clustering of vehicles occurring at intersections and propagating over the roads.

12.2.2 Traffic Stream Models

Traffic stream models look at vehicular mobility as a hydrodynamic phenomenon and try to relate the three fundamental variables of velocity $v(x, t)$ (measurable in km/h), density $\rho(x, t)$ (measurable in vehicles/km), and flow $q(x, t)$ (measurable in vehicles/h). All of these are functions of space x and time t, averaged over sufficiently large regions. Because traffic stream models consider vehicular traffic as a flow, they fall into the category of macroscopic descriptions.

The basic equation for traffic stream models comes from the idea that, given a road section, the number of vehicles on the section can only vary due to cars entering or leaving the section. This leads to the following continuity equation

$$\frac{\partial \rho}{\partial t} = -\frac{\partial q}{\partial x} = -\frac{\partial (\rho v)}{\partial x} \tag{12.5}$$

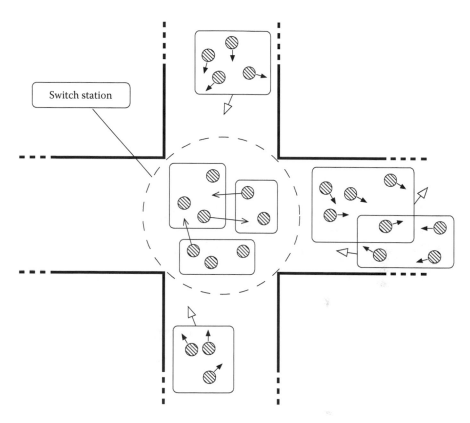

FIGURE 12.4 Example of Real Track motion at an intersection (switch station). Groups of cars moving according to the Reference Point Group Mobility model travel towards or away from the intersection, where vehicles merge into new groups.

that is, the density ρ measured over a small road section dx varies, during a small interval dt, according to the corresponding flow q. The outflow q can then be expressed as the product of density ρ and speed v. This formulation leads to the fundamental diagram of traffic flow depicted in Figure 12.5. The simplest model of this kind was proposed by Lighthill and Whitham,[13] assuming the velocity to be a function of the density

$$\frac{\partial \rho}{\partial t} = -\frac{d}{d\rho}(\rho v(\rho))\frac{\partial \rho}{\partial x} \qquad (12.6)$$

which is capable of modeling kinematic waves. This model has been widely used over the last few decades, but much more complex formulations, based on similar assumptions, can be found in traffic flow literature.

Given their macroscopic nature, traffic stream models can handle large quantities of vehicles, at the cost of precision. This makes them interesting for high-level analytical studies of traffic behavior. An example of such a macroscopic approach in the vehicular networking literature is found in Rudack et al.,[14] where vehicle-to-road

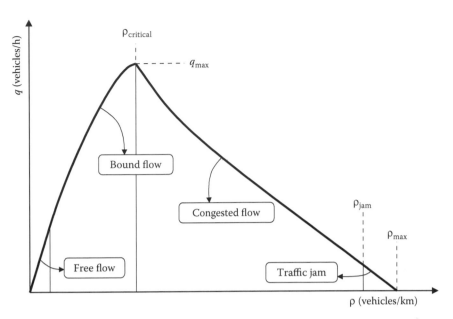

FIGURE 12.5 Qualitative representation of traffic flow fundamental diagram. Cars' outflow q is related to vehicular density ρ by a lambda-shaped curve. When the density is low, vehicles proceed at full speed, because traffic conditions are comparable to those in the absence of other cars, that is, in *free-flow* conditions. As the vehicular density increases, the presence of neighboring cars grows and cars are forced to progressively reduce their speed. However, the outflow still grows, as an effect of the increasing number of vehicles over the road. When a critical density $\rho_{critical}$ is reached, the impact of speed reduction becomes predominant, and cars start flowing out of the road at a reduced rate. This is referred to as congested flow. Finally, if the density grows over a congestion threshold ρ_{jam}, traffic jams are encountered.

infrastructure communication on a highway scenario is studied. In Ref. 14, the authors exploit common assumptions from traffic flow literature, considering the distribution of vehicular speed as normal[15] and the distribution of vehicles' interarrival times at the beginning of the considered road section as exponential.[16] As the goal of the paper is to determine the duration of connections between traveling vehicles and fixed gateways along the highway, no higher detail than the shape of these distributions is needed.

However, traditional traffic stream models cannot reproduce the independent motion of each vehicle, a fundamental aspect to account for in vehicular networking research, where it dramatically affects key communication factors such as network connectivity and link duration. As already stated, the consequence is that macroscopic-level models are rarely found in works on vehicular networking.

The *Fluid Traffic Motion* (FTM) model[17] represents an interesting exception to the above statement. As a matter of fact, it applies a traffic stream approach to individual mobile entities, exploiting macroscopic metrics on a microscopic scale, and thus gene-rating a meso-scopic description. It computes the speed of each car as a monotonically decreasing function of vehicular density, forcing a lower bound on the velocity when the traffic congestion reaches a critical state:

$$v_i(t + \Delta t) = \max\left[v_{min}, v_{max}\left(1 - \frac{\rho(x, t)}{\rho_{jam}}\right)\right] \qquad (12.7)$$

Equation 12.7 describes the model, where $\rho(x, t)$ is the current vehicular density on the road car i is traveling on, while ρ_{jam} is the vehicular density for which a traffic jam is detected. The Fluid Traffic Motion model computes $\rho(x, t)$ as *n/l*, where *n* is the number of cars on the same road of i and *l* is the length of the road segment itself. According to this formulation, cars traveling on very crowded streets are forced to slow down, possibly to the minimum speed, while the speed of cars is increased towards the maximum value when less congested roads are encountered. Figure 12.6 depicts the speed vs. vehicular density curves, for different settings of the model's parameter. The model divides a single road into multiple segments, in each of which the vehicular density is computed independently, in order to reproduce more detailed car traffic phenomena.

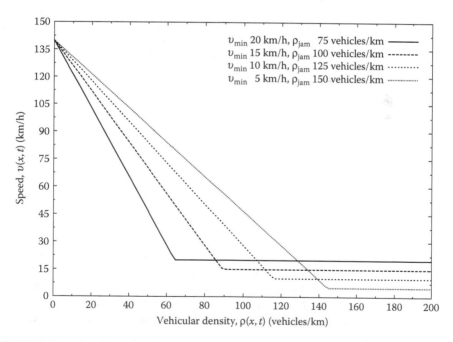

FIGURE 12.6 Speed vs. vehicular density curves obtained with the Fluid Traffic Motion model. Each curve represents a different setting of the parameters couple (v_{min}, ρ_{jam}), while v_{max} is forced to 140 km/h. Lower values of ρ_{jam} lead to steeper speed decrease and faster approach to traffic jam conditions. Lower values of v_{min} cause slower traffic under congestion conditions, which, however, become less likely to occur.

12.2.3 Car-Following Models

Car-following models describe the behavior of each driver in relation to its neighboring vehicles. As they regard each car as an independent entity, they fall into the category of microscopic-level descriptions. The first car-following models date back to the late 1950s and, since then, they have been one of the most popular methods to analytically represent vehicular traffic dynamics.

In this section, we will present several car-following descriptions, either well known in the traffic flow theory literature or employed in vehicular networking simulation. For a complete survey and comparison of car-following models, alongwith discussion of their implementation in traffic simulators, we refer the reader to Refs. 18 to 22.

12.2.3.1 Follow-the-Leader Models

Most car-following models determine the motion of a vehicle as a function of the state of a single neighboring car, typically the one in front. For this reason, they are also referred to as *follow-the-leader* models.

In such descriptions, the speed or acceleration depend on factors such as the distance from the front car and the absolute and relative speed or acceleration of both vehicles. With reference to the notation introduced in Figure 12.1, a general expression for a car-following model, in the form of a delayed differential equation formulation for a vehicle i, is

$$\frac{dv_i(t)}{dt} = -f[v_i(t),\, v_{i+1}(t),\, \Delta x_i(t)] \tag{12.8}$$

One of the best known expressions for the function f above is that proposed in the *GHR* model,[23] named after its authors, Gazis, Herman, and Rothery, and based on seminal work conducted during the 1950s.[24-27] With respect to previous car-following prototypes, the GHR model led to fundamental advances on the calibration of vehicular models. In particular, it was the first analytical description to introduce different calibrations for noncongested and congested traffic conditions. The GHR model is defined as

$$\frac{dv_i(t)}{dt} = k_1 v_i^{k_2}(t - \tau) \frac{\Delta v_i(t - \tau)}{\Delta x_i^{k_3}(t - \tau)} \tag{12.9}$$

where k_1, k_2, and k_3 are constants that are to be calibrated to adapt the model to specific drivers' behaviors or traffic scenarios. Their values vary the maximum instantaneous acceleration (k_1), the weight of vehicle i's absolute speed (k_2), and that of the bumper-to-bumper distance from the vehicle ahead (k_3). Note that the GHR formulation introduces a delay τ in the computation of the instantaneous acceleration, in order to account for the finite reaction time of drivers.

It should be said that, although very successful over at least three decades, the GHR model represents quite an inefficient car-following formulation. In fact, it has been progressively abandoned in favor of more recent car-following models that are easier to calibrate and account for many other factors, characterizing both cars' technical constraints and drivers' attitudes, increasing the level of realism of the motion description.

A close relative to the GHR model, the *linear* model by Helly[28] describes the acceleration of a vehicle as a linear function f of the same factors employed by the GHR formulation. It also introduces the concept of *desired* following distance $\Delta x_{des}(t)$, which represents the comfortable distance a driver would like to maintain with respect to the leading vehicle, considering its current speed and acceleration. The linear model computes the speed derivative for car i as from the following equations:

$$\frac{dv_i(t)}{dt} = k_1 \Delta v_i(t - \tau) + k_2[\Delta x_i(t - \tau) - \Delta x_{des}(t)] \tag{12.10}$$

$$\Delta x_{des}(t) = k_3 + k_4 v_i(t - \tau) + k_5 \frac{dv_i(t - \tau)}{dt} \tag{12.11}$$

where k_1, k_2, k_3, k_4, and k_5 are again constants that must be calibrated according to the traffic scenario. The linear model has several advantages over GHR, including a simpler formulation, a lower computational cost, a clearer physical meaning of the parameters' impact and, quite surprisingly, a higher degree of agreement with real-world traffic data. However, it still retains GHR's calibration problems.

The *Intelligent Driver Model* (IDM) by Treiber et al.[29] represents an evolution of the GHR/linear concept, and it is one of the most common car-following descriptions used in vehicular networking research.[5,30,31] This model characterizes drivers' behavior through the instantaneous acceleration of vehicles, calculated according to the following equations:

$$\frac{dv_i(t)}{dt} = a \left[1 - \left(\frac{v_i(t)}{v_{max}} \right)^4 - \left(\frac{\Delta x_{des}(t)}{\Delta x_i(t)} \right)^2 \right] \tag{12.12}$$

$$\Delta x_{des}(t) = \Delta x_{safe} + \left[v_i(t) \Delta t_{safe} - \frac{v_i(t) \Delta v_i(t)}{2\sqrt{ab}} \right] \tag{12.13}$$

In Equation 12.12, $\Delta x_{des}(t)$ is the so-called *desired dynamical* distance, whose significance is not dissimilar from the desired distance of the linear model. $\Delta x_{des}(t)$ is computed, in Equation 12.13, as a function of the minimum bumper-to-bumper distance Δx_{safe}, the minimum safe time headway Δt_{safe}, the speed difference with respect to the front vehicle $\Delta v_i(t)$, and the maximum acceleration and deceleration a and b. When combined, these formulae give the instantaneous acceleration of the car, divided into a desired acceleration $[1 - (v_i(t)/v_{max})^4]$ on a free road and a braking deceleration induced by the preceding vehicle $[\Delta x_{des}(t)/\Delta x_i(t)]^2$.

The *Krauss* model[32,33] is another variation of the GHR description that has often been used for research on vehicular networks.[34-36] Different from the GHR, linear, and IDM descriptions, Krauss proposes a discrete-time representation, modeling the vehicle's speed at each time step rather than its instantaneous acceleration. The Krauss model determines the speed of a vehicle i through the following formulation:

$$v_i^{safe}(t + \Delta t) = v_{i+1}(t) + \frac{\Delta x_i(t) - \tau v_{i+1}(t)}{[v_i(t) + v_{i+1}(t)]/2b + \tau} \tag{12.14}$$

$$v_i^{\text{des}}(t + \Delta t) = \min\{v_{\max}, v_i(t) + a\Delta t, v_i^{\text{safe}}(t + \Delta t)\} \tag{12.15}$$

$$v_i(t + \Delta t) = \max\{0, v_i^{\text{des}}(t + \Delta t) - k_1 a\Delta t \eta\} \tag{12.16}$$

Equation 12.14 computes the speed v_i^{safe} the vehicle is required not to exceed in order to maintain a safety distance from its leading vehicle. Equation 12.15 determines the desired new speed v_i^{des} of vehicle i, equal to the current speed plus the increment determined by a maximum uniform acceleration $a\Delta t$, with upper bounds represented by the maximum allowed speed and by the maximum safe speed v_i^{safe} computed above. Equation 12.16 determines the final value of the speed, introducing stochastic behavior into the model by means of a noise η, a random variable uniformly distributed in $[0, 1]$. Such randomness cannot exceed the measure of a maximum percentage k_1 of the highest achievable speed increment $a\Delta t$.

Vehicular motion formulations based on GHR represent the most common and straightforward way of extending car-following modeling. However, other descriptions, diverse from GHR, but still fitting the category of car-following models, are possible.

Collision avoidance car-following descriptions, also referred to as *safety distance* models, were first introduced in the late 1950s.[37] The rationale behind collision avoidance models is that a safety distance must be maintained between two vehicles to avoid contact. Thus, instead of modeling the acceleration, as in the GHR model, the original collision avoidance description determines the intervehicle distance as a function of the speed of cars as

$$\Delta x_i(t - \tau) = k_1 v_{i+1}^2(t - \tau) + k_2 v_i^2(t) + k_3 v_i(t) + k_4 \tag{12.17}$$

where k_1, k_2, k_3, and k_4 are constants to be calibrated according to the traffic scenario. $\Delta x_i(t - \tau)$ represents the minimum safety distance that car i has to keep at time $t - \tau$ in order to avoid a collision with its front vehicle $i + 1$ at time t. Because Equation 12.17 assumes that the front car will brake at the hardest rate b, this formulation guarantees that contact among vehicles is avoided at any time.

By trivial passages, Equation 12.17 can then be rewritten according to a generic car-following expression:

$$v_i(t) = -\frac{k_3}{2k_2} + \frac{1}{2k_2}\sqrt{k_3^2 - 4k_2[k_1 v_{i+1}^2(t - \tau) + k_4 - \Delta x_i(t - \tau)]} \tag{12.18}$$

or, more generally,

$$v_i(t) = F(v_i(t - \tau), v_{i+1}(t - \tau), \Delta_{x_i}(t - \tau)) \tag{12.19}$$

Here, vehicle i's speed at time t becomes a function of the speeds and intercar distance measured a driver's reaction time before t.

The well-known *Gipps* model[38] belongs to the category of collision avoidance car-following motion descriptions. In fact, Gipps determines the minimum safety distance at each time instant to be

$$\Delta x_i(t - \tau) = \frac{\tau}{2}\left[v_i(t - \tau) + v_i(t)\right] + \frac{v_i^2(t)}{2b} + \Delta t_{safe}v_i(t) - \frac{v_{i+1}^2(t - \tau)}{2b} \qquad (12.20)$$

In Equation 12.20, the minimum distance to avoid collision is obtained by applying a constant acceleration to vehicle *i*'s motion during $[t - \tau, t]$. $\Delta x_i(t - \tau)$ is expressed as a function of the same speeds already seen in Equation 12.17, with the introduction of a safety margin term, obtained via a headway delay Δt_{safe}. The meaning of $\Delta t_{safe}v_i(t)$ is thus similar to that of the safety distance Δx_{safe} introduced by other models, such as the Intelligent Driver Model.

As for the original collision avoidance model and also in the case of the Gipps formulation the velocity of vehicle *i* can be computed as the positive solution of the quadratic equation in $v_i(t)$, that is

$$v_i(t) = -b\left(\frac{\tau}{2} + \Delta t_{safe}\right) + \sqrt{b^2\left(\frac{\tau}{2} + \Delta t_{safe}\right)^2 + b\left[2\Delta x_i(t - \tau) + \frac{v_{i+1}^2(t - \tau)}{b} - \tau v_i(t - \tau)\right]}$$

$$(12.21)$$

from which we can note that the speed of vehicle *i* at time *t* is given by two contributions: an increasing function of the headway distance from the leading car $i + 1$ of its speed, $[2\Delta x_i(t - \tau) + (v_{i+1}^2(t - \tau))/b]$, and a decreasing function of its own speed, $-\tau v_i(t - \tau)$, all such terms being observed at time $t - \tau$.

Another approach to car-following modeling is based on behavioral thresholds referred to as *action points*, first proposed by Michaels.[39] The idea is that it is possible to identify space–time thresholds triggering different acceleration profile characterizations in a vehicle's driver. The evolution of such a concept led to the definition of the so-called *psycho-physical* models.

In detail, psycho-physical models apply to a bidimensional space, with axes representing the distance Δx_i and speed difference Δv_i of a vehicle *i* with respect to the car in front. Such space is divided into several areas, demarcated by the aforementioned action points, corresponding to different driver reactions.

An example of division of such bidimensional space into behavioral areas, with relative vehicle reactions, is that proposed by Wiedemann.[40] It detects the following four driving modes:

1. *No reaction.* No influence is exerted from the ahead vehicle, which is too distant or traveling at higher speed than the considered vehicle. The car under study is thus free to reach and keep the desired speed.
2. *Reaction.* The vehicle is approaching the car ahead, and thus has to reduce its speed to keep at a safe distance.

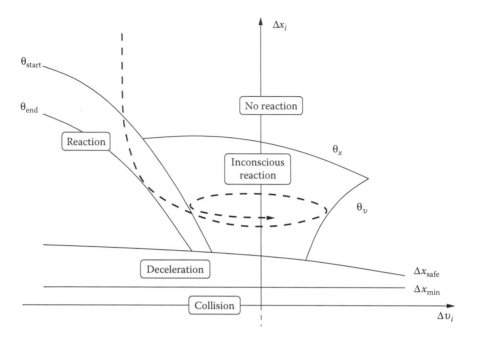

FIGURE 12.7 Example of Wiedermann's psycho-physical vehicular interaction. The Δx_i axis divides negative (vehicle approaching front car) and positive (front vehicle increasing its distance) speed differences Δv_i. A possible trajectory is shown by the arrow: a vehicle, initially in a free-flow condition (i.e., in the no reaction zone), starts approaching a vehicle in front ($\Delta v_i < 0$). As long as the distance is high enough, the driver ignores the obstacle ahead, keeping his/her pace. However, when the distance becomes too small with respect to the current speed (i.e., the reaction zone is entered) the driver starts slowing down. The braking process brings the driver to match the front car speed ($\Delta v_i = 0$). Any increase in the speed of the vehicle ahead ($\Delta v_i > 0$) is then matched by the back vehicle with some delay, due to the finite reaction time of the driver. For the same reason, any reduction in the front car's speed is matched by the back vehicle with delay as well (the last segment of the movement curve, in the $\Delta v_i < 0$ unconscious reaction area).

3. *Unconscious reaction.* The vehicle is a short distance from the car in front, the speed difference is small, and the two vehicles form a cluster.
4. *Deceleration.* The distance is too small if compared to the current speed difference, and the vehicle under consideration is forced to brake.

Figure 12.7 shows the space–time fragmentation proposed by Wiedemann.

As said before, the behavioral areas are separated by thresholds, which trigger consistent acceleration changes. In the example of Figure 12.7, the following thresholds can be identified:

1. Δx_{min} is the minimum distance between two still cars in a traffic jam. A Δx_i lower than this value determines an accident.

2. Δx_{safe} is the desired safety distance, and it is a function of Δx_{min} and of the speed difference, as more negative Δv_is, that is, higher approaching speeds, induce a larger safety spacing.
3. θ_{start} models the Δx_i at which the driver perceives that the distance from the leading vehicle is shortening. This is also referred to as *perception threshold*, and it is a function of the speed difference. Its calibration, that is, the shape of the θ_{start} curve, represents the core of car-following psycho-physical studies.
4. θ_{end} represents the distance at which the driver ends the approaching procedure through a comfortable deceleration. Points of the bidimensional space below these action points require (hard) braking by the driver.
5. θ_v is the speed difference threshold above which the driver is able to accelerate as in free-flow conditions. Thus, it also identifies the action points at which the driver notices that the front car has accelerated enough for him/her to start increasing speed as well, towards a $\Delta v_i = 0$ condition.
6. θ_x is the equivalent of θ_v in space, that is, the distance threshold beyond which the driver is free to accelerate as in free-flow conditions.

As could be expected, the main challenge of a psycho-physical representation of vehicular motion lies in the characterization of the different thresholds. Extensive investigations in that direction have been conducted in the past, especially during the 1970s,[41] and continue still today.[42-44]

Psycho-physical models have been recently employed in the implementation of tools for the study of vehicular networks.[45,46]

Another branch of car-following descriptions is that of *desired-spacing* models. These models are built upon the assumption that the objective of a driver is to maintain a determined distance from the vehicle in front. According to the original formulation by Hidas,[47] if we refer to such a desired distance as $\Delta x_{des}(t)$, this condition is modeled as

$$\Delta x_i(t) = \eta \Delta x_{des}(t) \tag{12.22}$$

where η is a random variable calibrated according to measurements on the driving scenario that has to be simulated, accounting for judgment errors of the driver. The objective distance, at least in the urban traffic conditions that the model addresses, can then be expressed as a linear function of the current speed

$$\Delta x_{des}(t) = k_1 v_i(t) + k_2 \tag{12.23}$$

where k_1 and k_2 are calibration constants. From Equation 12.22 and Equation 12.23, it is possible to express the acceleration of vehicle i as

$$\frac{dv_i(t)}{dt} = \frac{\tau}{\eta k_1 \tau + 1/2\tau^2} \Delta v_i(t-\tau) + \frac{1}{\eta k_1 \tau + 1/2\tau^2}(\Delta x_i(t-\tau) - \eta k_1 v_i(t-\tau) - \eta k_2)$$

$$+ \frac{1/2\tau^2}{\eta k_1 \tau + 1/2\tau^2} \frac{dv_{i+1}(t-\tau)}{dt} \tag{12.24}$$

Desired-spacing models have recently been employed in the field of vehicular networking, for the simulation of large-scale urban communication systems.[48]

12.2.3.2 Cellular Automata

During the 1990s, a novel approach to the analytical representation of traffic flows, based on *cellular automata*, was introduced.[49,50]

It should be said that considering cellular automata models as part of the car-following category would appear improper in traditional traffic flow theory literature. In fact, cellular automata are often intended as a class on their own, their mathematical approach to the problem being quite different with respect to that of the other car-following models we introduced above. However, the purpose of this chapter is to provide a high-level overview of diverse vehicular mobility models for network simulation purposes. Within such a context, assimilating cellular automata models to car-following descriptions is acceptable, as the basic principles behind the velocity update of cellular automata-based mobile entities are similar to those observed for standard car-following models.

Cellular automata models discretize not only time, but also space, which is fragmented into *cells*, each of which can host a single mobile entity at a time. Moreover, the possible *states* of each vehicle, that is, their instantaneous speed, must be finite, leading to a discretization of velocity as well. The movement of cars is thus described as a shift of finite states along a one-dimensional lattice of subsequent cells, as in Figure 12.8.

One of the first and most successful vehicular mobility representations based on cellular automata is the *Nagel–Schreckenberg model*.[51] It discretizes the time into slots of duration Δt and space into cells of length Δx. Because the space is discrete, we stress that the position of each car i, $x_i(t)$, becomes discrete as well, and coincides with the cell that i is lying in. In Ref. 51, Δt was set to 1 sec, to simplify the computational cost of the model, and Δx to 7.5 m, to allow each cell to host a single car. The speed of each vehicle, consistent with the finite-state constraint, is expressed as an integer in the range $[0, v_{max}]$.

The process of determining the movement of a generic vehicle i is divided into four phases:

1. *Acceleration.* The speed of the vehicle is incremented by one, unless the maximum speed has been reached. Formally

$$v_i(t + \Delta t) = \min\{v_i(t) + 1, v_{max}\} \tag{12.25}$$

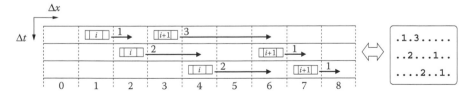

FIGURE 12.8 Cellular automata-based motion. The road is divided into cells, from 0 to 8, each accepting a single vehicle. Two cars move over time through subsequent cells, their discrete speed identified by the thick arrows. On the right, a simplified notation is employed to describe the same system.

2. *Deceleration.* To avoid a collision, the speed obtained from the first step must be upper bounded according to the distance from the front vehicle. Considering $x_{i+1}(t) - x_i(t) - 1$ as the distance, in cells, between i and its leading car, and $\Delta t = 1$, as in the original model formulation, then the deceleration step coincides with

$$v_i(t + \Delta t) = \min\{v_i(t + \Delta t), x_{i+1}(t) - x_i(t) - 1\} \tag{12.26}$$

3. *Randomization.* If the speed from the previous step is strictly positive, the model adds stochastic behavior to the process, randomly reducing the velocity by one. Denoting with η a binary random variable with probability distribution $P_\eta(1) = p$, $P_\eta(0) = 1 - p$, the randomization step turns into

$$v_i(t + \Delta t) = \max\{0, v_i(t + \Delta t) - \eta\} \tag{12.27}$$

4. *Movement.* The vehicle is moved forward according to the new speed value, that is,

$$x_i(t + \Delta t) = x_i(t) + v_i(t + \Delta t) \tag{12.28}$$

The first three phases can be collapsed to a single expression, which thus summarizes the speed evolution imposed by the model as

$$v_i(t + \Delta t) = \max\{0, \min\{v_i(t) + 1, v_{max}, x_{i+1}(t) - x_i(t) - 1\} - \eta\} \tag{12.29}$$

An example of vehicular motion described by the Nagel–Schreckenberg model is shown in Figure 12.9.

Cellular automata models trade some detail in the representation of vehicular motion with extremely low computational resource requirements. The enormous success these models have met over the last decade is mainly attributable to the fact that they allow for the simulation of, at a microscopic level, a large number of interacting vehicles even with limited processing power. For a complete survey of vehicular mobility modeling with cellular automata, the reader may refer to Ref. 52.

12.2.4 Flows-Interaction Models

The mobility descriptions presented in the previous sections are only representative of vehicular mobility over a single unidirectional lane. In other words, they regulate the inter-action of cars within the same *flow*, that is, cluster of cars moving along a common axis in a unidimensional motion. We refer to all such models as *car-interaction* representations.

The description of vehicular mobility provided by car-following representations is often insufficient to simulate real-world scenarios. As a matter of fact, when considering highway traffic it would be desirable that characterizing phenomena like overtakings between vehicles traveling on adjacent lanes or in-flow of cars at ramps be reproduced. Similarly, when simulating urban traffic the presence of intersections and roundabouts regulating the merging of flows from different roads cannot be neglected.

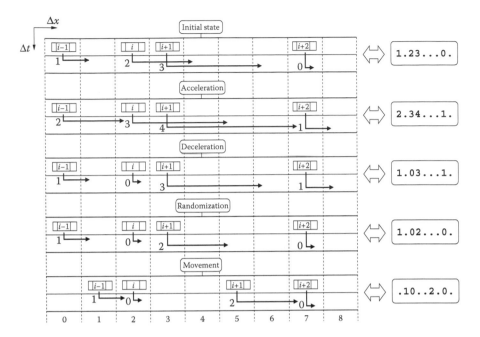

FIGURE 12.9 Cellular automata-based motion. Example of the four-step movement decision of the Nagel–Schreckenberg model. A possible choice of parameters fitting this example is $v_{max} = 4$, $p = 0.5$.

As a consequence, car-interaction descriptions, especially recent ones, are often coupled with lane-changing rules, which can be used to model overtakings and highway ramps. Also, extensions of the same models are possible, in order to cope with intersection management on metropolitan scenarios.

This section presents an overview of lane-changing models associated with some of the car-interaction representations described in the previous sections, and introduces the basic rationale behind the simulation of road junctions through car-interaction vehicular mobility descriptions.

12.2.4.1 Lane Changing

Studies of lane changes in microscopic models date back to the 1960s,[53] and have attracted growing interest from researchers in the field of vehicular traffic flow theory, some major results being those in Refs. 54 to 57.

It is commonly agreed that the modeling of lane changes represents a hard challenge, due to the high number of variables that come into play. In fact, the behavior of a driver in the presence of interacting vehicular flows cannot be described as a function of the state of the leading car anymore, but must also take into account the distance and speed of back and front vehicles on adjacent lanes.

The standard approach to the problem is to perform lane-changing decisions *half a time step* before updating vehicles' movement through car-following rules. In other words, the motion at each time step is divided into two subphases:

1. *Interlane movement.* Each vehicle decides if it has to move to a different lane. If so, the vehicle shifts to the new lane.
2. *Intralane movement.* Each vehicle changes its acceleration and speed according to the single-lane car-interaction model.

The lane-changing decision of the first phase above is usually driven by two criteria:

1. *A safety criterion.* The movement of a vehicle to a new lane must occur in a way that the safety of the vehicle itself as well as that of its new neighboring cars are not threatened. This normally translates to the requirement that a sufficiently large gap be present in the new lane for a vehicle to perform the lane change.
2. *An incentive criterion.* There must be a precise reason for a vehicle to perform the interlane movement. Generally, the motivation comes from an advantage in terms of acceleration or speed; that is, by changing lanes the vehicle is able to proceed faster. The incentive criterion often involves the use of hysteresis thresholds to avoid lane hopping in border conditions.

In the following, we will focus on lane-changing models designed to cooperate with some of the car-following descriptions previously introduced. We will employ a dedicated notation, with $|\cdot|_L$ and $|\cdot|_R$ and indicating values computed as if vehicle i had moved to the lane on its left and on its right, respectively.

The first lane-changing model we present is called *MOBIL,*[58] an acronym for *minimizing overall braking induced by lane-changes*, proposed by the same authors as the Intelligent Driver Model. It follows a game theoretical approach and allows a vehicle to move to an adjacent lane if its advantage, in terms of acceleration, is greater than the disadvantage of the back car in the new lane.

Under the MOBIL lane-change model, the safety criterion is expressed as

$$\left| \frac{dv_{j-1}(t + \Delta t)}{dt} \right|_L > -k_i b \tag{12.30}$$

and

$$\left| \frac{dv_{k-1}(t + \Delta t)}{dt} \right|_R > -k_i b \tag{12.31}$$

for movements of vehicle i to the lane to its left and right, respectively.

Referring to the notation depicted in Figure 12.1, Equation 12.30 states that the left lane change must not force the new back vehicle $j - 1$ to a deceleration higher (in modulus) than a maximum safe a deceleration. Such a deceleration is obtained as

k_1 times the standard deceleration b, with $k_1 \geq 1$. In other words, enough space must remain between the front of vehicle $j - 1$ and the back of vehicle i for $j - 1$ not to collide with i, after the latter has performed the interlane movement. A symmetric condition, in Equation 12.33, must be met for a movement to the right to occur.

The incentive criterion is based on a comparison of vehicle i's acceleration gain with the acceleration reduction suffered by the new back vehicle. In particular, the lane change is considered favorable if the first is greater than the latter, as in such case the overall acceleration of the system is increased.

$$\left.\left|\frac{dv_i(t + \Delta t)}{dt}\right|\right._{\text{L}} - \frac{dv_i(t + \Delta t)}{dt} + a_{\text{L}} \geq p\left(\frac{dv_{j-1}(t + \Delta t)}{dt} - \left|\frac{dv_{j-1}(t + \Delta t)}{dt}\right|_{\text{L}}\right) + k_2 a \qquad (12.32)$$

$$\left.\left|\frac{dv_i(t + \Delta t)}{dt}\right|\right._{\text{R}} - \frac{dv_i(t + \Delta t)}{dt} + a_{\text{R}} \geq p\left(\frac{dv_{k-1}(t + \Delta t)}{dt} - \left|\frac{dv_{k-1}(t + \Delta t)}{dt}\right|_{\text{R}}\right) + k_2 a \qquad (12.33)$$

Equation 12.32 expresses the incentive condition for a movement of vehicle i to its left lane. The left-hand term of the inequality represents the advantage, in terms of acceleration, of i after a potential move to the left, as the difference between the acceleration it would enjoy on the new lane and that experienced on the current lane. The right-hand term is instead the disadvantage of the new back car on the left lane $j - 1$, computed as the difference between the current acceleration and that induced by a lane change of i to its left lane.

Such a disadvantage is weighted by a *politeness factor p*. As the value of p grows over 1 the driver becomes more and altruistic, as it values others' disadvantage more than its own advantage. For $0 < p \leq 1$, MOBIL reproduces different realistic degrees of selfishness of the driver, as the disadvantage of other vehicles is taken into less consideration with respect to its own advantage. When $p = 0$ the driver completely ignores others' disadvantage, and performs lane changes whenever he/she finds it would bring a minimum gain. For $p < 0$ the driver's behavior becomes malicious, tending to change lane when this brings disadvantage to other vehicles, even at the cost of own acceleration loss.

Also, in order to avoid hopping between lanes in border conditions a *hysteresis threshold* is introduced. The acceleration employed as the threshold is a percentage k_2 of the comfortable acceleration a. Identical observations hold for lane changes to the right, in Equation 12.33.

Finally, the MOBIL model's incentive criterion introduces a *bias acceleration* that has different values for movements to the left a_{L} and to the right a_{R}. The bias acceleration is thus used to favor movements to one side. In European countries, this is typically used to skew traffic towards the right, to mimic the real-world tendency of drivers to stay on lane at their right. Therefore, in that case $a_{\text{L}} < a_{\text{R}}$. However, these values can also be adapted to left-biased traffic (e.g., U.K., India, Japan, and Australia) by inverting the inequality, $a_{\text{R}} < a_{\text{L}}$, or to nonskewed traffic (e.g., U.S.) by setting $a_{\text{R}} = a_{\text{L}}$.

An extended formulation of the incentive criterion in MOBIL also considers the advantage induced by the lane change on the current back vehicle $i - 1$. It thus adds a term

$$k_2 \left(\frac{dv_{i-1}(t + \Delta t)}{dt} - \left| \frac{dv_{i-1}(t + \Delta t)}{dt} \right|_R \right) \tag{12.34}$$

to the right-hand side of Equations 12.32 and 12.33. As a movement of vehicle i, no matter whether to the left or to the right, always opens up space in front of vehicle $i - 1$, the additional term in Equation 12.34 has a negative value, and thus increases the probability that the incentive criterion is verified. In other words, accounting for the gain of the current back vehicle mainly models the presence of "pushy" cars on the rear.

The MOBIL model has been employed for the simulation of vehicular networks in Refs. 5, 31, and 59.

In Ref. 33 a set of lane-changing rules is proposed within the framework of the *Krauss* model and is employed for vehicular network simulation in Refs. 35 and 36. As for the MOBIL model, it is possible to identify safety and incentive criteria also in the case of the Krauss overtaking model. The safety criterion has a similar meaning to that in MOBIL, but, consistent with the Krauss car-following formulation, it is expressed in terms of speed instead of acceleration. Recalling that v_i^{safe} represents the maximum speed at which vehicle i can travel to maintain a safety distance from its leading car (see Equation 12.14), the Krauss model safety criterion for a lane change to the left is

$$\left| v_{j-h}^{safe}(t + \Delta t) \right|_L \geq v_{j-h}(t) - b\Delta t, \quad \forall h \geq 1 \tag{12.35}$$

The inequality in Equation 12.35 guarantees that, if the lane change is performed, each back vehicle in the new lane is still able to brake with a comfortable deceleration b and avoid a collision with the vehicle ahead.

The incentive criterion for a left-handed lane changing by vehicle i is defined by the following conditions:

$$v_i^{safe}(t + \Delta t) < v_{max} \tag{12.36}$$

$$v_i^{safe}(t + \Delta t) \geq v_{jam} \tag{12.37}$$

$$\left| v_i^{safe}(t + \Delta t) \right|_L \geq v_{jam} \tag{12.38}$$

The Krauss lane-changing model states that, for the movement to the left to occur, the condition in Equation 12.36 must be verified; that is, vehicle i must not be able to reach the maximum desired speed on the current lane. Thus, the reason behind a lane change is the hope of approaching v_{max}. Also, the Krauss model prevents interlane movements when both lanes are experiencing traffic jams, identified by a speed of vehicles lower than a congestion threshold v_{jam}. Therefore, either Equation 12.37 or Equation 12.38 must hold; that is, at least one of the two lanes must not be congested.

On the other hand, when considering movements to the lane on the right, Krauss specifies different criteria, to account for the asymmetry of traffic in most European countries. The safety criterion for a right-handed lane change is similar to that observed before:

$$\left| v_{k-h}^{safe}(t + \Delta t) \right|_R \geq v_{k-h}(t) - b\Delta t, \quad \forall k \geq 1 \tag{12.39}$$

However, the incentive criterion is quite different, and it is described by the following set of inequalities:

$$v_i^{safe}(t + \Delta t) \geq v_{max} \tag{12.40}$$

$$\left| v_i^{safe}(t + \Delta t) \right|_R \geq v_{max} \tag{12.41}$$

According to Equations 12.40 and 12.41, a lane change to the right occurs only when free-flow conditions are observed on both lanes; that is, the maximum speed can be reached on the right lane as well, and occupancy of the left lane is unnecessary.

The incentive criterion of the Krauss model is completed by the possibility of random lane changes, with probability k_2. Therefore, a lane change to one of the two sides is performed if either the conditions described before are realized, or if the randomness inequality is verified:

$$\eta < k_2 \tag{12.42}$$

where η is a random variable uniformly distributed in [0, 1], and $0 \leq k_2 \leq 1$, with k_2 usually much smaller than 1.

As a final note on this model, its integration with the relative car-following model includes a rule to avoid overtakings on the right. Thus, a fourth speed update rule is added to those in Equations 12.14, 12.15, and 12.16, as follows

$$v_i(t + \Delta t) = \min\{v_i(t + \Delta t), \left| v_i^{safe}(t) \right|_L\} \tag{12.43}$$

that is, vehicle i upper bounds its speed not only on the safety speed evaluated with respect to the front vehicle on the lane it is currently traveling on, that is, $i + 1$, but also on that computed with respect to the front vehicle on its left lane, that is, $j + 1$. This way, the speed of vehicle i never exceeds the value it would have if i were traveling on the lane to its left, and no overtaking with respect to $j + 1$ can happen. In fact, Krauss applies such behavior only in the absence of traffic congestion, that is, when at least one of the inequalities in Equations 12.36 and 12.37 is verified. Instead, in the presence of traffic jams, passing on the right is allowed, because cars travel in large, slow clusters, and Equation 12.43 is not enforced.

Extensive studies on lane changing have also been conducted on models based on cellular automata. One of the best-known representations is that proposed by Nagel,[56] which is obviously compatible with the speed update model by the same author, presented in the previous section.

The safety criterion under the Nagel lane-changing model is satisfied if the two following inequalities hold:

$$x_{j+1}(t) - x_i(t) \geq v_i(t) \tag{12.44}$$

$$x(t) - x_{j-1}(t) \geq v_{max} \tag{12.45}$$

These equations refer to the case of a movement of vehicle i to the left lane. In the case of movement to the right, similar conditions must be met:

$$x_{k+1}(t) - x_i(t) \geq v_i(t) \tag{12.46}$$

$$x_i(t) - x_{k-1}(t) \geq v_{max} \tag{12.47}$$

Equations 12.44 and 12.46 refer to the fact that a *lead gap* (i.e., distance from the new front car) must exist in the new lane, sufficiently large to accommodate a movement of vehicle i equal to its current speed. Equations 12.45 and 12.47 impose that a *lag gap* (i.e., distance from the new back car) must exist in the new lane, sufficiently large to accommodate a movement equal to the maximum velocity the new back vehicle, either $j - 1$ or $k - 1$, can reach. The model thus assumes that the driver knows his/her own speed, but ignores that of other vehicles.

Before proceeding further, we stress that in a space-continuous system, according to our notation, the difference of positions includes the length of the front car. However, recall that the Nagel model, like all other cellular automata-based models, operates on a discretized space. As a consequence, the difference between the positions of cars is actually an integer number of cells. This explains why the length of the leading vehicle does not appear in the equations above.

As far as the incentive criterion is considered, the Nagel model once more distinguishes between movements to the left and to the right. Also, it separates behaviors reflecting different overtaking rules.

When considering European road regulations, left lane changes are constrained by

$$v_i(t) \geq \begin{cases} v_{i+1}(t) & \text{if } x_{i+1} - x_i \leq d_{max} \\ \infty & \text{otherwise} \end{cases} \tag{12.48}$$

$$v_i(t) \geq \begin{cases} v_{j+1}(t) & \text{if } x_{j+1} - x_i \leq d_{max} \\ \infty & \text{otherwise} \end{cases} \tag{12.49}$$

Nagel imposed that *at least one* of the conditions in Equations 12.48 and 12.49 be met for vehicle i to move to its left lane. Accordingly, i moves to the left if the current leading car $i + 1$ is slower ($v_i(t) \geq v_{i+1}(t)$), or if the front car on the left $j + 1$ is slower ($v_i(t) \geq v_{j+1}(t)$).

In the first case i tends to perform an overtaking; in the second it avoids passing on the right, which is prohibited in Europe.

Also note that the above conditions become systematically never met ($v_i(t) \leq v_{max} < \infty$) when the front vehicles $i + 1$ and $j + 1$ are farther than the *lookahead distance* d_{max}. That is, if a leading vehicle is too far ahead, free-flow conditions are assumed in the relative lane, and thus there is no reason for the car to change lanes.

The incentive criterion for lane changes to the right is represented by the logical negation of the criterion for movement to the left. Thus the following conditions must *both* be met:

$$v_i(t) < \begin{cases} v_{k+1}(t) & \text{if } x_{k+1} - x_i \leq d_{max} \\ \infty & \text{otherwise} \end{cases} \qquad (12.50)$$

$$v_i(t) < \begin{cases} v_{i+1}(t) & \text{if } x_{i+1} - x_i \leq d_{max} \\ \infty & \text{otherwise} \end{cases} \qquad (12.51)$$

In other words, the movement to the right is granted as soon as vehicles ahead are fast or distant enough.

Different rules describe the incentive criterion in the case of U.S. traffic, where passing on the right is not forbidden, but there is still a preference for overtaking on the left. For lane changes to the left *both* of the following conditions must hold

$$v_i(t) \geq \begin{cases} v_{i+1}(t) & \text{if } x_{i+1} - x_i \leq d_{max} \\ \infty & \text{otherwise} \end{cases} \qquad (12.52)$$

$$v_{j+1}(t) \geq \begin{cases} v_{i+1}(t) & \text{if } x_{i+1} - x_i \leq d_{max}, x_{j+1} - x_i \leq d_{max} \\ \infty & \text{if } x_{i+1} - x_i \leq d_{max}, x_{j+1} - x_i \leq d_{max} \\ 0 & \text{if } x_{j+1} - x_i > d_{max} \end{cases} \qquad (12.53)$$

Accordingly, left lane changes occur when the vehicle has a slower vehicle ahead (Equation 12.52), which is also slower than the front vehicle on the left lane (Equation 12.53). The first condition is never met when the current front vehicle is too far ahead, while the second is (1) never met when the current leading car is not within lookahead distance, while the left lane one is within lookahead distance, and (2) always met when the front car on the left lane is farther ahead than the lookahead distance.

As for European rules, in the case of U.S. regulations the movement to the right is also conditioned to dual properties with respect to those considered for the left lane change. Thus, vehicle i performs an interlane movement to its right if *at least one* of these conditions holds:

$$v_i(t) < \begin{cases} v_{k+1}(t) & \text{if } x_{k+1} - x_i \leq d_{max} \\ \infty & \text{otherwise} \end{cases} \qquad (12.54)$$

$$v_{i+1}(t) < \begin{cases} v_{k+1}(t) & \text{if } x_{k+1} - x_i \leq d_{\max}, x_{i+1} - x_i \leq d_{\max} \\ \infty & \text{if } x_{k+1} - x_i > d_{\max}, x_{i+1} - x_i \leq d_{\max} \\ 0 & \text{if } x_{i+1} - x_i > d_{\max} \end{cases} \qquad (12.55)$$

that is, vehicle i moves to the lane at its right if the front vehicle there, $k + 1$, is moving faster than i (Equation 12.54). or if the front vehicle on the right is traveling at a higher speed than the current leading car (Equation 12.55). The first always holds in the presence of an empty right lane, while the latter is (1) always met in the absence of a front car on the right lane, while there is a leading vehicle on the current lane, and (2) never met while vehicle i is enjoying free-flow conditions on the current lane.

A last incentive criterion proposed in Ref. 56 is the symmetric one, which has actually been shown to better fit real-world traffic in the United States. As its denomination suggests, this criterion does not differentiate between movements to the left or to the right, which are regulated by the following identical condition:

$$v_i(t) \geq \begin{cases} v_{i+1}(t) & \text{if } x_{i+1} - x_i \leq d_{\max} \\ \infty & \text{otherwise} \end{cases} \qquad (12.56)$$

As Equation 12.56 states, vehicle i's movement to a different lane is considered favorable when the current front vehicle is within lookahead distance and is traveling slower than i. The decision of the direction of movement, to the left or to the right, is then only determined by the safety criterion, and, if both lane changes are found to be safe, by a random choice.

A review of many other incentive criteria for cellular automata lane-changing models can be found in Ref. 56.

12.2.4.2 Intersection Management

Even with the addition of lane-changing criteria, car-following models can only reproduce vehicular systems on individual roads. At most, they can be employed to mimic ramps on highways, that is, merging between parallel flows.

However, for network simulation purposes, it is of great interest to study vehicular traffic in urban and suburban environments. In such cases, a proper modeling of the interaction among nonparallel traffic flows must be provided. Unsignalized road junctions, signalized intersections, and roundabouts all represent situations where nonparallel flows have to merge and then separate again. It is well known that such intersections largely dominate vehicular traffic dynamics in urban scenarios due to their capillary presence over the road topology and to the disruptive effect they have on the free-flow motion of vehicles.

Unfortunately, intersection management modeling within microscopic-level analytical frameworks often leads to intractable problems, due to the joint complexity of describing per-vehicle acceleration, lane changing, and nonparallel flows interaction. Moreover, traffic flow theory is mostly interested in determining the macroscopic effects of urban traffic, such as congestion levels and outflow rates at different road junctions of an urban topology, because that approach produces results that can then be employed to formulate optimal signal control strategies.

As a consequence, descriptions found in the literature of vehicular flow theory aim at modeling arrival, delay, and departure profiles at intersections, and their effect on large road topologies, more than defining microscopic car-to-car merging rules.[60] In particular, unsignalized intersections are modeled through stochastic analyses of capacity and delay, based on gap acceptance methods or queuing theory.[61] Signalized intersections are instead usually modeled as partly stochastic, partly deterministic systems,[62] where the stochastic phase accounts for randomness of vehicular traffic, for example, vehicles' interarrival times, while the deterministic phase reproduces periodic phenomena induced by traffic signals, for example, traffic lights cycling among colors.

Unfortunately, a macroscopic approach is hardly applicable to network simulation, where most applications require each mobile entity position to be modeled independently. Several solutions have been proposed in vehicular networking literature to cope with the issue of intersection modeling. In the following, we will outline such solutions, describing how stochastic and car-following motion representations can be extended to support intersection management to provide a methodology for microscopic-level interaction of flows in urban environments.

The simplest way to represent flows interaction at an intersection is to mimic the merging delay with a pause time. According to the *pause* model, all vehicles reaching a road junction stop for a certain amount of time, which can be deterministic or stochastic.

This minimum-cost solution provides very approximate results, as (1) it is independent from the inflow rate of vehicles, and thus does not generate delays proportional to the vehicular density, as encountered in unsignalized intersections, and (2) it lacks the temporization typical of signalized intersections. This solution can be employed within any graph-constrained mobility model, no matter whether it belongs to the stochastic, traffic stream, or car-following categories.

This trivial implementation has guaranteed to the pause model a non-negligible diffusion in vehicular networking research, especially during its early stages and typically in conjunction with stochastic models,[3,4] but not only with them.[63]

A correlation-based approach has been introduced within the *Constant Speed Motion* model[4] already discussed in Section 12.2.1.

As a matter of fact, Constant Speed Motion also includes the possibility of forcing pauses at intersections, that is, graph vertices, encountered within a trip. In that case, the model provides some rudimental flows-interaction mechanisms, as pauses of different cars at the same intersection are not independent, but correlated. In fact, considering a vehicle arriving at an intersection, two cases are possible:

1. If no other car is waiting at the intersection, it picks a random pause time uniformly distributed in a given range $[0, T_p]$.
2. If one or more other vehicles are already paused at the intersection, it forces its own pause time to match the residual pause time of the first car that arrived at the intersection under examination.

According to such rules, cars paused at a same road junction leave the intersection in the same instant, mimicking the clustering of vehicles leaving a crossroad, a typical effect in real-world traffic, due to the presence of semaphores.

However, it should be noted that this solution induces improper behaviors with increasing vehicular densities. Although one would expect delays to increase, the aforementioned rules tend to reduce the average waiting time at an intersection in the presence of higher densities, as the probability of finding already stopped vehicles at crossroads grows.

A stochastic approach is adopted in the *Probabilistic Traffic Sign Model* (PTSM).[63] Similarly to the Constant Speed Motion model above, PSTM distinguishes between vehicles arriving at a clear intersection and vehicles reaching a road junction where other cars are already waiting.

Thus, when a vehicle reaches an intersection:

1. If no other car is already stopped there, it directly crosses the intersection with probability p, while with probability $1 - p$ it picks a random pause time in $[0, T_p]$ and waits for such a time to expire, before traversing the intersection.
2. If other cars are waiting at the intersection, it stops as well, for a time equal to the residual pause time of the previous car that reached the same intersection, plus 1 sec. In other words, this behavior is identical to that of the Constant Speed Motion, but PSTM also adds to the pause time 1 sec of delay for each car already stopped.

With respect to the Constant Speed Motion model, PTSM increases the level of realism of the simulation, as it models the delay as a linear function of the density. However, it remains a stochastic solution that cannot reproduce the deterministic aspects of signalized intersection management.

Intersection handling capabilities have been proposed as an extension to the Intelligent Driver Model presented previously in this chapter, through the *Intelligent Driver Model with Intersection Management* (IDM-IM).[5] This model can be used to model road junctions ruled by stop signs or traffic lights, and, in both cases, IDM-IM only acts on the first vehicle approaching the intersection, as the car-following description in the IDM model automatically adapts the behavior of cars following the leading one.

The basic principle is to force the leading vehicle to believe that an obstacle (i.e., a still car) is present right before the intersection ahead if the vehicle should stop. In that case, the IDM car-following model induces a deceleration in the leading vehicle, in order to avoid a collision with the imaginary obstacle, and makes it stop right before the intersection. All following cars will queue up behind the first one, without need of any further intervention.

By removing the obstacle, the possibility of crossing the intersection is granted to the leading car. As an example, this occurs if the vehicle has right of way at a stop-sign-regulated intersection, or if the corresponding traffic light has turned green.

As this scheme is based on simple car-to-car interaction rules, its employment is not limited to the IDM description, but it can be adapted to any other car-following model.

In the IDM-IM model, the imaginary obstacle is implemented by tweaking, in the acceleration update equations of the first vehicle on lane i, the distance and speed difference with respect to its front car. As this front car does not actually exist, the standard values observed by i would be

$$\Delta x_i(t) = \infty \qquad (12.57)$$

$$\Delta v_i(t) = 0 \qquad (12.58)$$

which correspond to free-flow conditions (by looking at Equation 12.12, these settings cancel the deceleration term $(\Delta x_{\text{des}}(t)/\Delta x_i(t))^2$ due to the presence of a front car. Instead, IDM-IM imposes the following settings to i:

$$\Delta x_i(t) = x_{\text{stop}} - x_i(t) \qquad (12.59)$$

$$\Delta v_i(t) = - v_i(t) \qquad (12.60)$$

where x_{stop} is the position of the location vehicle i has to stop at. Thus, the settings mimic those that i would observe if a still $(v_{i+1}(t) = 0)$ front vehicle were located at a distance $x_{\text{stop}} - x_i(t)$. It is also desirable that x_{stop} be placed some distance from the actual vertex representing the intersection, so that i halts before reaching the "center" of the intersection.

Note that, through the formulation of the car-following model, this setting of $\Delta x_i(t)$ and $\Delta v_i(t)$ allows vehicles to freely accelerate when far from the next intersection and then to smoothly decelerate as they approach the road junction. The resulting deceleration profile is not completely accurate, as it has been shown that braking dynamics induced by intersections are different from those due to car-to-car interactions. However, such differences lead to errors of the order of a few meters or less and are thus sufficiently precise for network simulation purposes. For an accurate discussion of the issue, the reader may refer to Ref. 64.

The intersection management proposed by IDM-IM requires that a "smart" road topology determines when to create and remove the imaginary obstacles, thus regulating the flow of traffic from different directions. When modeling priority roads and roundabouts, the decision could be based on principles borrowed from gap acceptance theory,[54,61] while stop signs could be handled through priority queue techniques.[65,66] Traffic lights are easily realized by periodic switching between presence and absence states of the imaginary obstacle, reproducing red and green traffic light conditions, respectively.[5]

It should be said that such interaction between the "smart" road topology and the microscopic traffic model requires some attention. As a representative example, we consider the case of traffic light management. Let us assume that a vehicle, heading towards an intersection, finds the traffic light to be *red*. According to the IDM-IM rules above, when reaching a sufficiently small distance from the traffic light, it starts decelerating as if a car were halted right before the intersection. Now, let us imagine that the traffic light turns *green* when the vehicle is already braking, but has not come yet to a complete stop. In this case, the model behaves as one could expect in reality, as the sudden disappearance of the imaginary obstacle triggered by the red-to-green traffic light switch allows the vehicle to start accelerating again, as would happen in the real world.

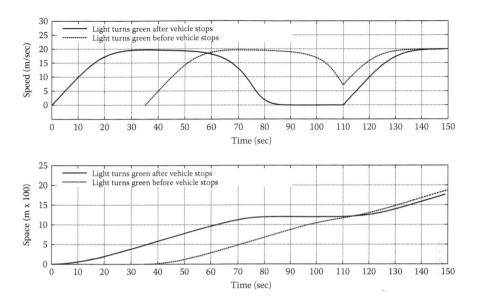

FIGURE 12.10 Traffic light *red-to-green* scenario under the IDM-IM model. A vehicle starts its movement with zero speed, accelerates and travels towards a red traffic light, which then turns green. The upper Figure shows the evolution of speed in time, while the lower one depicts the car movement on the road versus time (the upper curve can be seen as the time derivative of the lower one). [From Fiore, M. et al. Vehicular Mobility Simulation for VANETs, in SCS/IEEE Annual Simulation Symposium (ANSS), Norfolk, VA, March 2007. With permission.]

This case is illustrated in Figure 12.10. In the configuration represented by the solid curve, a vehicle starts moving at $t = 0$ sec, accelerates to the desired speed, decelerates as the traffic light becomes closer, and eventually comes to a full stop in front of the traffic light. The movement starts over again when the traffic light turns green at $t = 110$ sec. This can be easily observed in both figures. In a second configuration represented by the dashed curve, a vehicle starts its movement at $t = 35$ sec and thus at $t = 110$ sec, when the traffic light switches to green, it is still in the deceleration phase and has not yet halted. It then accelerates again. Thus, in this second configuration, the vehicle does not stop at the intersection and, as shown in the speed plot, the dashed curve is always greater than zero, while, in the image depicting the spatial evolution, the advantage in terms of speed experienced by the vehicle in the second scenario leads to an increased traveled space.

The opposite case is also possible, with a traffic light switching from *green* to *red* while a vehicle is approaching. Such a situation is not as straightforward as the previous one, as the behavior of the driver now strongly depends on the distance from the intersection when the traffic light color change occurs. In fact, in the real world, a driver reacting to a green-to-red switch would (1) brake and stop before the intersection, if far enough away when the traffic light state change happens, or (2) proceed at high speed and cross the intersection as fast as possible, if already too near to perform a safe deceleration.

In Ref. 5, the case of a green-to-red switch is addressed by defining a minimum safety braking distance Δx_{safe}^{g2r}. Then, if a vehicle i, under green-to-red conditions, finds that it is not possible to stop before the intersection, even by braking as hard as possible, that is, if

$$\Delta x_{safe}^{g2r} > x_{stop} - x_i(t) \tag{12.61}$$

then it crosses the intersection at its current speed. Otherwise, it stops by applying a strong enough deceleration. The safety braking distance is computed by means of simple kinematic formulae, as

$$\Delta x_{safe}^{g2r} = v_i(t)\frac{v_i(t)}{k_1 b} - \frac{k_1 b}{2}\left[\frac{v_i(t)}{k_1 b}\right]^2 = \frac{v_i^2(t)}{2k_1 b} \tag{12.62}$$

which describes the space needed to come to a full stop as a function of the current speed of the vehicle, $v_i(t)$, and of the maximum safe deceleration, $k_1 b$, that is, the IDM comfortable braking value b scaled by a factor $k_1 \geq 1$. In the expression above the terms $(v_i(t)/k_1 b)$ represent the time at which a zero velocity is reached by inducing a constant deceleration $k_1 b$ on the current speed $v_i(t)$.

Examples of driving behaviors in the presence of a green-to-red semaphore switch are shown in Figure 12.11. There, different curves represent different movement start locations, that is, different distances of the vehicle under study with respect to the traffic light when it switches from green to red (40, 100, 200, and 400 m, respectively). The 40 m case, represented by a solid curve, is an example of unsatisfied safety conditions, because 45 m $= \Delta x_{safe}^{g2r} > x_{stop} - x_i(t) = 40$ m. The car is too close to the traffic light when the color changes to red, thus the vehicle maintains its speed and does not stop. In the other cases, the safety condition is satisfied, and the vehicle comes to a complete stop in front of the semaphore, as shown in the spatial evolution. However, the deceleration starts at various distances from the traffic light, leaving different reaction margins to the driver. As proved by the speed profile plot, this results is a peculiar braking evolution, with more comfortable decelerations when the distance from the semaphore is larger, at the moment of the semaphore switch.

As a final comment on flows-interaction models, and on intersection management in particular, it should be said that realistic merging of vehicular flows is a necessary condition to modeling real-world traffic dynamics, but it is not a sufficient one. As a matter of fact, once an accurate microscopic description has been implemented, it must also be correctly employed. Therefore, in particular when considering urban environments, *traffic assignment* plays a major role in characterizing vehicular density distributions, system capacity profiles and occurrence of vehicular traffic phenomena. In brief, traffic assignment theory studies the mobility patterns of drivers, developing models aimed at a realistic representation of the route selection of vehicles over complex road topologies, and considering a number of factors, such as speed limitations, congestion levels,

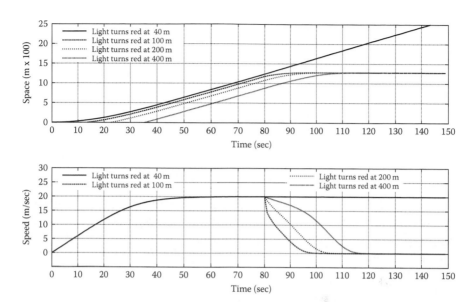

FIGURE 12.11 Traffic light *green-to-red* scenario under the IDM-IM model. A vehicle starts its movement with zero speed, accelerates and travels towards a green traffic light, which then turns red. The upper Figure shows the evolution of speed in time, while the lower one depicts the car movement on the road vs. time (the upper curve can be seen as the time derivative of the lower one). [From Fiore, M. et al. Vehicular Mobility Simulation for VANETs, in SCS/IEEE Annual Simulation Symposium (ANSS), Norfolk, VA, March 2007. With permission.]

time of the day, attraction and repulsion locations, and so on. It is out of the scope of this chapter to address traffic assignment in detail. Instead we invite the interested reader to refer to an introductory document on the topic by Gawron.[67]

12.3 Realism of Mobility Models

A fundamental issue to address when dealing with vehicular mobility models for networking studies is that of *realism*, intended as a model's ability to reproduce traffic phenomena as observed in the real world. In fact, in order to obtain reliable results from vehicular network simulations, it would be highly desirable that the mobility description employed be as near as possible to that encountered in the real world. In fact, it has been shown that diverse mobility representations affect in a dramatic way the performance of network protocols at all layers,[8,63,68–70] hence the necessity of a trustworthy modeling of car motion in vehicular network simulation.

Therefore, an important question is: *how to validate vehicular mobility representations?* In this section, we will overview some classic traffic flow theory benchmark tests employed to determine the level of realism of a vehicular mobility model.

12.3.1 Realism of Car-to-Car Interaction

As far as car-to-car interaction is concerned, we selected three well-known tests, which verify the capability of models to reproduce (1) speed/flow/density relationships, (2) reactions to a perturbation, and (3) shockwave effects, as they are observed in real-world measurements.

The first test verifies the capability of a model to correctly recreate the fundamental relationship between vehicular outflow, speed, and density. The correlation between outflow and density has already been discussed in Section 12.2.2, but here we add the vehicular speed as well. The behavior observed in everyday traffic involves the following dependences between these three parameters:

1. *Flow vs. density.* Given a straight road, as the inflow rate, and consequently the car density, increases, the outflow of vehicles is expected to grow linearly at first. When a critical vehicular density is reached, the road capacity does not sustain the arrival rate anymore, leading to queuing phenomena that slow down the system and reduce the outflow, up to traffic jam conditions (Figure 12.5).
2. *Speed vs. density.* The speed monotonically decreases with increasing density, because the initial free-flow conditions encountered with low vehicular density turn into more and more congested traffic as the inflow of cars grows. The more congested the traffic, the lower the speed of cars, constrained by reduced front and back gaps.
3. *Speed vs. flow.* Under free-flow conditions, high speed is coupled with low outflow, as the vehicular density is minimal. As the outflow grows, the speed is only slightly reduced, until the critical density is reached. Then the speed suffers from a dramatic decrease, coupled with reduced flow of cars, as traffic congestion occurs.

We test five of the models presented in Section 12.2, verifying their capability of reproducing the above relationship between density, flow, and velocity. The selected models are a stochastic representation (Manhattan model), a traffic stream description (Fluid Traffic Motion model, FTM), and three car-following formulations, one of which is based on cellular automata (IDM, Krauss model, and Nagel–Schreckenberg model). We do not include flows-interaction models in the list, as no merging among flows is present in this test. As already stated, flows-interaction models are based on car-interaction descriptions; thus, in this test, any flows-interaction model would perform exactly as the car-interaction representation it is based on.

Results from the tests are depicted in Figure 12.12, where it appears evident that car-following models behave as expected, as the curves match the real-world observations outlined before.

The FTM only approximates the desired result, due to the nontrivial calibration of three parameters of the model: the minimum speed v_{min}, the traffic jam density ρ_{jam}, and the length of segments the road has to be divided into. Also, a limitation of the FTM model is quite evident from the results: due to its formulation (see Equation 12.7), this model cannot reproduce zero-speed conditions, that is, it must be $v_{min} > 0$, to avoid the model reaching a deadlock state. As a matter of fact, a density $\rho(x, t)$ inducing a zero speed on vehicles would

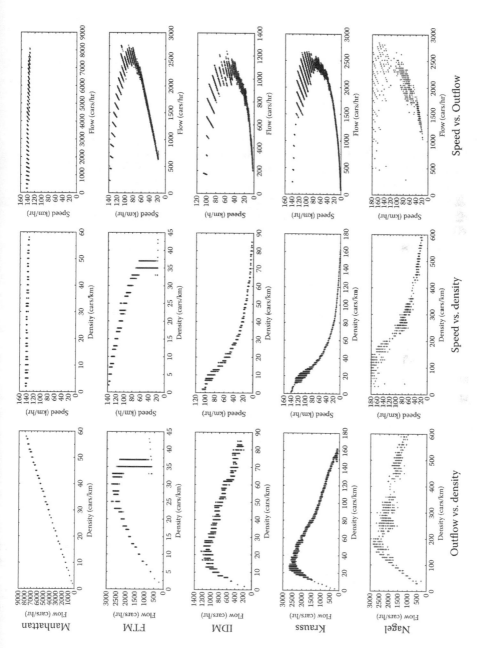

FIGURE 12.12 Fundamental diagrams of vehicular outflow, density, and speed, as generated by the Manhattan, FTM, IDM, Krauss, and Nagel–Schreckenberg models.

force cars to stop, a situation that, in turn, would prevent the density from being updated. Thus, $\rho(x, t)$ would stay constant, and the whole system would freeze.

On the other hand, the Manhattan model generates an unrealistic, persistently linear relationship between the three variables. This is a clear proof of the stochastic nature of the model that, even if introducing some car-following management, lacks the detail necessary to reproduce even the most basic phenomena observed in everyday vehicular traffic.

As a second test, we study the reaction of models to a mild perturbation, monitoring the behavior of a flow of cars traveling on a single-lane road and encountering a slow vehicle ahead. A real-world behavior would involve the vehicles slowing down (each with different dynamics, as the first vehicle brakes the hardest, while the following cars experience progressively smoother decelerations) and forming a queue as they approach the obstacle. Then, when the obstacle is removed, cars should start accelerating again, with a speed increment that is propagated along the queue.

Four models are employed for this test, again selected in a way to cover the different categories of representation distinguished before. So, we considered a stochastic representation (Manhattan model), a traffic stream description (FTM), and two car-following formulations (IDM and Krauss model). For models belonging to the flows-interaction class, the same considerations made for the first test apply also to this case.

Results are shown in Figure 12.13, depicting the speed profile of 20 vehicles as they approach a stopped car ahead. At 60 sec, the stopped car starts accelerating and moves away. This is considered a mild perturbation, as the obstacle is removed before the approaching vehicles come to a full stop.

The plots referring to the IDM and Krauss models in Figure 12.13 reproduce the expected behaviors. The speed profile of the vehicles can be separated into three phases: an initial free-flow phase, a deceleration phase induced by the stopped car in front, and an acceleration phase due to the removal of the obstacle. Different cars result in different profiles, as each reacts to the behavior of its own leading vehicle, introducing a delay proportional to the position of the car within the row. Thus, the leading car in the row is the one that starts braking first, reaching the lowest speed, and it is also the first to start

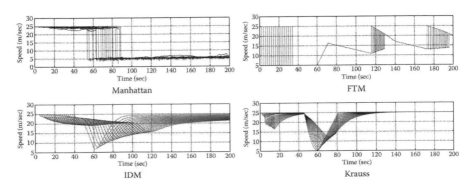

FIGURE 12.13 Evolution of speed for the first 20 vehicles belonging to a queue of cars encountering a mild perturbation.

accelerating after the obstacle has moved away. The two models, IDM and Krauss, present identical behaviors over different timescales, an effect of the different parameter settings employed in the two tests.

When following the Manhattan model, vehicles do not smoothly decelerate, but suddenly slow down to the minimum speed v_{min} when they arrive next to the obstacle. After the stopped front car is removed, vehicles do not accelerate back to full speed, but tend to keep the minimum speed. The reason for the lack of realistic deceleration is the collision-avoidance trigger implemented by the Manhattan model, inducing a very hard braking only when very near to the front car. The lack of acceleration is due to the fact that the model misses the concept of desired speed, and only implements a random acceleration, so that vehicles do not have any incentive to accelerate back to full speed.

The FTM model only generates a very rough approximation of the correct behavior, with vehicles abruptly stopping one after the other (vertical lines in the first part of the plot) in the presence of the obstacle. Once the impediment is removed, all the vehicles begin a periodic, growing speed evolution, with speed increment and decrements, due to the movement of cars between subsequent road segments where the vehicular density is computed independently.

Traffic congestion is also known to produce typical slow-speed waves,[71-73] which move backwards with respect to the direction of vehicular motion, as time progresses. Although common, especially over long, straight roads such as freeways and highways, this phenomenon is quite counterintuitive and requires some explanation.

An example off a process generating a traffic wave is depicted in Figure 12.14. There, all vehicles are initially traveling at a constant, high speed (denoted by the white color of the cars), and are uniformly spaced, thus realizing optimal traffic conditions. At some point in time, one of the vehicles starts decelerating to a full stop (notice the darker and darker color of the car, indicating a decreasing speed), due to some event not shown in the Figure (possible examples encountered in the real world are a front vehicle entering the road from a ramp or lateral street, a traffic light switching to red, or a human error). The deceleration introduces a perturbation on the back vehicles, which are progressively forced to stop behind the braking car. The traffic wave is generated when the vehicle that first started the deceleration process resumes its movement (its color becoming lighter). In fact, the finite response time of any human driver leads to a delay in the reaction of the back vehicles. Such delay prevents the other already stopped cars from moving again before other vehicles join the slow-speed queue. As a consequence, when the cars that had to slow down due to the original perturbation can resume a high-speed movement, the queue of stopped vehicles has already grown, and other back drivers are now stuck. As time passes, this process interests cars that are increasingly distant from the origin of the perturbation, in a backward direction with respect to that of vehicular movement. We stress that traffic wave generation is highly dependent on the density of traffic, as in congested driving conditions even a short braking by a vehicle, not necessarily leading to a full stop, can lead to the generation of slow-speed waves interesting to tens or even hundreds of cars.

In a third test, we thus verify the capability of each model to recreate the traffic wave phenomenon.

The models under study are, in this case, a stochastic representation (Manhattan model), a traffic stream description (FTM), and two car-following formulations, one of

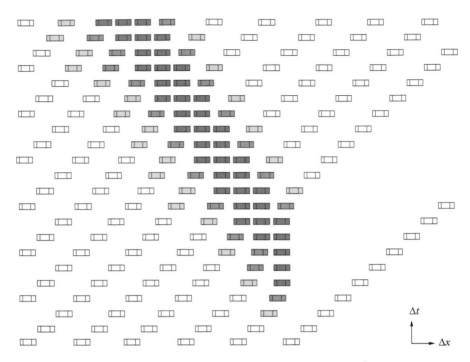

FIGURE 12.14   Traffic wave due to a perturbation induced by a stopping vehicle.

which is based on cellular automata (IDM and Nagel–Schreckenberg model). As far as flows-interaction models are concerned, the same considerations made before also apply in this case.

The plots in Figure 12.15 show the space–time evolution of the speed of vehicles in congested traffic, as obtained with different models. Once more, the car-following description, that is, the IDM and Nagel–Schreckenberg models, successfully represent

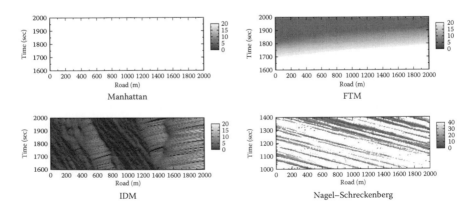

FIGURE 12.15   Speed vs. time and space, in the presence of increasing car inflow.

the real-world phenomenon of traffic shockwaves. As a matter of fact, dark low-speed waves can be observed in the relative plots. Such waves move against traffic direction (which is assumed to proceed towards positive abscissae) with time. The discrete-space nature of the Nagel–Schreckenberg model and the diverse settings of the parameters lead to sensibly different profiles of the waves under the two models. In particular, as already observed in the previous test, the IDM setting seems to lead to a slow reaction of cars and, consequently, to lower global speeds and longer delays in waves propagation.

The FTM model produces an inaccurate result, as the relative plot shows a transition to lower speeds in time, but no traffic shockwaves. This is attributable to the per-segment computation of speed, which forces all vehicles along a same segment (in this case, the whole road represented in the plot) to maintain an identical speed.

Finally, the Manhattan model completely fails the test, as vehicles movements are unaffected by the increasing density, consistent with the results from the first test. Thus, even in the presence of congestion conditions, the Manhattan model generates a constant high-speed profile, resulting in a completely *white* plot.

12.3.2 Realism of Flows Interaction

The previous analysis concerns car-interaction properties of the models, which is the reason why flows-interaction models were not considered for testing. Obviously, it would be desirable to validate the capability of different models to reproduce realistic merging of flows at road intersections and highway ramps.

We therefore consider two flows-interaction tests, based on (1) speed profiling of vehicles approaching and leaving an intersection, and (2) average vehicular density monitoring in a simple road topology.

In the first test, we record the speed dynamics of vehicles approaching a stop-sign-ruled intersection. A real-world behavior would involve vehicles decelerating to a complete stop, waiting for their turn, and then crossing the road junction, accelerating to full speed again. As we assume scarce traffic at the intersection, vehicles are expected to resume their motion after a very short pause.

We tested one model for each category, namely a stochastic representation (Manhattan model), a traffic stream description (FTM), a car-following formulation (IDM), and a flows-interaction model (IDM-IM). This last model is set to consider a stop sign right before the intersection.

In Figure 12.16, the average speed of vehicles, recorded with the aforementioned models, is depicted as a function of the distance to the center of the intersection. The result matches intuition, as the only model that can reproduce a realistic speed profile is obviously the flows-interaction one. As a matter of fact, IDM-IM forces vehicles to decelerate, stop, and then accelerate again.

The Manhattan model completely ignores the intersection, as it does the FTM description. In the second case, as the intersection joins two different road segments, traversing it causes FTM to change the speed of cars according to the vehicular density variation.

The IDM description also neglects the presence of the intersection. However, a slightly lower speed is observed *exactly* at the center of the intersection. This is an artifact of the IDM implementation. The car-following model's formulation does not allow vehicles to

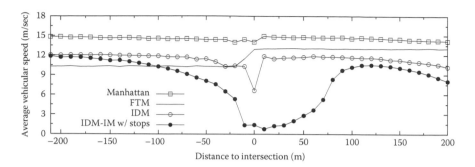

FIGURE 12.16   Speed profile of the in- and outflow of cars.

overlap; however, it cannot prevent such a situation from occurring in the presence of multiple independent flows merging at an intersection, When this occurs, and the overlapping vehicles are directed to the same outbound lane, the only solution is to force them to stop, and then make them resume their motion one after the other. When considering the speed profile, such a behavior leads to zero-speed contributions from vehicles stopped right at the middle of the intersection, and, therefore, to a lower average speed with respect to that recorded along road segments.

The importance of the flows-interaction representation is even more evident in the second test we conducted. Here, we measured the average vehicular density over a simple road topology, as obtained with different mobility models. The road layout is a 3 × 3-block regular grid, thus featuring four intersections, with street segments of 250 m connecting the crossways. A traffic load of 20 vehicles/km/lane is uniformly distributed over the road topology. Such a value represents average traffic conditions, and should thus not lead to congestion situations. In the real world, we would expect higher densities of cars around intersections, where queuing and clustering phenomena occur, and lower densities along the streets, where vehicles can travel at high speeds.

In this test, in addition to representatives for the stochastic [Constant Speed Motion model (CSM) and Manhattan model], traffic stream (FTM), and car-following (IDM) classes, we tested multiple models from the flows-interaction category, because the first test above showed that car-interaction models produce unrealistic results even in a single intersection scenario. Thus, we also considered the following:

1. The flows-interaction CSM (CSM w/ pauses), with pauses characterized as detailed in Section 12.2.4.
2. Different flavors of the IDM-IM, considering (i) intersections regulated by stop signs (IDM-IM w/ stops), (ii) intersections regulated by traffic lights (IDM-IM w/ lights), and (iii) intersections regulated by traffic lights, plus multiple lanes for each road segment (IDM-IM multilane).
3. The MOBIL model, built upon the IDM-IM multilane description [which we dubbed IDM with Lane Changes (IDM-LC)]. Thus, while IDM-IM multilane simply forces each vehicle to move over a randomly picked lane, IDM-LC also involves the possibility of vehicles changing lanes, if MOBIL's safety and incentive criteria are satisfied.

We stress that, although all the other models only employ a single lane on each road segment, IDM-IM multilane and IDM-LC are the only two descriptions that use multilanes streets.

The average vehicular densities generated by the diverse models listed before are depicted in Figure 12.17. There, car-interaction and flows-interaction models can be clearly distinguished, as the first generate quasi-uniform distributions of cars over the road topology, while the latter determine peaks of vehicular density at or in the proximity of road junctions.

By considering intersections as a generic location of the road topology, not different from any other point over the road segments, the CSM, Manhattan, FTM, and IDM models cause vehicles to traverse crossroads as if they were traveling over unobstructed streets. Thus, no semantic difference in the vehicular density is observed at intersections with respect to any other location of the road layout, but a doubled value attributable to the superposing of the densities belonging to the two overlapping streets. The only exception to this scheme is presented by the IDM, which produces slightly higher densities right at the center of each intersection. The cause is again the IDM implementation issue discussed for the previous test, which, under particular circumstances, forces overlapping vehicles to stop at the center of an intersection, and wait until overlying cars move away to safety distance. The presence of such still vehicles obviously increases the average density recorded at intersections.

As far as flows-interaction models are concerned, all of them generate a gathering of cars at or around intersection points, as expected. However, the shape of the high-density peaks at crossroads is quite different for the diverse models. In particular, CSM w/ pauses determines very high-density peaks (the vehicular density scale of the CSM w/ pauses plot is three times larger than that of all other plots in Figure 12.17), concentrated at the very center of each road junction. The reason for this is that CSM w/ pauses is based on a stochastic car-interaction description, and thus does not create queues of cars in the proximity of intersections, but simply "stacks" vehicles paused at streets' junction points. The resulting density is obviously not very realistic.

When considering IDM-IM-based models, things change, as the car-following properties of the underlying car-to-car interaction model allow for the reproduction of queuing of cars around intersections. Nevertheless, differences among these models can be noticed as well. In fact, the IDM-IM w/ stops tends to produce higher vehicular densities, spread over larger areas near intersections. This denotes higher congestion of the road network, and it is what one could expect in reality, as employing stop signs slows down the traffic flow, reducing the overall system capacity.

On the other hand, traffic-light-based models speed up the crossing of intersections, leading to shorter queues, lower delays, and reduced vehicular densities. Also, we can observe that introducing multiple lanes (IDM-IM multilane and IDM-LC) further shortens the queues of vehicles, as density peaks are less spread away from intersection points. Again, this can be observed in everyday traffic, because roads with multiple lanes increase the capacity of the system, allowing more than one vehicle at a time to approach the same intersection from the same direction. Instead, we do not observe any particular difference when comparing multilane models without (IDM-IM multilane) and with (IDM-LC) lane-changing capabilities. This is due to the fact that, given

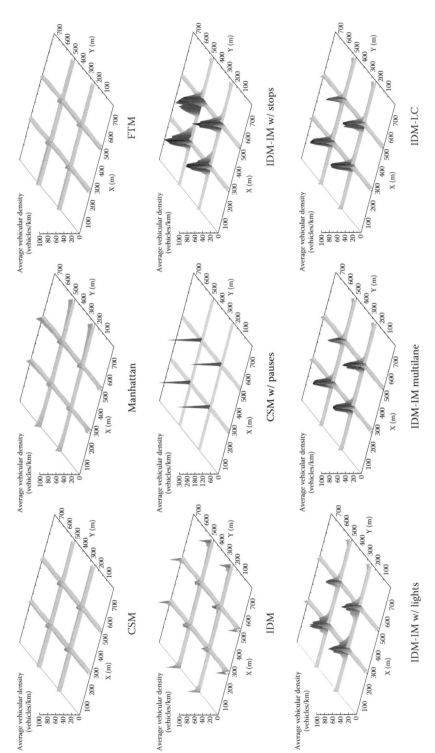

FIGURE 12.17   Average vehicular density distribution for different mobility models.

the length of road segments and the typical speeds of vehicles in urban scenarios, lane changes occur with low frequency, and therefore do not significantly affect the outcome of the simulation.

As a final note on the problem of determining the level of realism of flows-interaction models, we underline that the tests employed above are aimed at making evident how different models reproduce intuitive real-world traffic dynamics at road junctions.

In traffic flow theory, the actual validation of flows-interaction models is usually performed by considering existing road junctions, recording the real-world inflow rates of vehicles, and measuring the resulting delays and outflow rates. The inflow rates are then used as an input parameter for the model, whose output delay and outflow rates are compared with those observed in reality. The same applies to lane-changing models, whose output is checked against real-world measurements of different lane usage, for example.

12.3.3 Discussion of Results

From the results presented above, we can conclude that, although car-following models provide a faithful representation of real-world vehicular dynamics, stochastic models fail all the tests and cannot be considered realistic reproductions of vehicles' behavior in everyday traffic. Traffic stream models, on the other hand, achieve partial success, because they are intended to reproduce large-scale phenomena, but are inadequate when microscopic car interactions are taken into account. Moreover, a realistic intersection management built on car-following models is required to recreate traffic dynamics in the presence of road junctions, an aspect of particular importance in urban environments.

12.4 Impact of Mobility on Network Simulation

In Section 12.3 we have clarified the level of realism of models belonging to the different classes of vehicular mobility descriptions introduced in Section 12.2. However, an interesting problem is that of understanding what level of detail in the mobility representation is actually needed for the simulation of vehicular networks.

The goal of this section is thus to understand if, how, and why different mobility models affect network simulation. In particular, our approach is intended to be as general as possible, and therefore we are not interested in evaluating the performance of one or more specific protocol(s) under varying vehicular mobility. Instead our study focuses on physical-level network connectivity, in a way such that its results are applicable to any protocol stack configuration. To this end, we define several metrics through which to derive network topological properties of interest, and show that different analytical representations of traffic motion have a non-negligible impact on such metrics.

As our focus is on nodes' mobility, we avoid biases that could be introduced by complex road topologies and nonuniform traffic, by studying a simple grid-like street layout.

12.4.1 Network Connectivity Metrics

As stated before, our analysis requires metrics capable of capturing the dynamics of the physical topology of a potential network built over moving vehicles. For the formal definition of such metrics, we model the network topology at time t as a graph $G(t) = \{V, E(t)\}$, where vehicles (also referred to as nodes in the following) correspond to the set of vertices $V = \{v_i\}$ and communication links to the set of time-dependent edges $E(t) = \{e_{ij}(t)\}$. An edge $e_{ij}(t)$ exists if there is a direct wireless communication link from v_i to v_j at time t, with $i \neq j$. Using this notation, we define the following metrics:

1. *Link duration.* This is the time span between the instant at which a vehicle enters within transmission range of another vehicle, and the instant at which the physical connection is lost, due to the relative movement of nodes. Formally, the duration of the link from v_i to v_j at time t is defined as

$$l_{ij}(t) = \begin{cases} 0, & \text{if } \nexists e_{ij}(t) \\ t_f - t_0, & \text{if } \exists e_{ij}(t), \end{cases} \tag{12.63}$$

   with t_0, t_f such that $\nexists e_{ij}(t_0-)$, $\nexists e_{ij}(t_f+)$ and $\exists e_{ij}(\tau)$, $\forall \tau \in [t_0, t_f]$. The link duration measures how stable a connection is over time.

2. *Nodal degree.* This is the number of vehicles within the transmission range of a node, that is, the number of neighbors of the node. If the set of neighbors, both symmetric and asymmetric, of v_i at time t is defined as

$$N_i(t) = \{v_j \mid \exists e_{ij}(t)\}, \tag{12.64}$$

   then the degree of v_i at the same time is given by $d_i(t) = |N_i(t)|$. The nodal degree determines how dense the network is, from the physical connectivity point of view.

3. *Cluster number.* This is the number of coexisting, nonconnected clusters of nodes at a given instant. We define a cluster as a group of vehicles that is *logically* fully connected, that is, within which a bidirectional (multihop) route exists between whichever couple of nodes. Formally, the existence of a unidirectional path from v_i to v_j at time t can be represented as the binary variable

$$p_{ij}(t) = \begin{cases} 1, & \text{if } \exists e_{ij}(t) \text{ or } \exists v_k \mid \exists e_{ik}(t), p_{kj}(t) = 1 \\ 0, & \text{otherwise.} \end{cases} \tag{12.65}$$

   The cluster in which v_i lies at time t can be defined as

$$C_i(t) = v_i \cup \{v_j \mid p_{ij}(t) = 1, p_{ji}(t) = 1\}, \tag{12.66}$$

   and the set of unique clusters in the network at t as

$$C(t) = \{C_i(t) \mid C_i(t) \cap C_j(t) = \varnothing, \ \forall j < i\}. \tag{12.67}$$

The cluster number corresponds to the cardinality of this last set, $c(t) = |C(t)|$, and provides information on the degree of fragmentation of the overall simulated network, in terms of the number of mutually isolated groups of nodes.

4. *Normalized cluster size.* This is the number of vehicles in a cluster at a given instant, normalized over the number of simulated vehicles. For a cluster $C_i(t)$, it can be defined as

$$s_{C_i}(t) = |C_i(t)|/|V|, \tag{12.68}$$

The normalized cluster size characterizes the distribution of nodes into clusters, distinguishing clusters of different cardinality, and it is independent from the simulated network size.

5. *Clustering coefficient.* This is an index of the connectivity of nodes within a cluster. Formally, if we define the set of links within a cluster $C_i(t)$ as

$$E_{C_i}(t) = \{e_{jk}(t)|v_j, v_k \in C_i(t)\}, \tag{12.69}$$

then the clustering coefficient of the same cluster is

$$k_{C_i}(t) = \frac{|E_{C_i}(t)|}{|C_i(t)|(|C_i(t)| - 1)}, \tag{12.70}$$

which is the ratio between the number of existing links and the maximum number of unidirectional links that could exist in the cluster. According to this definition, the clustering coefficient has a maximum value 1 if the cluster is a clique, while it tends to zero in a linear topology, with a number of nodes drifting to infinity.

12.4.2 Mobility Models Analysis

Mobility models determine the way nodes move in a vehicle-based network and, as a consequence, the connectivity properties of the resulting system. Thus, they represent the first aspect that has to be taken into consideration when exploring the relationship between vehicular motion and network topology.

In order to determine the impact of vehicular mobility modeling on network connectivity, we employ a deliberately simple scenario. The rationale behind this choice is that a plain environment evidences at best differences that are solely attributable to the mobility description, avoiding any bias that complex road topologies and activity models could induce into the system.

Thus, we simulate a regular 3×3-block grid, each block measuring 250 m on it side. Vehicles enter/leave the scenario from the borders of the topology and randomly select their trips. A typical vehicular density of 20 vehicles/km/lane is considered, and a transmission range of 100 m is assumed.

Nine different models were selected as representative of the diverse classes introduced in Section 12.2. The models are the same as those employed in the flows-interaction vehicular density test conducted in Section 12.3.2. We list them, along with their parameter settings, in Table 12.2. An in-depth discussion of models settings calibration is

TABLE 12.2 Mobility Models' Parameter Settings

		Parameters					
Model	Type	v_{min} (m/sec)	v_{max} (m/sec)	a (m/sec^2)	b (m/sec^2)	Δt_{safe} (sec)	Δx_{safe} (m)
CSM (w/ pauses)	Stochastic	5.0	15.0	—	—	—	—
Manhattan	Stochastic	5.0	15.0	0.6	—	—	1.0
FTM	Traffic stream	2.0	15.0	—	—	—	—
IDM	Car-following	10.0	15.0	0.6	0.9	0.5	1.0
IDM-IM	Flows-interaction	10.0	15.0	0.6	0.9	0.5	1.0
IDM-LC	Flows-interaction	10.0	15.0	0.6	0.9	0.5	1.0

		Parameters				
Model	Type	p	a_L (m/sec^2)	a_R (m/sec^2)	ρ_{jam} (cars/km/lane)	T_p (sec)
CSM (w/ pauses)	Stochastic	—	—	—	—	30
Manhattan	Stochastic	—	—	—	—	—
FTM	Traffic stream	—	—	—	0.125	—
IDM	Car-following	—	—	—	—	—
IDM-IM	Flows-interaction	—	—	—	—	—
IDM-LC	Flows-interaction	0.5	−0.2	0.2	—	—

beyond the scope of this chapter, as the interested reader can refer to the models' references for details. We just stress that the selected parameters fit real-world values and that we calibrated them according to the simulated urban scenario.

12.4.2.1 Network-Level Analysis

We start our analysis from the clustering metrics, which reflect network-wide connectivity properties. Figure 12.18 reports the cluster number, normalized cluster size, and clustering coefficient, averaged over space and time, relative to different mobility models.

As a first observation, it is evident that diverse motion descriptions generate dissimilar network clustering dynamics and can be ordered by increasing average cluster number, which coincides with increasing clustering coefficient and decreasing normalized cluster size. This relationship between the clustering metrics is expected, because the number of simulated vehicles is the same for all the models, and a higher number of noncommunicating clusters implies a smaller size of each cluster. Also, clusters of fewer vehicles tend to experience a slightly higher degree of internal connectivity, as their spatial extension is reduced, and the probability that two vehicles in the cluster are out of transmission range is lower.

Secondly, by looking at the individual models, it is clear that the sequence follows a precise scheme: models that neglect flows interaction (CSM, Manhattan, FTM, and IDM) come first, generating fewer, larger clusters, that is, a more globally connected network, in which the probability that two nodes over the road topology cannot communicate is low. On the other hand, motion descriptions that consider intersection

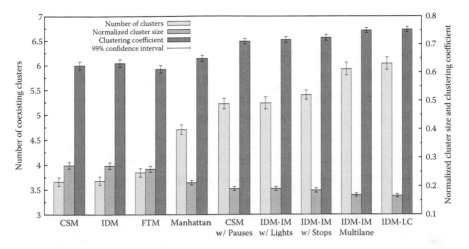

FIGURE 12.18 Cluster statistics for different mobility models. (From Fiore, M. and Härri, J., The Networking Shape of Vehicular Mobility, in ACM International Symposium on Mobile Ad Hoc Networking and Computing, Hong kong, May 2008. With permission.)

management and employ multiple-lane roads (IDM-IM multilane and IDM-LC) determine the highest cluster numbers, fragmenting the network into smaller and more isolated groups of nodes. Models featuring flows-interaction handling but forcing the vehicular traffic over single-lane roads (CSM w/ pauses, IDM-IM w/ stops, and IDM-IM w/ lights) result in an intermediate behavior.

To understand why the analytical representations of movement have such an impact on the network clustering, a spatial analysis is required. To this end, we need to look back at Figure 12.17, showing the vehicular density measured over the simulated road topology, averaged over time, for the diverse models.

As already noticed when first discussing the distributions of vehicles depicted in Figure 12.17, two main trends can be identified: models neglecting flows interaction tend to produce quasi-uniform distributions of vehicles, while models accounting for intersection management result in peaks of vehicular density at crossways, which smooth down to lower density elsewhere.

The first behavior is symptomatic of a presence of cars spread over the road topology, at each instant of simulation. From the network connectivity point of view, this leads to the creation of a few, large clusters, as it is hard to find gaps in the vehicular instantaneous distribution that are larger than the nodes' transmission range. The higher average cluster number obtained with the Manhattan model with respect to CSM, FTM, or IDM, even in the presence of a similar vehicular distribution, is attributable to the lack of desired speed of the model, which, coupled with its nonovertaking rules, creates small clusters of cars traveling at low speed over the road segments (see the results in Figure 12.13 for an example of such behavior).

On the other hand, when considering stop signs or traffic lights at road junctions, the accumulation of vehicles waiting at crossways tends to create clusters around these locations, while the lower density over roads eases the absence of communication paths along the streets that join the intersections. This effect is more evident when employing stop signs than traffic lights, as the first are slower in granting passage to vehicles and thus produce a higher congestion at intersections. Moreover, the presence of multiple lanes magnifies the clustering effect at intersections, as they allow vehicles to gather nearer to the road junctions, forming shorter, denser queues. The latter is the reason why IDM-IM multilane and IDM-LC result in a higher average cluster number with respect to the other flows-interaction models, as seen in Figure 12.18.

CSM w/ pauses falls between the uniformly distributed and intersection clustering behaviors identified above, as it adds unrealistically abrupt and high (again, we recall of the different density scale in Figure 12.17 for the CSM w/ pauses plot) peaks at intersections, while a perfectly uniform distribution of vehicles is measured along the roads.

12.4.2.2 Link-Level Analysis

The diverse vehicular distributions also result in noticeable differences when looking at communication link properties, as demonstrated in Figure 12.19, showing the average nodal degree and link duration obtained with the various mobility models. Models producing quasi-uniform densities have comparable performance in terms of nodal degree and link duration, and the same is true for models considering traffic-lights-ruled intersections (IDM-IM w/ lights, IDM-IM multilane, and IDM-LC). In the last case, the link duration and the nodal degree are almost doubled with respect to the first group of models. Evidently, the presence of flows-interaction management at road junctions forces vehicles to stop, increasing the average duration of links, and creates high-density spots, where vehicles enjoy an elevated number of neighbors. CSM w/ pauses performs between the two groups above, as it provides an approximate form of intersection handling,

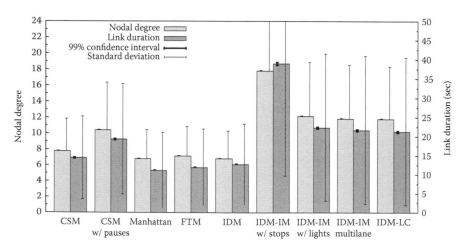

FIGURE 12.19 Link statistics for different mobility models. (From Fiore, M. and Härri, J., The Networking Shape of Vehicular Mobility, in ACM International Symposium on Mobile Ad Hoc Networking and Computing, Hong Kong, May 2009. With permission.)

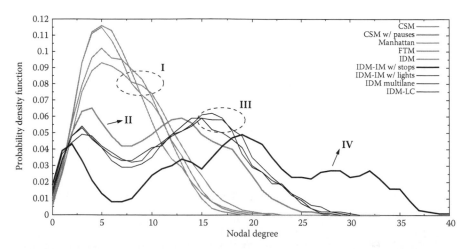

FIGURE 12.20 Nodal degree PDF for different mobility models. (From Fiore, M. and Härri, J., The Networking Shape of Vehicular Mobility, in ACM International Symposium on Mobile Ad Hoc Networking and Computing, Hong Kong, May 2010. With permission.)

whereas IDM-IM w/ stops results in a much higher degree and link duration than the other models. As already stated, stop signs slow down flows merging at crossways, which in turn emphasizes the effects described for the traffic lights case.

Averages provide an immediate picture of the effect of mobility modeling on nodal degree and link duration; however, a deeper understanding can be achieved considering the respective PDFs. In Figure 12.20, the nodal degree of the network topology obtained from models neglecting flows interaction is characterized by a very regular, single-peaked PDF (marked as I). On the other hand, models that assume traffic-light-managed cross-ways produce almost overlapping PDFs (III), which are more spread and present two local maxima, instead of one. CSM w/ pauses (II) generates a PDF between those of group I and group III, while IDM-IM w/ stops further spreads the PDF, reducing its regularity.

The regular shape of the degree PDF obtained with car-interaction models (I) is a result of the uniform distribution of vehicles they produce. In fact, in the absence of intersection characterization, cars limit themselves to traveling at high speed through the road topology, and the degree is just a function of the instantaneous number of neighboring vehicles. This is true even when considering car-following models, such as IDM, as they have been observed to produce similar spreading of cars with respect to less realistic mobility descriptions.

On the other hand, the nodal-degree PDF produced by models accounting for traffic lights at intersections (III) is characterized by two peaks. The first, corresponding to lower degrees, corresponds to neighbors traveling either at high speed in the same lane or in the opposite direction with respect to the vehicle whose degree is under study. Thus, such a peak in the PDF occurs when nodes are traveling over the road segments. The second, high-degree peak is instead due to neighbors (1) queued along with the car under study at an intersection, or (2) vehicles crossing a road junction, while the vehicle under study is paused at a red traffic light. Therefore, the second peak is produced by traffic at intersections.

The overall distribution in the case of models belonging to group III is thus the result of the superposition of two single-peaked PDFs, corresponding to diverse driving situations over the road topology: a low-degree connectivity encountered while cars are traveling at high speed on road segments, and a high-degree connectivity recorded when they are waiting at intersections. The first effect is that producing the PDFs of models belonging to group I, while the second is characteristic of the flows-interaction description added by group III representations.

The degree PDF generated by IDM-IM w/ stops (IV) resembles that obtained with models of group III, but it is distinguished by a much higher probability of encountering queued neighbors at road junctions, due to the slower traffic management provided by stop signs.

The same approach can be applied to the link duration, whose average values were depicted in Figure 12.19. The corresponding PDFs are shown in Figure 12.21. As for the previous metric, the link duration distribution is also strongly correlated to the mobility model employed, and four main behaviors can be identified, matching the same groups of models already distinguished before. Once more, models disregarding flows interaction (I) generate regular distributions, whereas traffic-light-managed models (III) produce very peculiar double-peaked PDFs. CSM w/ pauses (II) approximates the behavior of models of group III, with lower precision with respect to the nodal degree metric case. Finally, the presence of stop signs (IV) spreads the probability over a much larger range than any other model, almost canceling the low-duration peak.

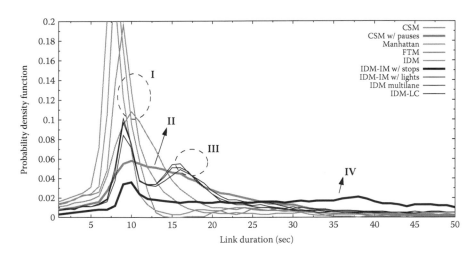

FIGURE 12.21 Link duration PDF for different mobility models. (From Fiore, M. and Härri, J., The Networking Shape of Vehicular Mobility, in ACM International Symposium on Mobile Ad Hoc Networking and Computing, Hong Kong, May 2011. With permission.)

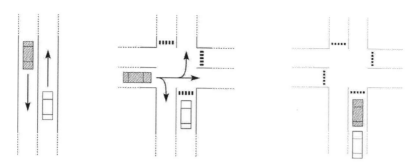

FIGURE 12.22 Driving situations influencing the duration of links established by the *white* vehicle: neighbor car driving in opposite direction on the same road (left); neighbor car crossing an intersection while the white vehicle is stopped (center); and neighbor car queued with the white vehicle (right). (From Fiore, M. and Härri, J., The Networking Shape of Vehicular Mobility, in ACM International Symposium on Mobile Ad Hoc Networking and Computing, Hong Kong, May 2012.)

The regular shape of the distribution recorded with models belonging to group I is easily explained through the same considerations made for the degree PDF before. That of CSM w/ pauses (II) is just an approximation of a more realistic behavior, induced by more detailed intersection management (III and IV).

In the case of group III, two peaks in the PDF can be clearly distinguished again. However, the first one is this time due to links established by cars traveling over parallel road segments in the *opposite* direction only. As vehicles move at very high relative speeds, the resulting links are necessarily short-lived. Also, such a driving situation has a high probability of occurring, as cars driving in the opposite direction are encountered with a much higher frequency than vehicles moving in the same direction or paused at intersections. The second peak, characterized by higher link durations, is due to links established with cars passing through road junctions, while the vehicle under examination is stopped. The occurrence of this kind of link is quite high, as many cars are seen passing by while waiting at an intersection. For the same reason, a low number of cars is observed traveling in the same direction, or stopped at a same intersection, of the vehicle under study: such cars tend to stay in the same location for a long time, so the number of recorded samples is much lower than that measured in other driving situations, leading to a low probability in the link duration PDF. However, such interactions also determine long-lasting links, building up the tail of the distribution in Figure 12.21. Examples of the three driving situations outlined above are depicted in Figure 12.22. The conclusions, as far as models from group III are concerned are that (1) there is a high probability of establishing short-lived links with vehicles traveling in the opposite direction, (2) links built while queued at intersections may also not be very reliable, if neighbors belong to different vehicular flows, and (3) only nodes within the same traffic flow enjoy long-lasting links.

IDM-IM w/ stops magnifies the third behavior just outlined above, as the probability of establishing links with other queued vehicles noticeably increases with the employment of stop signs in place of traffic lights.

12.4.3 Discussion of Results

The results of the network connectivity analysis shown before invite reflection upon the need for realism in vehicular networking. We can state the fllowing:

1. Car-to-car interaction plays a minor role in urban scenarios, when networking metrics are taken into account. As a matter of fact, stochastic, traffic stream, and car-following models, although embodying different levels of realism, generate comparable results in the absence of intersection management.
2. Realistic flows interaction at intersections has a dramatic impact on networking metrics, producing unique connectivity Figures, and is thus required to produce simulation results that are consistent with reality. Pseudorandom pauses at intersections cannot recreate the same dynamics. We stress that a realistic flows interaction is only built on top of a car-following, or similarly realistic, model.
3. Employing multiple lanes does not introduce noticeable differences in the link-level connectivity metrics, as the relative results coincide with those obtained with single-lane roads. However, they have an impact on the network-level clustering, and thus their presence cannot be neglected.
4. Lane changes do not have any noticeable effect on the network topology; urban scenarios, with roads whose lengths are in the order of hundreds of meters, do not leave a lot of chances for performing overtakings, as would happen in highway traffic.

Thus, the way mobility is modeled has a major impact on the topological properties of a vehicular network, and, apart from overtakings, all of the motion features (flows merging and the car-to-car interaction it is built upon, traffic light management, and multiple lanes) proved to generate unique cluster- and link-level effects, motivating the employment of realistic motion representations instead of approximate ones.

12.5 Conclusions

In this chapter, we addressed the issue of vehicular mobility modeling from a networking perspective.

We first presented a classification of analytical representations of vehicular motion employed in network simulation, distinguishing among stochastic, traffic stream, car-following, and flows-interaction models. We introduced formal definitions for several models belonging to each category.

Secondly, we addressed the problem of determining the level of realism of diverse car mobility descriptions, showing how detailed models from traffic flow theory clearly outperform the stochastic representations widely employed for vehicular network simulation.

Finally, we verified the impact of different mobility descriptions on the connectivity of a network built on moving vehicles. We demonstrated that, even in basic scenarios, such a impact is dramatic, and strongly motivates the adoption of realistic flows-interaction models for network simulation.

References

1. Einstein, A., Investigations on the Theory of the Brownian Motion, in *Collected Papers*, R. Fürth, 1926, pp. 170–182.
2. Johnson, D.B., Maltz, D.A., and Broch, J., Dsr: The dynamic source routing protocol for multi-hop wireless ad hoc networks, in *Ad Hoc Networking*, (ed. Perkins, C.). Addison-Wesley, 2001, pp. 139–172.
3. Davies, V., Evaluating Mobility Models Within an Ad Hoc Network, Master's Thesis, Colorado School of Mines, 2000.
4. Canu Project, available at http://canu.informatik.uni-stuttgart.de.
5. Fiore, M., Härri, J., Filali, F., and Bonnet, C., Vehicular Mobility Simulation for VANETS, in SCS/IEEE Annual Simulation Symposium (ANSS), Norfolk, VA, USA, March 2007.
6. Saha, A.K. and Johnson, D.B., Modeling Mobility for Vehicular Ad Hoc Networks, in ACM Workshop on Vehicular Ad Hoc Networks (VANET), Philadelphia, PA, USA, October 2004.
7. U.S. Census Bureau. Topologically Integrated Geographic Encoding and Referencing (Tiger) System, available at http://www.census.gov/geo/www/tiger.
8. Bai, F., Sadagopan, N., and Helmy, A., The important framework for analyzing the impact of mobility on performance of routing protocols for adhoc networks, *Elsevier Ad Hoc Networks*, 1, 383–403, 2003.
9. ETSI, Universal Mobile Telecommunication System (UMTS), Selection Procedures for the Choice of Radio Transmission Technologies of the UMTS, 1998–2004, UMTS 30.03 Version 3.2.0, available at http://www.3gpp.org/ftp/Specs/html-info/3003U.htm.
10. Zhou, B., Xu, K., and Gerla, M., Group and Swarm Mobility Models for Ad Hoc Network Scenarios Using Virtual Tracks, in IEEE Military Communications Conference (MILCOM), Monterey, CA, USA, October 2004.
11. Nandan, A., Tewari, S., Das, S., Gerla, M., and Kleinrock, L., Adtorrent: Delivering Location Cognizant Advertisements to Car Networks, in IEEE/IFIP Annual Conference on Wireless On-demand Network Systems and Services (WONS), Les Ménuires, France, January 2006.
12. Hong, X., Gerla, M., Pei, G., and Chiang, C.-C., A Group Mobility Model for Ad Hoc Wireless Networks, in ACM/IEEE International Symposium on Modeling, Analysis and Simulation of Wireless and Mobile Systems (MSWiM), Seattle, WA, USA, August 1999.
13. Lighthill, M.J. and Whitham, G.B., On kinematic waves ii: A theory of traffic flow on long crowded roads, *Royal Society A*, 229, 317–345, 1955.
14. Rudack, M., Meincke, M., and Lott, M., On the Dynamics of Ad Hoc Networks for Inter-Vehicle Communications (ivc), in International Conference on Wireless Networks (ICWN), Las Vegas, NV, USA, June 2002.
15. Schnabel, W. and Lohse, D., *Grundlagen der Straßenverkehrstechnik und der Verkehrsplanung*, Verlag für Bauwesen, Berlin, Germany, 1997.
16. Leutzbach, W., *Introduction to the Theory of Traffic Flow*, Springer, Berlin, Germany, 1988.

17. Seskar, I., Marie, S., Holtzman, J., and Wasserman, J., Rate of Location Area Updates in Cellular Systems, in IEEE Vehicular Technology Conference (VTC), Denver, CO, USA, May 1992.

18. McDonald, M. and Brackstone, M., Car-following: A historical review, *Transportation Research, Part F: Traffic Psychology and Behaviour*, 2(4), 181–196, 1999.

19. Panwai, S. and Dia, H., Comparative evaluation of microscopic car-following behavior, *IEEE Transactions on Intelligent Transportation Systems*, 6(3), 314–325, 2005.

20. Helbing, D., Traffic and related self-driven many-particle systems, *Reviews of Modern Physics*, 73(4), 1067–1141, 2001.

21. Aycin, M., and Benekohal, R., Comparison of car-following models for simulation, *Transportation Research Record*, 1678, 116–127, 1999.

22. Zhang, Y., Scalability of Car-Following and Lane-Changing Models in Microscopic Traffic Simulation Systems, Master's Thesis, Louisiana State University, 2002.

23. Gazis, D.C., Herman, R., and Rothery, R.W., Nonlinear follow-the-leader models of traffic flow, *Operations Research*, 9, 545–567, 1961.

24. Reuschel, A., Fahrzeugbewegungen in der kolonne bei gleichförmig beschleunigtem oder verzögertem leitfahrzeug, *Z. Öslerr, Ing. Arch. Vereines*, 95, 59–62, 73–77, 1950.

25. Pipes, L.A., An operational analysis of traffic dynamics, *Journal of Applied Physics*, 24, 274–281, 1953.

26. Chandler, R.E., Herman, R., and Montroll, E.W., Traffic dynamics: Studies in car following, *Operations Research*, 6(2), 165–184, 1958.

27. Kometani, E. and Sasaki, T., On the stability of traffic flow, *Journal of Operations Research Japan*, 2, 11–26, 1958.

28. Helly, W., Simulation of Bottlenecks in Single Lane Traffic Flow, in Symposium on Theory of Traffic Flow, 207–238, Research Laboratories, General Motors, New York, NY, USA, 1959.

29. Treiber, M., Hennecke, A., and Helbing, D., Congested traffic states in empirical observations and microscopic simulations, *Physical Review E*, 62(2), 1805–1824, 2000.

30. Jaap, S., Bechler, M., and Wolf, L., Evaluation of Routing Protocols for Vehicular Ad Hoc Networks in City Traffic Scenarios, in IEEE International Conference on Intelligent Transportation Systems (ITS), Vienna, Austria, September 2005.

31. Sommer, P., Design and Analysis of Realistic Mobility Model for Wireless Mesh Networks, Master's Thesis, ETH Zurich, Switzerland, 2007.

32. Krauss, S., Wagner, P., and Gawron, C., Metastable states in a microscopic model of traffic flow, *Physical Review E*, 55(304), 55–97, 1997.

33. Krauss, S., Microscopic Modeling of Traffic Flow: Investigation of Collision Free Vehicle Dynamics, PhD Thesis, Universität zu Köln, 1998.

34. Breisemeister, L., Group Membership and Communication in Highly Mobile Ad Hoc Networks, PhD Thesis, Technical University of Berlin, 2001.

35. Krajzewicz, D., Hertkorn, G., Rossel, C., and Wagner, P., Sumo (simulation of urban mobility): An Open-Source Traffic Simulation, in SCS Middle East Simulation Multiconference (MESM), Sharjah, UAE, October 2002.

36. Lan, K.-C., Karnadi, F., Mo, Z., Rapid Generation of Realistic Mobility Models for VANET, in IEEE Wireless Communications and Networking Conference (WCNC), Hong Kong, March 2007.

37. Kometani, E. and Sasaki, T., Dynamic Behaviour of Traffic with a Non-Linear Spacing–Speed Relationship, in Symposium on Theory of Traffic Flow, Research Laboratories, General Motors, New York, NY, 1959, pp. 105–109.
38. Gipps, P.G., A behavioural car following model for computer simulation, *Transportation Research B*, 15, 105–111, 1981.
39. Michaels, R.M., Perceptual Factors in Car Following, in International Symposium on Theory of Road Traffic Flow, Paris, France, 1963, pp. 44–59.
40. Wiedemann, R., Simulation des Straßenverkehrsflusses, Schriftenreihe des Instituts fur Verkehrswesen 8, Universität Karlsruhe, 1974.
41. Evans, L., and Rothery, R., Perceptual thresholds in car following—a recent comparison. *Transportation Science*, 11(1), 60–72, 1977.
42. Wiedemann, R., Modelling of RTI-Elements on Multi-Lane Roads, Telematics in Road Transport, Commission of the European Community, DG XIII, Brussels, Belgium, 1991.
43. Fliess, T. and Schulze, T., Urban Traffic Simulation with Psycho-Physical Vehicle-Following Models, in Winter Simulation Conference (WSC), Atlanta, GA, USA, December 1997.
44. Fellendorf, M., and Vortisch, P., Validation of the Microscopic Traffic Flow Model VISSIM in Different Real-World Situations. UC Berkeley Transportation Library, 2001.
45. Gorgorin, C., Gradinescu, V., Diaconescu, R., Cristea, V., and Ifode, L, An Integrated Vehicular and Network Simulator for Vehicular Ad-Hoc Networks, in European Simulation and Modelling Conference (ESM), Toulouse, France, October 2006.
46. Bononi, L., Di Felice, M., Bertini, M., and Croci, E., Parallel and Distributed Simulation of Wireless Vehicular Ad Hoc Networks, in ACM/IEEE International Symposium on Modeling, Analysis and Simulation of Wireless and Mobile Systems (MSWiM), Torremolinos, Spain, October 2006.
47. Hidas, P., A car-following model for urban traffic simulation, *Traffic Engineering and Control*, 39(5), 300–305, 1998.
48. Kim, J., Sridhara, V., and Bohacek, S., Realistic Simulation of Urban Mesh Networks—Part i: Urban Mobility, Technical Report, Univeristy of Delaware, 2006.
49. Cremer, M., and Ludwig, J., A fast simulation model for traffic flow on the basis of boolean operations, *Mathematics and Computers in Simulation*, 28(4), 297–303, 1986.
50. Schütt, H., Entwicklung und erprobung eines sehr schnellen, bitorientierten verkehrssimulationssystems für strassennetze, Schriftenreihe der Arbeitsgruppe Automatisierungstechnik 6, Technische Universität Hamburg-Harburg, 1990.
51. Nagel, K. and Schreckenberg, M., A cellular automaton model for freeway traffic, *Journal de Physique I*, 1992(2), 2221–2229, 1992.
52. Chowdhury, D., Santen, L., and Schadschneider, A., Statistical physics of vehicular traffic and some related systems, *Physics Reports*, 329(4), 199–329, 2000.
53. Fox, P., and Lehmann, F.G., A digital simulation of car following and overtaking, *Highway Research Record*, 199, 33–41, 1961.
54. Gipps, P.G., A model for the structure of lane-changing decisions, *Transportation Research B*, 20, 403–414, 1986.
55. Helbing, D., *Verkehrsdynamik*, Springer, Berlin, Germany, 1997.

56. Nagel, K., Wolf, D.E., Wagner, P., and Simon, P., Two-lane traffic rules for cellular automata: A systematic approach, *Physical Review E*, 58, 1425–1437, 1998.
57. Ahmed, K.I., Modeling Drivers' Acceleration and Lane Changing Behavior, PhD Thesis, Massachusetts Institute of Technology, 1999.
58. Treiber, M. and Helbing, D., Realistische Mikrosimulation von Strassenverkehr mit Einem Einfachen Modell, in Arbeitsgemeinschaft Simulation (ASIM), Rostock, Germany, September 2002.
59. Jaap, S., Bechler, M., and Wolf, L., Evaluation of Routing Protocols for Vehicular Ad Hoc Networks in Typical Road Traffic Scenarios, in Open European Summer School on Networked Applications (EUNICE), Colmenarejo, Spain, July 2005.
60. Papageorgiou, M., Traffic control, in *Handbook of Transportation Science*, (ed. Hall, R.W.), Kluwer Academic Publishers, Boston, 1999, pp. 233–267.
61. Troutbeck, R.J. and Brilon, W., Unsignalized intersection theory, in Revised Monograph on Traffic Flow Theory, Turner-Fairbank Highway Research Center, 2001.
62. Rouphail, N., Tarko, A., and Li, J., Traffic Flow at Signalized Intersections, in Revised Monograph on Traffic Flow Theory, Turner-Fairbank Highway Research Center, 2001.
63. Mahajan, A., Potnis, N., Gopalan, K., and Wang, A., Urban Mobility Models for VANETS, in IEEE Workshop on Next Generation Wireless Networks (WoNGeN), Bangalore, India, December 2006.
64. Akçelik, R., and Besley, M., Acceleration and Deceleration Models, in Conference of Australian Institutes of Transport Research (CAITR), Melbourne, Australia, December 2001.
65. Gawron., C., An iterative algorithm to determine the dynamic user equilibrium in a traffic simulation model, *International Journal of Modern Physics C*, 9(3), 393–207, 1998.
66. Cetin, N., Burri, A., and Nagel, K., A Large-Scale Multi-Agent Traffic Microsimulation Based on Queue Model, in Swiss Transport Research Conference (STRC), Monte Verità, Ascona, Switzerland, March 2003.
67. Gawron, C., Simulation-Based Traffic Assignment, PhD Thesis, University of Köln, 1998.
68. Jardosh, A., Belding-Royer, E.M., Almeroth, K.C., and Suri, S., Towards Realistic Mobility Models for Mobile Ad Hoc Networks, in ACM Annual International Conference on Mobile Computing and Networking (MobiCom), San Diego, CA, USA, September 2003.
69. Musolesi, M. and Mascolo, C., A Community Based Mobility Model for Ad Hoc Network Research, in ACM International Workshop on Multi-hop Ad Hoc Networks: from Theory to Reality (REALMAN), Florence, Italy, May 2006.
70. Fiore, M., Härri, J., Filali, F., and Bonnet, C. Understanding Vehicular Mobility in Network Simulation, in IEEE International Workshop on Mobile Vehicular Networks (MoVeNet), Pisa, Italy, October 2007.
71. Triterer, J., Investigation of Traffic Dynamics by Aerial Photogrammetry Techniques, Technical Report 278, Transportation Research Center, Ohio State University, 1975.

72. Smith, S., Freeway Data Collection for Studying Vehicle Interactions, technical report FHWA/RD-85/108, U.S. Department of Transportation, Federal Highway Administration, Office of Research, Washington, DC, USA, 1985.

73. Coifman, B., Time Space Diagrams for Thirteen Shock Waves. PATH Technical Report UCBITS- PWP-97-1, University of California, Berkeley, CA, USA, 1997.

74. Eritico Geographic Data Files, available at http://www.ertico.com/en/links/links/gdf_-_geographic_data_files.htm.

75. Fiore, M., and Härri, J., The Networking Shape of Vehicular Mobility, in ACM International Symposium on Mobile Ad Hoc Networking and Computing (MobiHoc), Hong Kong, China, May 2008.

76. Fiore, M., and Härri, J., The Networking Shape of Vehicular Mobility, in ACM International Symposium on Mobile Ad Hoc Networking and Computing (MobiHoc), Hong Kong, China, May 2009.

77. Fiore, M., and Härri, J., The Networking Shape of Vehicular Mobility, in ACM International Symposium on Mobile Ad Hoc Networking and Computing (MobiHoc), Hong Kong, China, May 2010.

78. Fiore, M., and Härri, J., The Networking Shape of Vehicular Mobility, in ACM International Symposium on Mobile Ad Hoc Networking and Computing (MobiHoc), Hong Kong, China, May 2011.

79. Fiore, M., and Härri, J., The Networking Shape of Vehicular Mobility, in ACM International Symposium on Mobile Ad Hoc Networking and Computing (MobiHoc), Hong Kong, China, May 2012.

13

Vehicular Network Simulators

Gongjun Yan,
Khaled Ibrahim, and
Michele C. Weigle
Department of Computer Science
Old Dominion University

13.1 Introduction

Vehicular networking is an emerging area of interest in the wireless networking community as well as in the transportation research community. The potential of vehicular networks to provide vital services, from real-time traffic information to advance collision warning, makes this an important area of study. Vehicular networking can comprise vehicle-to-vehicle (V2V) communication, vehicle-to-infrastructure (V2I) communication, or a combination of both. Typically, networks formed without infrastructure support are termed *ad hoc networks*; thus, vehicular networks with only V2V communication have been called *vehicular ad hoc networks* (VANETs). VANETs have much in common with the well-studied mobile ad hoc networks (MANETs). Both are ad hoc networks of mobile nodes that are capable of wireless communication. Much of the similarity ends there, however. Because VANET nodes are vehicles rather than hand-held devices, there is little concern with energy consumption, storage capacity, or computation power. Additionally, because vehicles move much faster than humans, VANETs have the added complication of a quickly changing topology. In order to evaluate VANETs*, researchers almost always must resort to simulation as the expense of actual deployment is too high. Unfortunately, there is no standard vehicular networks simulator. Currently, most researchers generate a mobility trace using a vehicular mobility simulator and input this trace to a standard networking simulator. The choice of the mobility simulator is important as performance in vehicular networks depends highly on the connectivity of the nodes, and the manner in which nodes move greatly affects this connectivity.

The use of simulation in evaluating MANETs has been well documented in recent years,[1–3] but little attention has been paid to how simulation is used in VANETs. The work of Harri et al.[4,5] is an exception, with the authors focusing on classifying the available options for vehicular mobility simulators. This chapter presents an overview of vehicular network simulators. Section 13.2 discusses popular simulators used in papers presented at the ACM VANET workshop and categorizes simulators as either *loosely integrated* or *tightly integrated*. Section 13.3 describes several vehicular mobility simulators, while Section 13.4 describes several popular network simulators. Tightly integrated simulators, those that contain bothnetwork and mobility components, are described in Section 13.5.

13.2 An Overview of Vehicular Network Simulators

VANET simulations require both a networking component and a mobility component. In most cases, these components are provided by two separate simulators. Researchers build a topology and produce a trace of vehicle movements using a mobility simulator. This movement trace, representing network node movements, is then fed to the networking simulator. Recently there has been work on developing integrated simulators that contain both networking and vehicular mobility components, allowing feedback between the two components (e.g., TraNS[6] and ASH[7]), but so far, most research has been performed using two separate simulators.

Table 13.1 shows the usage of simulators in papers presented at the ACM VANET workshop from 2004 to 2007. In these four years of the workshop, there were a total of

TABLE 13.1　Simulator Usage Summary from Survey of 69 Papers in ACM VANET (2004–2007)

Item	Papers (%)
Used simulation	51 (74%)
Specified simulator	36 (70%)
Self-developed simulator	8 (16%)
Did not specify simulator	7 (14%)
Did not use simulation	18 (26%)

* In this chapter, the terms *VANET* and *vehicular network* are used interchangeably and can refer to V2V or V2I communication.

TABLE 13.2   Mobility Simulators Used
in ACM VANET (2004–2007)

Simulator	Papers
SHIFT/SmartAHS	3
CORSIM	2
VanetMobiSim	1
GrooveSim	1
VISSIM	1
MicroApplet	1
PARAMICS	1

69 papers and posters presented. Of these, 74% used some type of simulation as part of their evaluation. Out of these, 70% named the specific simulator that was used, while 16% used a self-developed simulator.

As mentioned earlier, most researchers use a combination of a mobility simulator and a network simulator. Table 13.2 shows the number of papers using particular mobility simulators, and Table 13.3 shows the papers using particular network simulators. As these network simulators allow for node mobility, researchers can use a built-in, but often simplified, mobility model instead of generating a trace with a separate mobility simulator. Because of this, there are many more network simulators named in ACM VANET papers than mobility simulators.

Vehicular network simulators can be classified as either *loosely integrated* or *tightly integrated*. Loosely-integrated simulators use separate mobility simulators and network simulators. The mobility simulator generates the mobility of vehicles and records the vehicular movements into trace files. The network simulator imports these trace files, but there is no direct interaction between the two simulators. Tightly integrated simulators do not use trace files but rather embed the mobility simulator and network simulator into a single vehicular network simulator. In some cases, the mobility model and network model can communicate, providing feedback from the network simulator to adjust parameters of vehicular movement. For example, in traffic congestion notification systems, the receipt of certain network messages may cause a vehicle to change its path (i.e., take an early exit). In collision avoidance systems, network messages may

TABLE 13.3   Network Simulators Used
in ACM VANET (2004–2007)

Simulator	Papers
ns-2	18
QualNet	4
SWANS	2
J-Sim	1
NAB	1

cause a vehicle to slow down to avoid an accident. This type of feedback is not supported by loose integration of mobility and network simulators.

As loosely integrated simulation is currently more prevalent, separate mobility and network simulators will be discussed first, followed by a description of several tightly integrated simulators.

13.3 Mobility Simulators

Vehicular mobility simulators are used to generate travel paths of vehicles. In most cases, the movement traces can be saved and imported into a network simulator to study how VANET applications might perform. As mentioned earlier, network connectivity affects network performance, and connectivity is determined by these movement traces. Therefore, it is important to generate realistic movement traces in order to rigorously evaluate VANET protocols. For most VANET studies, which focus on communication between individual vehicles, the most appropriate level of mobility is microscopic, which describes traffic based on the behavior of individual vehicles. This section presents an overview of several popular microscopic vehicular mobility simulators, someof which are commercial products and some are freely available.

13.3.1 TSIS-CORSIM

TSIS-CORSIM (Traffic Software Integrated System–Corridor Simulation)[8] is a powerful commercial traffic simulation package, developed at the University of Florida and funded by the Federal Highway Administration (FHWA). CORSIM is a microscopic simulation model that is especially designed to simulate highways and surface streets and thus includes traffic signals and stop signs. CORSIM consists of two main components, NETSIM for simulating surface streets, and FRESIM for simulating freeways. CORSIM is a featured tool of the FHWA[9] and has been widely used in the transportation research community. TSIS-CORSIM requires Microsoft Windows and Internet Explorer.

CORSIM can simulate very complex traffic scenarios and thus can take a large amount of information as input. A graphical editor, TRAFED, is included with TSIS-CORSIM to allow the traffic scenario to be described. The description of highways, or freeways, includes intersections, road segment lengths, number of lanes, merge lane lengths, adding or removing lanes, free flow speeds, roadway curvature, and roadway grade. The description of surface streets, or arterial roadways, includes road segment lengths, lane utilization, free flow speeds, and signal timings. CORSIM also allows the traffic volume to be specified.

The output of CORSIM is a MOE (Measure Of Effectiveness) file. The output is processed into tables that summarize the information. For freeway models, the key metrics produced are throughput, speed, and density. The information for arterial models includes throughput, control delay, and maximum queues. The tables also highlight problem areas that affect arterial and freeway performance, such as ramp intersections. The full TSIS-CORSIM package also includes TRAFVU, which uses CORSIM input and output files to produce animations of the traffic network that has been simulated.

13.3.2 VISSIM

VISSIM[10] is a commercial microscopic traffic simulator developed by PTV America. VISSIM allows users to simulate a large array of traffic scenarios and complex traffic situations. The simulator allows for many different types of objects in the simulation, from cars and trucks to buses to pedestrians. Interaction with VISSIM is done through a sophisticated graphical user interface (GUI). Users can perform 3D modeling and generate 3D animations. VISSIM requires Microsoft Windows.

13.3.3 PARAMICS

The PARAMICS[11] suite, developed by Quadstone Ltd, is a set of high-performance micro-simulation software tools. PARAMICS can simulate intersections, highways, urban areas, work zones, and so on. It also includes a 3D visualization tool. The PARAMICS suite consists of eight modules: Modeller, Analyser, Processor, Estimator, Designer, Converter, Programmer, and Monitor. Like TSIS-CORSIM and VISSIM, PARAMICS requires Microsoft Windows.

These three traffic simulation tools (CORSIM, VISSIM, and PARAMICS) are widely used by traffic engineers as they are very powerful and allow for the simulation of complex roadway systems. Choa et al.[12] performed a comparison of these commercial simulators, evaluating how they performed at simulating a particular freeway and interchange improvement project. Although all three of these simulators have been used for some VANET studies, most VANET researchers are interested more in evaluating network protocols in a realistic, but limited, traffic environment with open-source tools than in modeling a complex traffic scenario with closed systems.

13.3.4 SmartAHS

SmartAHS[13–15] is an automated highway system simulator developed as part of the California PATH project[16] at UC-Berkeley. It was originally built to simulate automated vehicles, although a human driver model[17] based on the cognitive driver model COSMODRIVE[18] was later added. SmartAHS, based on the specialized programming language Shift,[19] is available for free download from PATH.[20]

SmartAHS simulations are first coded using the Shift language, then translated to C, which is compiled into a Unix executable. The highway topology is built using the Smart PATH highway description.[21] Vehicles are added by creating a file that describes their locations on the roadway. The TkShift debugger[22] allows the user to graph simulation data and animate the simulation.

13.3.5 Microscopic Traffic Applet

The Microscopic Traffic Applet[23] is a Java applet used to demonstrate the Intelligent Driver Model (IDM) car-following model[24] and the MOBIL lane change model[25] in six different scenarios. There are two basic topologies: a ring (circular) road and an oval-shaped road. In most scenarios, both cars and trucks are simulated, with trucks having a slower desired speed and deceleration.

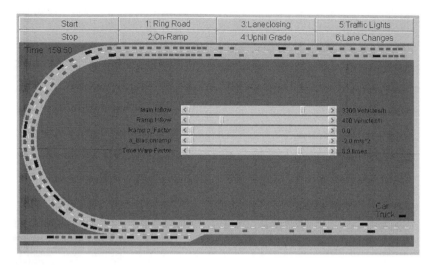

FIGURE 13.1 Screen shot of microscopic traffic applet. (From Treiber, M., Microsimulation of road traffic, available at http://www.traffic-simulation.de/. With permission.)

In the basic ring road scenario, the user can set the average density of vehicles and the percentage of vehicles that are trucks rather than cars. In the other ring road scenario, demonstrating lane changes, there are obstacles at certain positions on the road. The user can set the politeness factor and the lane-changing threshold in the lane-changing algorithm.

The remaining scenarios take place on the oval-shaped road. The first oval scenario shows the impact of an on-ramp, pictured in Figure 13.1. The user can adjust the rate at which new vehicles enter the main road, the rate at which vehicles enter the on-ramp, and the politeness factor. The next oval scenario illustrates the effect of a lane closure. The user can adjust the rate at which cars enter the road, the percentage of trucks, and the imposed speed limit. These are the same parameters that are available for the uphill grade scenario, where vehicles, and especially trucks, slow down as they begin the incline. The final scenario is an oval road with a traffic light. Again, the user can set the inflow rate, the truck percentage, and the speed limit.

As this is an applet designed to illustrate IDM, it does not include any method to import maps from other sources, set a path from source to destination, or output a trace file for input into a network simulator.

13.3.6 CanuMobiSim and VanetMobiSim

CanuMobiSim[26] is a Java-based simulator designed for mobility simulation by the CANU research group at the University of Stuttgart. It implements several mobility models and can generate mobility traces suitable for use in various network simulators, including ns-2 and GloMoSim. CanuMobiSim allows maps formatted according to the GDF (Geographical Data Files) standard[27] to be imported.

Since CanuMobiSim was originally designed for MANET, the mobility models are more suitable for MANETs than VANETs. The mobility models implemented in the CanuMobiSim simulator include the following:

1. Brownian motion
2. Gauss–Markov random motion[28]
3. Incrementally changed random motion[29]
4. Random waypoint movement[30]
5. Graph-based mobility model[31]
6. Constant speed motion
7. Fluid traffic model[32]
8. Intelligent driver model[24]
9. Smooth motion[33]

VanetMobiSim[34,35] is an extension to CanuMobiSim that implements the IDM-IM (IDM with Intersection Management) and IDM-LC (IDM with Lane Changes) mobility models and allows users to import maps from the U.S. Census Bureau's TIGER/Line database.[36] In addition, VanetMobiSim has been validated against the commercial traffic generator CORSIM. Both CanuMobiSim and VanetMobiSim have visualization components for node movements and topology, an example of which is shown in Figure 13.2.

The inputs to both CanuMobiSim and VanetMobiSim are an XML configuration file and a map description. The XML configuration file specifies the simulation scenario

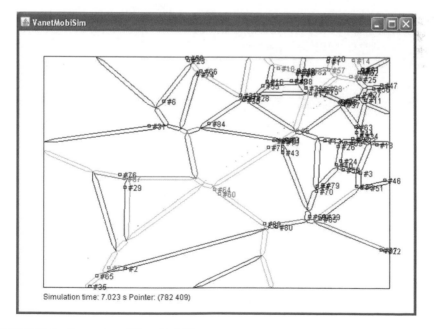

FIGURE 13.2 Screenshot of VanetMobiSim.

including globally specified extensions (e.g., maps, trip models, outputs) and node-specific extensions (e.g., mobility model).

In addition to importing GDF maps, both CanuMobiSim and VanetMobiSim allow the user to manually specify a topology. In this case, the graph's vertices and edges must be specified using the <vertex> and <edge> tags. The XML configuration file also contains globally specified extensions, such as nodes' initial positions, nodes' trips, and so on. Each node has a start and destination. The path can be manually specified or can be generated using either the Dijkstra shortest-path algorithm or Dial's STOCH algorithm.[37]

CanuMobiSim and VanetMobiSim provide three types of output: *LoaderOutput* to record the processing of the simulation scenario, *DebugOutput* to track simulation debug information, and a trace file for a network simulator (ns-2 or GloMoSim). The following is an example of a trace file generated for ns-2:

```
$node_(39) set X_ 600.000001
$node_(39) set Y_ 1.0E-6
$node_(39) set Z_ 0.0
. . .
$node_(44) set X_ 1300.000001
$node_(44) set Y_ 2000.000001
$node_(44) set Z_ 0.0
. . .
$ns_ at 65.0 "$node_(22) setdest 1834.9902902243118
  807.000001 0.8378965"
$ns_ at 65.0 "$node_(23) setdest 593.000001
  1588.8521043171565 18.421394"
$ns_ at 65.0 "$node_(24) setdest 593.000001
  1687.0351099518518 17.74068"
. . .
$ns_ at 83.0 "$node_(52) setdest 1293.000001
  1710.677801020203 17.030092"
$ns_ at 83.0 "$node_(53) setdest 593.000001
  1913.3284389187972 7.75004"
$ns_ at 83.0 "$node_(54) setdest 593.000001 1940.000001
  5.794535"
. . .
```

This sets initial positions, destinations, and speeds for various nodes in the simulation.

13.3.7 SUMO

SUMO (Simulation of Urban Mobility)[38,39] is an open-source mobility simulator written in C++ that uses Random Waypoint path movement and the Krauß car-following model. SUMO supports maps from TIGER/Line and ESRI.[40] MOVE[41] is an extension to SUMO that adds a GUI for describing maps and defining vehicle movement and allows the user to import Google Earth maps. MOVE also includes a visualization tool that allows users to view the generated mobility trace.

The input files in SUMO are output files from two other programs, `netconvert` and `duarouter`. Figure 13.3 shows the relationship among these programs and their input and output files. The user would employ this series of programs to generate the input files for SUMO. `traceExporter` is used to export the mobility trace generated by SUMO to ns-2, or another network simulator.

The first program, `netconvert`, converts the user-defined map (specified in two XML files, nodes.xml and edges.xml) into a road-network file, net.xml, one of the two input files to `duarouter`. nodes.xml includes three fields for each node, or road intersection: id, x-coordinate and y-coordinate. An example is `<node id="0" x="0" y="800">`. edges.xml includes edge information: origin node, destination node, edge id, and number of lanes. An example of an edge is `<edge fromnode="8" id="edge1" tonode="10" nolanes="4">`.

`duarouter` is used to create routes for vehicles and generates routes.xml. It takes two input files: flow.xml and net.xml (output from `netconvert`). flow.xml includes id, origin-edge, destination-edge, start-time, end-time, and number of vehicles. An example

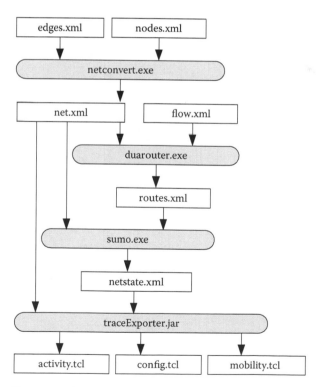

FIGURE 13.3 The input and output files for SUMO.

of a route is <flow id="flow0" from="edge1" to="edge13" begin="0" end="600" no="30">.

sumo takes two input files, net.xml (output from netconvert) and routes.xml (output from duarouter), and generates a simulation record file, netstate.xml.

traceExporter parses netstate.xml from sumo and net.xml from netconvert to create files for input into a network simulator. For example, if the network simulator chosen is ns-2, traceExporter will generate ns-2 activity, mobility and configuration files.

13.3.8 Summary

This section has provided an overview of prominent vehicular mobility simulators. The first set of simulators, CORSIM, VISSIM, and PARAMICS, are powerful commercial microscopic traffic simulators. The advantages of these simulators are that they are widely used and validated by traffic engineers and provide user-friendly graphical interfaces. The disadvantages are that they are not freely available and are often too detailed for most VANET evaluation purposes.

The second set of simulators presented are freely available, and many of them are open-source. Unfortunately, because most of these have been developed at universities, they may have limited support. In addition, only a few of these simulators, notably VanetMobiSim, have been validated against the commercial tools.

13.4 Network Simulators

Because vehicular networks involve solely wireless communications, all of the networking simulators described here support performing simulations with mobile wireless nodes. As described earlier, a mobility simulator is generally used to produce node movement traces that are then fed to the network simulator. The network simulator then controls the communications between the mobile nodes. As these network simulators support wireless communication, most of them include at least a simple node mobility model.

13.4.1 OPNET

OPNET[42] is a commercial network simulator mainly used for simulations of wired and wireless communications networks. It supports a wide range of wireless technologies such as MANETs, IEEE 802.11 wireless LANs, WiMAX, Bluetooth, and satellite networks. OPNET provides a graphical editor interface to build models for various network entities from physical layer modulator to application processes, and includes graphical packages and libraries for presenting simulation scenarios and results.

There are three basic phases of the OPNET deployment process. First, choose and configure node models to use in simulations, such as a wireless node, trajectory, and so on. Second, build and organize the network by setting up connections for different entities. Third, select the desired statistics (local or global) to collect during the simulation.

OPNET includes simple mobility options with trajectories:

1. *Random Drunken Model*: At each intersection, the vehicle randomly selects from four directions or chooses to remain stationary.

2. *Random Waypoint Model*: The vehicle randomly picks a destination. Upon reaching the destination, the vehicle pauses a random amount of time and then picks a new destination.
3. *Trace*: Vehicle movement is based on an imported trace file.

Each mobile node, or vehicle in the case of VANETs, is assigned a trajectory, which can be generated from a microscopic traffic simulator (using the trace option). The scenario file describes network parameters, mobility parameters, and global parameters (e.g., simulation time). A configuration file converter retrieves specific elements from the network and node configuration files and reorders these elements into a standard (default) configuration format. The network configuration file specifies all of the network-related parameters, for example, nodes, routers, applications, protocols, radios, and so on. The node configuration file includes the nodes' identities (such as IP addresses). The global parameters define the simulation time, coordinate system, random number seed, file-based node placement, protocol stack, statistic filter, and so on. In addition to the GUI-based output, the simulation produces a statistics file, which contains all the statistics generated from the simulation run. Statistics can also be exported directly from the generated plots.

13.4.2 GloMoSim and QualNet

GloMoSim[43,44] was developed at UCLA in 1999 to support MANET simulations. It includes a large set of wireless and MANET routing protocols, as well as physical layer implementations. GloMoSim also supports Random Waypoint, Random Drunken, and Trace-based mobility models. GloMoSim is still available for download only for educational purposes, but its last release was in December 2000.

QualNet,[45] developed by Scalable Network Technologies, is the commercial version of GloMoSim. QualNet includes a sophisticated GUI for setting up and running simulations, and provides a large set of wireless physical and MAC layer models. It also includes the following mobility models: group mobility, pedestrian mobility, and Random Waypoint. QualNet is available for both Windows and Unix/Linux platforms.

GloMoSim's input files are `nodes.input`, `mobility.in`, `app.conf`, and `config.in`. The main configuration parameters for setting up a scenario are defined in `config.in`, which specifies the mobility, simulation time, radio-related parameters, all layers' protocols, and the application configuration. The file `node.input` specifies the nodes' topology (coordinates of each node). The file `app.conf` specifies the applications that generate traffic, and `mobility.in` specifies the `trace`, or trip, of nodes. An output file `GLOMO.STAT` is generated at the end of a simulation, which contains statistical information for each node at a certain layer. For example, the information provided for a sender is the number of unicast packets, the number of broadcast packets, the total number of bytes, and the total number of packets.

The inputs to QualNet are similar to the ones for GloMoSim, but the format is slightly different. In addition, because QualNet is a popular network simulator in the research community, several mobility simulators output movement traces suitable for import into QualNet. The output files in QualNet include the ones in GloMoSim, for example `GLOMO.STAT`, but there is also animation output as well as report tables and figures.

13.4.3 ns-2

ns-2[46,47] is an open-source discrete event network simulator that supports both wired and wireless networks, including many MANET routing protocols and an implementation of the IEEE 802.11 MAC layer. There are implementations of several mobility models available for ns-2, including Random Trip Mobility[48] and Semi-Markov Smooth Mobility.[49] ns-2 simulates the wireless physical layer and the important parameters that influence its behavior (e.g., channel fading). ns-2 is the most widely used simulator for academic networking research.

The core of ns-2 is written in C++, but users interact with ns-2 by writing TCL scripts. This script should contain all of the commands needed for specifying the simulation (e.g., setting up the topology, specifying wireless parameters, and so on). As with QualNet, several of the mobility simulators can generate node descriptions and movement traces suitable for use in ns-2.

A typical wireless ns-2 simulation produces an event trace file and an animation trace file, used by the included utility nam to provide animation of the simulation. The event trace file includes packet enqueue (transmission), packet dequeue (forwarding), packet drops, and packet reception. An example of a wireless event trace file is given below:

```
N -t 20.222848 -n 34 -e 9999.128810
N -t 20.222848 -n 0 -e 9999.013416
. . .
r 20.223646686 _33_ MAC --- 0 AODV 48 [0 ffffffff 12 800]
    [energy 9999.012389 ei 0.681 es 0.000 et 0.001 er 0.306]
    ----- [18:255 -1:255 28 0] [0x2 3 1 [3 0] [2 4]] (REQUEST)
r 20.223646827 _38_ MAC --- 0 AODV 48 [0 ffffffff 12 800]
    [energy 9999.012709 ei 0.681 es 0.000 et 0.001 er 0.306]
    ----- [18:255 -1:255 28 0] [0x2 3 1 [3 0] [2 4]] (REQUEST)
. . .
r 20.223671686 _33_ RTR --- 0 AODV 48 [0 ffffffff 12 800]
    [energy 9999.012389 ei 0.681 es 0.000 et 0.001 er 0.306]
    ----- [18:255 -1:255 28 0] [0x2 3 1 [3 0] [2 4]] (REQUEST)
r 20.223671827 _38_ RTR --- 0 AODV 48 [0 ffffffff 12 800]
    [energy 9999.012709 ei 0.681 es 0.000 et 0.001 er 0.306]
    ----- [18:255 -1:255 28 0] [0x2 3 1 [3 0] [2 4]] (REQUEST)
. . .
s 20.269883256 _7_ AGT --- 64 cbr 512 [0 0 0 0] [energy
    9998.931443 ei 0.681 es 0.000 et 0.211 er 0.176] -----
    [7:2 9:0 32 0] [63] 0 2
```

The first two lines (starting with N) indicate the initial state of two nodes (numbers 34 and 0), including the time they enter the simulation (after the -t tag) and their initial energy (after the -e tag). The remaining lines indicate packet reception (lines starting with r) and packet transmission (lines starting with s). The trace file includes information such as the time the event occurred, the node associated with the event, the packet protocol type, packet size, time-to-live (TTL), and sender–receiver pair of the packet.

13.4.4 J-Sim

J-Sim[50] is an open-source simulation environment, developed entirely in Java. J-Sim provides two mobility models: trajectory-based and random waypoint. J-Sim is presented as an alternative to ns-2, because it is designed to be easier to use. In J-Sim, applications are built as a set of components that can be designed and tested separately. J-Sim can take a TCL file as input, similar to ns-2, but with a different format. Like ns-2, J-Sim produces an event trace file and an animation file, suitable for use in nam.

13.4.5 OMNeT++

OMNeT++[51] is an open-source simulation environment. The primary simulation applications are Internet simulations, mobility, and ad hoc simulations. OMNeT++ has a component-based design, meaning that new features and protocols can be supported through modules. OMNeT++ supports network and mobility models through the independently developed Mobility Framework and INET Framework modules. Simulation design in OMNeT++ is GUI-based, and output data can be plotted through the GUI as well. OMNEST[52] is the commercial version of OMNeT++, offered by Simulcraft, Inc.

13.4.6 SWANS

SWANS (Scalable Wireless Ad hoc Network Simulator)[53] was developed to be a scalable alternative to ns-2 for simulating wireless networks. Based on comparisons of SWANS, GloMoSim, and ns-2,[54] SWANS was determined to be the most scalable and the most efficient in memory usage with the fastest runtime. Along with better performance, SWANS delivered similar results as ns-2, at least for the network components that were implemented in both.

The input for SWANS is a Java file that creates the nodes and specifies how these nodes should move (the node movement scenario) and how they should communicate (the communication scenario). The user can select any of the ready-made applications in SWANS and associate it with any node(s) to execute it at the node application layer. Also, SWANS gives the user the flexibility to build a custom application and execute it at the application layer of any node.

13.5 Tightly Integrated Simulators

As mentioned before, the simulation of VANET applications not only requires simulating the wireless communication between the vehicles, but also requires simulating the mobility of the vehicles. Unfortunately, these two aspects of VANET simulation have often been decoupled. Both vehicular mobility and wireless communication have large communities concerned with their modeling and simulation, so high-quality simulators exist in each of these areas, as discussed in the previous sections.

The problem is how to merge the two types of simulators (network simulator and mobility simulator). A simple method to achieve this merge and create a tightly integrated simulation is to implement mobility models in a network simulator, but without allowing the network messages to feed back to the mobility model. This kind of simulation is called *one-way communication* (from mobility model to network). These types of simulators are suitable for simulating infotainment-related VANET applications, including Internet connectivity, multimedia applications, and peer-to-peer applications, where the communication does not affect vehicles' movements.

In contrast, tightly integrated simulators that offer *two-way communication* usually consist of two subsimulators (network and mobility) that can communicate with each other. These simulators are more appropriate for safety-related and traffic information applications that assume that feedback from the network will affect vehicles' movements. In these types of applications, the traffic simulator feeds the network simulator with position information, speed, acceleration, direction, and so on. The VANET application that runs at the top level of the network simulator incorporates this information with surrounding vehicles' information in order to notify the driver of upcoming congestion or a possible collision. Based on this notification, driving decisions (i.e., vehicle mobility) may be affected. For example, in a congestion notification system, the driver may choose to change lanes or take a different path. These decisions need to propagate back to the mobility simulator to be reflected in the vehicle mobility information.

Usually, any simulator has an events queue to store the events that should be executed according to their scheduled execution time. In the case of two-way communication simulators, each subsimulator has its own events queue. These two events queues can be combined into one events queue, or they can be kept separate, which implies that extra overhead will be needed for synchronization. Based on this decision, the two-way communication simulators are separated into two categories: those with a single events queue and those with two events queues.

Having a single events queue can be achieved through implementing one of the subsimulators in the other. Often, the vehicular mobility subsimulator is implemented in the network subsimulator, as in ASH,[7] or the two simulators can be highly integrated together, as in NCTUns.[55] The advantages of having a single events queue are that the vehicles' mobility events and the network events will be inserted in the same queue, which removes the burden of synchronizing the two types of events. In addition, the simulation will be more efficient from the execution time and memory consumption perspectives. The main disadvantage of having a single events queue is that the process of maintaining and extending such simulators is not easy.

With two events queues, two-way communication is achieved through an interface that is implemented between the network subsimulator and the mobility subsimulator. The main function of that interface is to update each subsimulator with the recent events in the other subsimulator. Also, it synchronizes the event execution in each of the events queues. The main disadvantages are that these types of simulators consume more memory and execution time.

In this section, tightly integrated simulators with one-way communication will be discussed first. However, only those simulators that attempt to package the mobility and network simulator in a single program are included, rather than those that manually feed the mobility simulator's movement trace to the network simulator. Next will come an overview of tightly integrated simulators with two-way communication and two events queues, and finally tightly integrated simulators with two-way communication and a single events queue.

13.5.1 SWANS++

SWANS++[56] extends the network simulator SWANS by adding a GUI to visualize the scenario and a mobility model, STRAW (STreet RAndom Waypoint),[57] for the vehicles' movement in street scenarios. STRAW uses the simple random waypoint mobility model, but it restricts the vehicles' movement to real street boundaries, loaded from TIGER/Line data files.

STRAW consists mainly of three components: intrasegment mobility, intersegment mobility, and route management and execution. In intersegment mobility, the vehicles move according to a car-following model and change their speed only in certain situations:

1. When the vehicle arrives at an intersection and the next segment is full, the vehicle stops until the next segment has a free slot.
2. When the vehicle has a vehicle in front of it, the vehicle adjusts its speed accordingly in order to maintain a certain distance in between.
3. When the vehicle arrives at a traffic control or stop sign.
4. When the vehicle makes a turn.

For intersegment mobility, according to the system design, there is either a traffic control sign or a stop sign at each intersection that forces the vehicle to alter its speed. The mobility model implemented in STRAW (and therefore, SWANS++) does not support lane changing. The route management and execution (RME) module is responsible for determining the vehicles' routes during the simulation. The RME has two techniques to fulfill its task. The first technique is simple intersegment mobility (simple STRAW) at which the vehicle's next segment is determined stochastically, while the second technique is origin–destination mobility (STRAW OD), in which the vehicle's route is predetermined based on the shortest path between the origin and destination. SWANS++ is a tightly integrated simulator, but it does not provide feedback between the mobility and networking modules.

13.5.2 GrooveNet

GrooveNet[58,59] (originally known as GrooveSim) is an integrated network and mobility simulator that allows communication between real and simulated vehicles. Originally, GrooveNet extended the open-source simulator roadnav[60] by adding a network model and a GUI based on Qt.[61] GrooveNet can load real street maps from the TIGER/Line database in order to simulate vehicles' mobility on real roads. GrooveNet includes fixed mobility, street speed, uniform speed, and car-following mobility models. GrooveNet supports many operational modes, including drive mode, simulation mode, playback mode, hybrid simulation mode, and test generation mode. These modes. GrooveNet with capabilities such as communicating with surrounding vehicles to get real traffic information, executing a specific scenario, mixing between the previous two modes, and finally playing back the log file generated during any of the modes' operation for further analysis. GrooveNet's unique ability to integrate simulated vehicles with real vehicles allows it to function as testbed software as well as a simulator.

13.5.3 TraNS

TraNS (Traffic and Network Simulation Environment)[6] can be called the first VANET simulator. It was the first work to combine a network simulator, ns-2, with a vehicular mobility simulator, SUMO, and to provide feedback from the network simulator to the mobility simulator. TraNS can operate in two modes: network-centric mode and application-centric mode. In the network-centric mode, there is no feedback provided from ns-2 to SUMO, so the vehicles' mobility trace file can be pregenerated and fed to the network simulator later. The link between the two simulators in this case is done through a parser that analyzes the mobility trace file generated by SUMO and converts it to a suitable format for ns-2. In the application-centric mode, the feedback between ns-2 and SUMO is provided through an interface called TraCI.[62] In this mode the two simulators (SUMO and ns-2) must run simultaneously. TraCI achieves the link between ns-2 and SUMO by converting the mobility commands coming from ns-2 to a sequence of mobility primitive commands such as stop, change lane, change speed, and so on that can be sent to SUMO. As both simulators are running separately at the same time, the two-way communication in application-centric mode uses two separate events queues.

13.5.4 Veins

Veins (Vehicles in Network Simulation)[63] is another simulator that couples a mobility simulator with a network simulator. In Veins, SUMO is paired with OMNeT++ by extending SUMO to allow it to communicate with OMNeT++ through a TCP connection. In order to create a bidirectional communication between the two simulators, OMNeT++ has also been extended by adding a module that allows all participating nodes (vehicles) to send commands via the established TCP connection to SUMO. In this case, the two extensions represent the interface between the network simulator and the mobility simulator. Thus, the network simulator can react to the received mobility trace from the mobility simulator by introducing new nodes, by deleting nodes that have reached their destination, and by moving nodes according to the instructions from the mobility simulator.

In Veins, there is a manager module that is responsible for synchronizing the two simulators. At regular intervals, the manager module triggers the execution of one timestep of the traffic simulation, receives the resulting mobility trace, and triggers position updates for all modules it had instantiated. Thus, as with TraNS, this simulator has two separate events queues.

13.5.5 NCTUns

NCTUns 4.0 (National Chiao Tung University Network Simulation 4.0)[55] implements two-way communication with a single events queue. NCTUns 1.0 was developed only as a network simulator, but the most recent version, NCTUns 4.0, integrates some traffic simulation capabilities, such as designing maps and controlling vehicles' mobility. A large variety of maps can be designed using different types of supported road segments (e.g., single-lane roads, multilane roads, crossroads, T-shape roads, and lane-merging roads). Also, NCTUns includes a GUI to aid in the design process of the maps.

The supported vehicular movement has two modes, prespecified and autopilot. In the prespecified movement mode, the scenario designer specifies the moving path and the speed for each vehicle. In autopilot mode, the scenario designer specifies the following parameters for each vehicle: initial speed, maximum speed, initial acceleration, maximum acceleration, maximum deceleration, and so on. Then, the autopilot selects the best route to navigate in the map. Autopilot mode is also capable of performing car following, lane changing, overtaking, turning, and traffic light obeying. The main disadvantage of NCTUns 4.0 is that its network stack (network subsimulator) is not validated against any well-known network simulator. Although the logic behind the vehicles' movements considers the road conditions and the traffic light signal states ahead of the vehicle, it does not match any known vehicular mobility model. The code for the vehicles' movement logic is integrated with the network simulation code, which make itdifficult to extend.

13.5.6 Gorgorin et al.

Gorgorin et al.[64] developed an integrated vehicular network simulator that allows feedback between the network and mobility modules using a single event queue. Map information is imported from the TIGER/Line database, but because the information in these maps does not include the number of lanes or traffic control information, the authors augment the maps with this missing information heuristically.

For the vehicular mobility model, the authors implemented a microscopic traffic simulator that is based on the driver's behavior, assuming that the driver will be in one of the following four modes: free driving, approaching, following, or braking. The driver's behavior is determined based on the distance to the preceding vehicle. Not all drivers have the same personality, so the simulator allows for different driver profiles (i.e., aggressive, regular, and calm). The mobility model implemented has been validated against traces taken from a German freeway and a U.S. freeway.[65]

For the network simulator, the authors implemented a physical layer that depends on both cumulative noise calculation and signal to noise ratio (SNR) to determine whether to accept or drop the received packet. Also, they have implemented CSMA/CA to represent the MAC layer. For the routing layer, they have implemented a geographical routing protocol. The new feature in this simulator is that the authors implemented a model to compute the fuel consumption and pollutant emissions in order to find the relationship between these measurements and the vehicle's speed and acceleration.

13.5.7 ASH

ASH (Application-aware SWANS with Highway mobility)[7] is an extension of the wireless network simulator SWANS that implements the IDM vehicular mobility model and

MOBIL lane changing. ASH supports feedback between the vehicular mobility subsimulator and the network subsimulator, making it one of the two-way communication simulators with a single events queue. ASH allows users to design a simple highway segment and customize it by specifying the directions (one-way or two-way), the number of lanes, the number of entries and exits and their corresponding locations along the segment.

In addition to adding highway mobility models to SWANS, ASH extends the node types available:

1. *Mobile Communicating Node.* This represents a participating vehicle that should execute a user-defined application, which specifies how the vehicle should behave.
2. *Mobile Silent Node.* This represents a nonparticipating vehicle that should execute a null application so that it will not be able to send or receive any messages.
3. *Static Communicating Node.* This represents road-side infrastructure that should execute a user-defined application, which specifies how the road-side unit should behave. Also, this kind of node may have different physical layer characteristics (e.g., transmission power) than the mobile communicating nodes.
4. *Static Silent Node.* This represents a road obstacle that should execute a null application.

In particular, the addition of the mobile silent node is important to allow testing of protocols under different penetration rates, where not all vehicles are equipped with communication devices. The location of the static silent node can either be predetermined before running the simulation or can be determined at runtime, in order to simulate an accident, for example.

Because most VANET applications use flooding-based techniques to disseminate data, ASH also implements the Inter-Vehicle Geocast protocol (IVG).[66] Moreover, it supports a probabilistic version of IVG[67] to take the surrounding traffic density into account. ASH also implements statistical and logging utilities to support simulations. The utilities provide information about the simulation entities at different granularity. This information can be retrieved at the simulation level, lane level, vehicle level, or message type level. The statistics utility provides statistics about all possible events that can occur in the simulation.

ASH accepts a configuration file for the highway scenario. The nodes' creation and the communication scenario should be specified in a Java file as done in SWANS.

13.6 Conclusion

There is currently no standard, or even preferred simulator for evaluating vehicular networking research. Most often, researchers have combined existing mobility simulators with existing network simulators by generating mobility traces that are fed to the network simulator. This loose integration allows no feedback between the two simulators and is not suitable for simulating congestion notification or collision warning applications where vehicle movements will be affected by received messages. More recently, tightly integrated simulators have been developed that combine both mobility and networking components into a single application. More advanced simulators allow for two-way communication between the mobility and networking components to allow for feedback between the network and vehicle movement. This chapter has provided an overview of the current landscape of vehicular network simulation. New developments in simulators providing feedback between realistic mobility models and realistic networking models will help researchers to better evaluate their emerging systems.

References

1. Camp, T., Boleng, J., and Davies, V., A survey of mobility models for ad hoc network research, *Wireless Communications and Mobile Computing*, 2(5), 483–502, 2002.
2. Kurkowski, S., Camp, T., and Colagrosso, M., MANET simulation studies: The incredibles, *ACM SIGMOBILE Mobile Computing and Communications Review*, 9(4), 50–61, 2005.
3. Hogiea, L., Bouvry, P., and Guinand, F., An overview of MANETs simulation, Electronic Notes in Theoretical Computer Science, 150(1), 81–101, 2006.
4. Häerri, J., Filali, F., and Bonnet, C., Mobility models for vehicular ad hoc networks: A survey and taxonomy, IEEE Communications Surveys and Tutorials, 2009 (to be published).
5. Häerri, J., Fiore, M., Filali, F., Bonnet, C., Casetti, C., and Chiasserini, C.F., A realistic mobility simulator for vehicular ad hoc networks, in Proceedings of the Workshop on Wireless Communications (EWCOM), in conjunction with IEEE ICC, Istanbul, Turkey, June 2006.
6. Piorkowski, M., Raya, M., Lezama Lugo, A., Papadimitratos, P., Grossglauser, M., and Hubaux, J.-P., TraNS: Realistic joint traffic and network simulator for VANETs, *ACM SIGMOBILE Mobile Computing and Communications Review Special Issue*, April 2008.
7. Ibrahim, K. and Weigle, M.C., ASH: Application-aware SWANS with highway mobility, in Proceedings of IEEE INFOCOM Workshop on MObile Networking for Vehicular Environments (MOVE), April 2008.
8. Center for Microcomputers in Transporation (McTrans), Traffic software integrated system—corridor simulation (TSIS-CORSIM), available at http://mctrans.ce.ufl.edu/featured/tsis/.
9. Federal Highway Administration, Traffic analysis tools—CORSIM, available at http://ops.fhwa.dot.gov/trafficanalysistools/corsim.htm.
10. PTV America, VISSIM, available at http://www.ptvamerica.com/vissim.html.
11. Quadstone, Inc., The PARAMICS transportation modeling suite, available at http://www.paramicsonline.com/.
12. Choa, F., Milam, R., and Stanek, D., CORSIM, PARAMICS, and VISSIM: what the manuals never told you, in Proceedings of the TRB Conference on the Application of Transportation Planning Methods, Baton Rouge, LA, April 2003, pp. 392–402.
13. Gollu, A. and Varaiya, P., SmartAHS: A simulation framework for automated vehicles and highway systems, *Mathematical and Computer Modeling*, 27(9), 103–128, 1998.
14. Antoniotti, M. and Göllü, A., SHIFT and SmartAHS: A language for hybrid systems engineering, modeling, and simulation, in Proceedings of the USENIX Conference of Domain Specific Languages, Santa Barbara, CA, October 1997, pp. 171–182.

15. Kourjanski, M., Gollu, A., and Hertschuh, F., Implementation of the smartahs using shift simulation environment, in Proc. of SPIE Conference on Intelligent Systems and Advanced Manufacturing, Vol. 3207, Pittsburgh, PA, October 1997, pp. 192–202.

16. California partners for advanced transit and highways (PATH), available at http://www.path.berkeley.edu.

17. Delorme, D. and Song, B., Human driver model for SmartAHS, Technical report, California PATH, University of California, Berkeley, April 2001.

18. Bellet, T., Modélisation et simulation cognitive de l'opérateur humain: une application à la conduite automobile, Ph.D. thesis, Paris Université Paris V, 1998.

19. Shift: The hybrid system simulation programming language, available at http://path.berkeley.edu/ shift.

20. Smart AHS, available at http://path.berkeley.edu/smart-ahs/index.html.

21. Kourjanski, M. and Hertschuh, F., Highway description compiler manual, available at /urlhttp://path.berkeley.edu/smart-ahs/doc/hwyc.ps, March 1998.

22. TkShift, available at http://path.berkeley.edu/shift/TkShift/TkShift.html.

23. Treiber, M., Microsimulation of road traffic, available at http://www.traffic-simulation.de/.

24. Treiber, M., Hennecke, A., and Helbing, D., Congested traffic states in empirical observations and microscopic simulations, *Physical Review E*, 62(2), 1805–1824, 2000.

25. Kesting, A., Treiber, M., and Helbing, D., MOBIL: General lane changing model for car following models, *Transportation Research Record: Journal of the Transportation Research Board*, 86–94, January 2007.

26. CanuMobiSim, available at http://canu.informatik.uni-stuttgart.de.

27. International Organization for Standardization (ISO), Intelligent transport systems—geographic data files (GDF)—overall data specification, ISO 14825:2004, 2004.

28. Sanchez, M., Manzoni, P., and Haas, Z.J., Determination of critical transmission range in ad-hoc networks, in Proceedings of Multiaccess Mobility and Teletraffic for Wireless Communications Workshop (MMT'99), Venice, Italy, October 1999.

29. Haas, Z.J. and Pearlman, M.R., The performance of a new routing protocol for the reconfigurable wireless networks, in Proceedings of IEEE International Conference on Communications (ICC '98), 1998, pp. 156–160.

30. Johnson, D.B., and Maltz, D.A., Dynamic Source Routing in Ad Hoc Wireless Networks, in Mobile Computing Vol. 353, Kluwer Academic Publishers, 1996, pp. 153–181.

31. Tian, J., Häehner, J., Becker, C., Stepanov, I., and Rothermel, K., Graph based mobility model for mobile ad hoc network simulation, in Proceedings of the 35th Annual Simulation Symposium, Washington, DC, 2002, p. 337.

32. Seskar, I., Marie, S.V., Holtzman, J., and Wasserman, J., Rate of location area updates in cellular systems, in Proceedings of IEEE Vehicular Technology Conference (VTC'92), Denver, CO, May 1992.

33. Bettstetter, C., Smooth is better than sharp: A random mobility model for simulation of wireless networks, in Proceedings of the ACM International Workshop on Modeling, Analysis, and Simulation of Wireless and Mobile Systems (MSWiM), Rome, Italy, June 2001.

34. VanetMobiSim project, available at http://vanet.eurecom.fr.
35. Härri, J., Fiore, M., Fethi, F., and Bonnet, C., Vanetmobisim: Generating realistic mobility patterns for vanets, in Proceedings of the 3rd International Workshop on Vehicular ad hoc Networks, Los Angeles, CA, September 29, 2006, p. 96.
36. U.S. Census Bureau, Topologically Integrated Geographic Encoding and Referencing System (TIGER), available at http://www.census.gov/geo/www/tiger.
37. Dial, R.B., A probabilistic multipath traffic assignment model which obviates path enumeration, *Transportation Research*, 5(2) 83–111, 1971.
38. SUMO, available at http://sumo.sourceforge.net/.
39. Krajzewicz, D., Bonert, M., and Wagner, P., The open source traffic simulation package SUMO, in RoboCup 2006 Infrastructure Simulation Competition, Bremen, Germany, 2006.
40. ESRI, Data and maps, available at http://www.esri.com/data/data-maps/overview.html.
41. Karnadi, F.K., Mo, Z.H., and Lan, K.-C., Rapid generation of realistic mobility models for VANET, in Proceedings of the IEEE Wireless Communications and Networking Conference (WCNC), March 2007, pp. 2506–2511.
42. OPNET Technologies, available at OPNET. http://www.opnet.com/.
43. GloMoSim, available at http://pcl.cs.ucla.edu/projects/glomosim/.
44. Zeng, X., Bagrodia, R., and Gerla, M., GloMoSim: A library for parallel simulation of large-scale wireless networks, in Proceedings of the Workshop on Parallel and Distributed Simulation, 1998, pp. 154–161.
45. Scalable Network Technologies, Qualnet, available at http://www.scalable-networks.com/index. php.
46. McCanne, S. and Floyd, S., ns Network Simulator, available at http://www.isi.edu/nsnam/ns/.
47. Breslau, L., Estrin, D., Fall, K., Floyd, S., Heidemann, J., Helmy, A., Huang, P., McCanne, S., Varadhan, K., Xu, Y., and Yu, H., Advances in network simulation, *IEEE Computer*, 33(5), 59–67, 2000.
48. PalChaudhuri, S., Le Boudec, J.-Y., and Vojnovic, M., Perfect simulations for random trip mobility models, in Proceedings of the 38th Annual Simulation Symposium, San Diego, CA, April 2005.
49. Zhao, M. and Wang, W., A novel semi-markov smooth mobility model for mobile ad hoc networks, in Proceedings of IEEE Globecom, San Francisco, CA, November 2006.
50. J-Sim, available at http://www.j-sim.org/.
51. OMNeT++, available at http://www.omnetpp.org/.
52. OMNEST, available at http://www.omnest.com.
53. Barr, R., Haas, Z., and van Renesse, R., Scalable wireless ad hoc network simulation, in *Handbook on Theoretical and Algorithmic Aspects of Sensor, Ad hoc Wireless, and Peer-to-Peer Networks*, CRC Press, 2005, pp. 297–311.
54. Kargl, F. and Schoch, E., Simulation of MANETs: A qualitative comparison between JiST/SWANS and ns-2, in Proceedings of the International Workshop on System Evaluation for Mobile Platforms (MobiEval), San Juan, Puerto Rico, 2007, pp. 41–46.

55. Wang, S.Y., Chou, C.L., Chiu, Y.H., Tzeng, Y.S., Hsu, M.S., Cheng, Y.W., Liu, W.L., and Ho, T.W., NCTUns 4.0: An integrated simulation platform for vehicular traffic, communication, and network researches, in Proceedings of IEEE VTC-Fall, 2007, pp. 2081–2085.
56. SWANS++, available at http://www.aqualab.cs.northwestern.edu/projects/swans++/.
57. Choffnes, D.R. and Bustamante, F.E., An integrated mobility and traffic model for vehicular wireless networks, in Proceedings of ACM VANET, Cologne, Germany, 2005, pp. 69–78.
58. Mangharam, R., Weller, D.S., Stancil, D.D., Rajkumar, R., and Parikh, J.S., GrooveSim: a topography-accurate simulator for geographic routing in vehicular networks, in Proceedings of ACM VANET, Cologne, Germany, 2005, pp. 59–68.
59. Mangharam, R., Weller, D.S., Rajkumar, R., Mudalige, P., and Bai, F., GrooveNet: A hybrid simulator for vehicle-to-vehicle networks, in Proceedings of the International Workshop on Vehicle-to-Vehicle Communications (V2VCOM), San Jose, CA, July 2006.
60. Roadnav, available at http://roadnav.sourceforge.net/.
61. Qt cross-platform GUI development, available at http://trolltech.com.
62. Wegener, A., Piorkowski, M., Raya, M., Hellbrck, H., Fischer, S., and Hubaux, J.-P., TraCI: An Interface for Coupling Road Traffic and Network Simulators, in 11th Communications and Networking Simulation Symposium (CNS'08), 2008.
63. German, R., Sommer, C., Yao, Z., and Dressler, F., Simulating the influence of IVC on road traffic using bidirectionally coupled simulators, in Proceedings of IEEE INFOCOM Workshop on MObile Networking for Vehicular Environments (MOVE), April 2008.
64. Gorgorin, C., Gradinescu, V., Diaconescu, R., Cristea, V., and Ifode, L., An integrated vehicular and network simulator for vehicular ad-hoc networks, in Proceedings of the 20th European Simulation and Modelling Conference, 2006.
65. Fellendorf, M. and Vortisch, P., Validation of the microscopic traffic flow model VISSIM in different real-world situations, in Proceedings of the Transportation Research Board Meeting, 2001.
66. Bachir, A. and Benslimane, A., A multicast protocol in ad hoc networks inter-vehicle geocast, in Proceedings of IEEE VTC-Spring, April 22–25, 2003, pp. 2456–2460.
67. Ibrahim, K. and Weigle, M.C., CASCADE: Cluster-based accurate syntactic compression of aggregated data in VANETs, in Proceedings of IEEE MASS, 2008.

VI

Human Factors

14

Mental Workload and Driver Distraction with In-Vehicle Displays

Carryl L. Baldwin
Department of Psychology
George Mason University

14.1 Introduction

Emerging technologies present the opportunity to provide drivers with a vast amount of information, much of which is intended to keep the driver safe and to increase transportation efficiency. For example, collision avoidance warnings have the potential to augment the sensory capabilities of the driver, serving as a second set of eyes or ears in heavy traffic or when the driver is fatigued. In-vehicle navigation systems can assist drivers with not only getting from one place to another but also in finding restaurants, gas stations, shops, and other points of interest. Well-designed navigation systems can help alleviate the anxiety some drivers have about getting lost and can reduce navigational inefficiency, resulting in tremendous economical and ecological savings. Adaptive speed control systems and lane departure warnings relieve the driver of some of the mental workload of vehicular control and monitoring. Although most of these systems were designed to increase safety by decreasing the mental effort involved in the driving, navigating, and collision avoidance tasks, there is ample reason to question their potential to distract, confuse, and disorient drivers. Driver distraction and the mental workload involved in the use of in-vehicle displays with different characteristics (i.e., modality, complexity) are the focus of this chapter. Particular emphasis is placed on auditory displays, which have the greatest potential to minimize both visual distraction and mental workload, but which may also divert attention from more important visual tasks. First, we will look at the prevalence of driver distraction and its impact on certain types of drivers most susceptible to variations in workload and distraction due to in-vehicle displays. Attention will then be given to the many types of emerging displays and their relative potential to affect the driving task. Finally, we will conclude with recommended design guidelines to mitigate the potential negative impact of these displays.

14.2 Driver Distraction and Its Impact

Driver distraction involves diverting the driver's attention away from the task of driving. So, before defining driver distraction and its prevalence, a more detailed description of the task of driving is warranted. Driving can be thought of as predominantly involving three subtasks of hierarchical priority. Listed in descending order, starting with the task of highest priority, the driver must maintain vehicular control, detect potential hazards, and navigate.[1] Many in-vehicle systems and their displays are designed to assist drivers with one of these three tasks and can therefore be thought of in a similar hierarchical priority fashion. Cruise control, adaptive cruise control, assisted braking, and lane departure warning systems are some of the many systems included in the first category. Collision avoidance systems, back-up assist, and enhanced night vision detection systems are examples of the second category. The third category includes the wide range of commercially available factory-installed and after-market navigational aids. An additional category of systems has no direct relevance to the task of driving but rather is designed to entertain the driver. This fourth class may be referred to as *infotainment* systems and includes devices ranging from a simple radio or compact disk player to wireless communications such as mobile phones, email, and other Internet functions. The potential for distraction is present with systems at all levels of the hierarchy,[2] as the following indicators demonstrate.

14.2.1 Prevalence

Driver distraction is a real and persistent threat to traffic safety. Distraction, a specific type of driver inattention, refers to a situation where some person, object, event, or interface has the driver's attention rather than the driver being just "lost in thought."[3] Estimates of the number of collisions caused by distraction vary. It has been estimated that roughly 25 to 30% of all crashes that come to the attention of police involve driver inattention or distraction.[3] The actual number was always suspected to be even higher, because people are often reluctant to admit they were distracted prior to a crash where they are determined to be at fault. Imagine, particularly in the litigious U.S. culture, the reluctance that someone may have to admitting that he was reaching for a mobile phone or reading the headlines on the newspapers lying next to him as he rear-ended the car ahead of him.

Confirmation that these numbers were underestimates comes from the recent 100-Car Naturalistic Driving Study conducted by research scientists at Virginia Tech.[4]

They admittedly broadened the definition of driver inattention to include even driving-related but noncritical tasks such as checking rearview mirrors and instrument gauges. However, they observed that driver inattention of one category or another was a factor in nearly 80% of all observed crashes and a substantial proportion of near crashes.

As reviewed in several previous publications,[3,5] drivers devote their attention to a host of tasks other than driving. Nearly all drivers are distracted at least part of the time and the majority of drivers on U.S. roads are engaged in distracting activities roughly 30% of the time they are driving.[5,6] A majority of this distraction comes from conversing with other passengers and eating. Drivers on U.S. roadways engage in a host of activities while driving. They eat, drink, smoke cigarettes, groom themselves, reach for objects in the car (sometimes even in the back seat) and engage in conversations with people both in the vehicle and over mobile phone devices.

The increasing prevalence of in-vehicle devices and displays has the potential to increase driver distraction in at least two significant ways. First, in-vehicle displays may directly distract drivers by diverting their attention away from other more relevant aspects of driving. Second, the increased automation afforded by sophisticated in-vehicle interfaces and systems may reduce mental workload to such an extent that the driver no longer perceives the need to stay directly engaged in the task of driving.[7] Parasuraman and Riley[8] forewarned of this overuse of automation. Certain types of drivers may be more prone to an overreliance on automated systems than others. For example, the two highest crash risk age groups—teens and seniors—are even more susceptible to driver distraction than are their middle-aged counterparts.

The 100-Car naturalistic driving study[4] has provided the largest sample to date of the everyday driving habits of people in the U.S. This extensive project allowed determination of "average" driving episodes or baseline driving situations as well as documentation of the driver status and events surrounding critical situations such as crashes and near-crashes. They found that approximately half of the near-crash episodes were associated with the driver taking his or her eyes and attention off the forward roadway. The majority of these episodes involved driver distraction (attention diverted to a secondary source). Using a cell phone was the most prevalent cause of driver distraction, playing a part in approximately 15% of the near-crashes. Another 10% of these episodes involved the driver being engaged with some other device in the vehicle in a task unrelated to the main task of driving.

The estimate of distraction caused by cell phones was found to be much lower in previous studies. Stutts et al.,[3] using the National Crashworthiness Database of police-reported crashes, estimated the number of distraction incidents that could be attributed to cell phone use to be only 1.5%. In a large-scale observational study in which 70 drivers were unobtrusively videotaped, Stutts et al.[5] found slightly higher rates of cell-phone-related distraction. Roughly 34% of drivers used their cell phones at some point while driving and these drivers were engaged in cell phone use about 3.8% of the time the vehicle was moving. The Stutts et al. data was collected during a year-long span over the period 2000 to 2001. Comparing their figures to those collected by Dingus et al.[4] several years later may suggest a trend toward increased cell phone usage and cell-phone-induced distraction.

The tendency to engage in distracting behaviors varies as a function of age. Stutts et al.[6] found that drivers under age 50 were much more likely to engage in a wide range of distracting behaviors compared to drivers over age 60 years. Drivers between the ages of 18 and 29 years had the highest rate of engagement in potentially distracting behaviors overall.

It is clear that in-vehicle technologies have the potential to impact the safety of drivers. Now we turn to the issue of whether or not these systems impact all drivers equally or whether some drivers may potentially benefit more or conversely be at greater risk of the distracting effects or potentially excessive mental workload stemming from in-vehicle displays. The relative plethora of increasingly inexpensive in-vehicle devices (i.e., the new Ford SYNC system) means that more system functionalities are available to a wider range of drivers than ever before. No longer the isolated instance of being a novel device in a few high-end luxury cars, advanced in-vehicle devices are making their way into the vehicles of young novice drivers and older drivers—two groups of drivers that are at even greater risk of crashes and, as we shall see, distraction also.

14.2.2 High-Risk Drivers

Crash risk can be calculated in numerous ways. One method is to examine the absolute number of crashes per number of licensed drivers stratified by age, gender, or geographic location involved. This method is a frequent choice of automobile insurance companies. As insurance rates indicate, when calculated using this method, young novice drivers— and males, in particular—are at highest risk of being involved in a severe automobile crash.[9] When calculated in terms of the number of licensed drivers, older drivers have no higher crash risk than their middle-aged cohorts.[10] Other calculation methods show a different picture.

An alternative method of calculating crash risk adjusts crash rates in terms of the number of vehicle miles traveled (VMT). This method takes into account the fact that although seniors do not tend to drive as often, when they do get behind the wheel, they are nearly as likely to be involved in a crash as a teenage male. When adjusting for the number of VMT, we see the highest crash risk in drivers under the age of 25 (young drivers) and drivers over the age of 70 (older drivers). As illustrated in Figure 14.1, when crash risk is calculated in terms of VMT traveled, both young and old drivers are at greater risk of being fatally injured than their middle-aged cohorts. Males maintain a higher traffic fidelity rate across the lifespan.

The highest crash risk, calculated both in terms of per trip and per VMT, is observed among the youngest drivers, at age 16 years.[11] The types of crashes that younger and older drivers have differ, although they all usually involve failures of attention at critical times. Both older and younger drivers have perceptual-attentional issues that make them more susceptible to driver distraction. The causes of their issues are different, but they share the commonality of an increased risk of crash and a greater likelihood that the driver will be at fault due to not paying attention (or not having adequate attentional capacity to pay attention) to the most relevant aspects of the driving task at the time of the incident.

14.2.2.1 Novice Drivers

The crashes of young drivers most frequently involve some type of risk-taking behavior.[12] Risk-taking behaviors engaged in by teens include driving at excessive speeds, following too close, and passing when there is little time to do so.[13] Of particular concern to teen

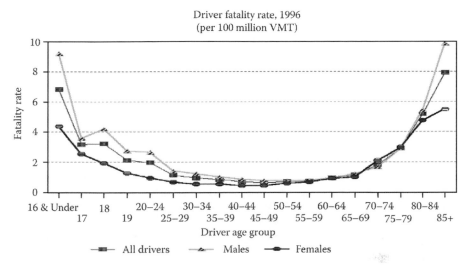

FIGURE 14.1 Driver fatality rate as a function of the number of vehicle miles traveled. (From Cerrelli, E., Crash data and rates for age–sex groups of drivers, 1996, Research Note, National Highway Traffic Safety Administration, Washington, DC, 1998.)

drivers is the potential for collision avoidance warnings to promote more risky driving behavior. If these young risk-taking drivers become over confident in the capabilities of their collision avoidance systems they may feel even more confident making risky maneuvers and/or pay even less attention to the task of driving. Drivers under the age of 20 are overrepresented in the distraction-related crashes.[3] Their most frequent type of distraction while driving involves some type of infotainment device—such as adjusting the radio, CD player, or other form of electronic media. Young drivers are more likely to use in-vehicle devices and are much more likely to use after-market devices that are not integrated with the automobile's other systems. Integrated systems ensure a priority system such that a cell phone ring does not interrupt collision avoidance system warning. After-market systems typically have no such integration.

Younger drivers are also much less likely to believe that their performance is disrupted by the use of an in-vehicle device. This more easily distracted, less experienced, and more risk-taking group of drivers poses a particular challenge to the design of safe in-vehicle devices.

14.2.2.2 Older Drivers

Older drivers are more likely to be distracted by a person or object outside the vehicle.[3] They are particularly challenged by such tasks as navigating in an unfamiliar area[14] and making unprotected left-hand turns, and other tasks that require judging the speed and distance of other vehicles at junctions.[15] Older drivers' accidents frequently involve looking but not seeing critical vehicles and other roadway users.[16] As Merat, Anttila, and Luoma[17] discuss, the accidents of older drivers are generally not due to careless or aggressive behaviors, but rather difficulty in handling complex traffic situations.

In-vehicle driver assistance systems may be particularly beneficial to older drivers if they are designed in accordance with the drivers' perceptual-cognitive capabilities.[18] For example, older drivers benefit from auditory route guidance directions alone or in combination with visual displays.[19] However, the auditory display must be well above masked threshold.[20] In other words, auditory warnings or navigational instructions must not only be audible, they must be at least 10 to 15 dB above the background noise in the vehicle. Baldwin et al. observed that when verbal messages were presented at loudness levels close to the background noise present in a simulated vehicle, older drivers' speed keeping became more variable. These near-threshold levels also resulted in older drivers misinterpreting a significantly higher number of messages relative to when the messages were presented at least 10 dB above ambient background noise levels. Additionally, older drivers need information regarding navigational turn points earlier than their younger counterparts.[21] All drivers benefit from the inclusion of a salient landmark in the navigational instruction.[22] For example, providing the instruction, "Turn Left on Baker Street at the Fire Station" results in less mental demand and fewer navigational errors than simply presenting, "Turn Left on Baker Street." Including salient landmarks are expected to particularly benefit older drivers. They reduce the need to search for street signs, which are frequently difficult to both find and read. Providing older drivers with salient landmarks is expected to be particularly advantageous during complex driving situations.

During periods of low to moderate driving demand, there is generally very little difference between the driving performance of older and younger individuals.[17,23] But when something unexpected happens or when the driving situation becomes particularly complex (as is often the case in laboratory experiments) large performance differences are often observed. This fact complicates our understanding of the potential impact of in-vehicle technologies on older drivers. Driving performance as a function of the use of an in-vehicle task may differ in real-world driving relative to that observed during simulated driving. This mismatch between the performance of older drivers in a driving simulator vs. on the actual road was documented in one recent investigation.[17] Merat et al. observed that older drivers performing in-vehicle tasks in a driving simulator maintained longer distance headways between their car and a lead vehicle and they drove slower. However, when performing the same in-vehicle tasks while driving a real car, older drivers tended to drive closer to a lead vehicle and exhibit greater speed variation than they did when driving the car without performing the additional tasks.

Driver distraction is not limited to high-crash-risk driving populations. The distracting effects of mobile phone conversations to drivers of all ages are now well documented.[24–28] Given the prior attention to this form of distraction, it will not be reviewed here. Of relevance to the current aim though, is the fact that these distracted drivers are frequently not aware of how impaired their driving actually is.[29–31] This mismatch between drivers' perception of their performance and their *actual* performance is a critical issue of concern for the safe implementation and use of all in-vehicle systems. Most drivers appear to have little awareness of the many different sources of distraction they face and the impact that these various forms have on vehicular safety.

14.2.3 Sources of Distraction

Distraction can come from many sources, and definitions of exactly what constitutes distraction differ. Anything that takes a driver's attention away from the primary tasks of maintaining control of the vehicle, avoiding obstacles, and navigating to a desired location can be considered a source of distraction. Broadly, distraction can be classified as either exogenous (external to the driver; i.e., in-vehicle displays, flashing lights along the roadside) or endogenous (internal or within the driver; i.e., ruminations over a recent argument with a spouse or colleague, preoccupation with a complex task to be performed).[32]

14.2.3.1 Exogenous and Endogenous Forms

Endogenous distraction can be extremely detrimental to the driving task and is difficult to assess because it generally cannot be directly seen. Unlike exogenous distraction, which usually captures the drivers' visual attention, when a driver is endogenously distracted he or she may be looking at the roadway or hazard but not seeing the information.[33,34] These instances of looking but not seeing—or inattentional blindness—are frequently brought on by excessive mental or cognitive load and are a key source of distraction-related crashes.[27] As pointed out in a recent report on driver distraction compiled by the National Highway Traffic Safety Administration,[35] cognitive distraction may also include mental preoccupation or emotionally upset drivers. However, while recognizing that endogenous distraction can affect driver performance and safety, the topic of exogenous in-vehicle displays are the primary focus of the current discussion. Exogenous forms of distraction include any event, task, activity, or informational display (visual, auditory, or haptic) that diverts the driver's attention away from the task of driving.

In a large-scale observational study of 70 drivers across a broad age range, Stutts et al.[5] found that drivers engaged in a wide variety of distracting behaviors. Drivers were eating and drinking or preparing to eat and drink nearly 5% of the time they were driving and were conversing with a passenger over 15% of the time.[5] It should be noted, however, that at least one simulator investigation has provided evidence that conversing with a passenger may actually increase safe driving behavior.[36] Passengers can sometimes serve as an extra set of eyes and they tend to moderate their conversation based on the complexity of the traffic situation. Similar patterns of moderation are not observed for cell phone conversations.

In addition to eating, drinking, and conversing, Stutts et al.[5,6] found that drivers engage in a host of other activities including but not limited to grooming (shaving, applying makeup, combing or brushing their hair, etc.), reaching for objects inside the vehicle, turning the radio or CD player on and surfing through selections, and reading and writing. Sometimes drivers were engaged in more than one of these distracting activities at a time.

Clearly drivers engage in a wide range of distracting activities and sometimes are even distracted by their own internal thoughts and emotions. However, coverage of all these sources of distraction is well beyond the scope of the current aim. Focus in this chapter will be on those exogenous forms of distraction stemming from in-vehicle displays and information systems.

14.2.3.1.1 In-Vehicle Systems

In-vehicle technologies are becoming increasingly common in the modern automobile. A hierarchical categorization scheme can be used to classify these systems as related to the following:

1. Vehicular control
2. Hazard identification
3. Navigation and information
4. Infotainment

Among the many devices that can be included in the first category are adaptive speed control, active steering, and assistive braking devices. Systems in the second category include a host of collision avoidance systems and backup or reverse alerts. Global positioning systems (GPS) utilizing satellite technology have enabled the development of increasingly sophisticated and increasingly popular mobile navigation and information systems. These GPS-based navigation systems can provide directions to specific points of interest and landmarks, suggest nearby restaurants of particular food types, or provide the location and directions to the nearest gas station or repair shop. Many systems can also warn drivers of traffic congestion or areas of temporary road construction so that alternative routes may be planned.

Last but not least, multimedia entertainment and communications systems are increasingly found and used in the modern automobile. Cell phone use is on the rise, and some parents say they would no longer take a family trip in a vehicle that was not equipped to play DVDs to entertain children in the back seat. Many young people do not leave home without their iPods or MP3 players, and often the vehicles they drive are equipped for this form of mobile music interface.

These various forms of in-vehicle technologies are highly desirable for many motorists. Many of them were specifically designed to aid the driver in the tasks of vehicle control, hazard detection, and navigation. However, as the previous section pointed out, they also have the potential to distract drivers. We next turn to methods of assessing the impact of these systems and devices.

14.3 Methods and Tools for Assessment

It is clear that driver distraction is a persistent threat to automotive safety. Determining the effects of distraction can be studied in numerous ways. In this section, major methods and theory-based tools for assessing distraction potential are presented. First, different forms of investigation along with their strengths and weaknesses are discussed.

The four major forms of investigation for driving research include the following:[35]

1. Surveys
2. Observational studies
3. Crash-based studies
4. Experimental studies of driving performance

Discussion of each of these forms will then be followed by key theoretical constructs that are essential to understanding the issue of driver distraction, regardless of the form of investigation utilized.

14.3.1 Forms of Investigation

14.3.1.1 Surveys

Surveys provide an efficient method of obtaining information regarding motorists' attitudes and opinions about in-vehicle systems. Ultimately, the preferences of consumers will drive much of the emerging automotive technology, regardless of whether it conforms to acceptable standards of safety and human factors principles. Surveys are easy to administer, require relatively little training on the part of the experimenter or administrator, and can provide certain forms of information quickly and inexpensively. They are, however, only as good as the psychometric properties they are based on and the honesty of the people responding to them. Given these limitations, surveys have been constructed and administered either alone or in conjunction with other forms of investigation to examine a number of in-vehicle systems' issues.

Several survey instruments have been developed and utilized to estimate various aspects of motorist behavior, frequently related to crash involvement for different segments of the driving population. For instance the Driver Behavior Questionnaire (DBQ) has been used to examine factors related to the crashes of older drivers[37] and to determine if causal factors differ between different types of driving violations.[38]

Surveys were instrumental in the initial development of many in-vehicle systems. For example, surveys were used during the initial development of in-vehicle navigation system interfaces for the purpose of assessing commuters' willingness to use different forms of in-vehicle navigational systems[39] and their willingness to use alternative routes when commuting to work if suggested by a navigational system. Surveys are also frequently used in conjunction with experimental investigations to determine user preferences for different in-vehicle display formats.[40–42]

14.3.1.2 Observational Studies

Observational studies have been and are frequently carried out to determine the prevalence rates of various motorists' behaviors. They include investigations in which the experimenter or observer cannot or does not randomly assign individuals to treatment vs. control conditions. Random assignment is frequently not an option, nor is it a feasible method, of studying many forms of motorist behavior. Observational studies can help determine the roadways motorists use most often, what time of day they are used, and whether or not motorists follow posted regulatory signs, such as stopping at stop signs or observing speed limits. Advances in technology have enabled many forms of observational studies to be conducted relatively unobtrusively, such as the red-light-running cameras increasingly being installed at major intersections in some parts of the country. These camera systems are generally unobtrusive to the driver, at least until the ticket comes in the mail.

Observational studies include both site-based investigations and naturalistic in-vehicle observations. As will be discussed, advances in technology have dramatically increased research capabilities for naturalistic observation by allowing sophisticated camera, steering, and control sensors and other instrumentation to be inexpensively installed in personal automobiles. However, first we look at the more traditional site-based form of observational study.

14.3.1.2.1 Site-Based Methods

Site-based methods have been used for several decades to determine the types and frequencies of behaviors drivers engage in. Although not necessarily aimed at distraction-related issues per se, a large-scale site-based survey of driving behavior carried out each year is the National Occupant Protection Use Survey, or NOPUS. Individual states conduct statewide observations of the rate of seatbelt use among motorists traveling along various types of roads in an effort to gather a nationwide estimate of seatbelt usage.[43]

14.3.1.2.2 Naturalistic Observation

Naturalistic in-vehicle observational investigations are becoming more common with the advent of powerful, unobtrusive instrumentation. One recently completed large-scale investigation of this type is the 100-Car Naturalistic Driving Study conducted by researchers at Virginia Tech (Dingus et al., 2006). This investigation involved collecting data from 100 instrumented cars over the course of a one-year period and has provided a vast amount of information regarding driving distraction. In fact, the amount of data collected from this investigation is so large that much of it will still be being analyzed and reported for several years to come. Similar studies, though of smaller magnitude, are currently being conducted elsewhere across the nation.

14.3.1.3 Crash-Based Studies

Crash-based studies have the advantage of providing direct evidence of the impact of distraction. However, they are limited by the honesty of the people being interviewed, who may be faced with legal and economic pressures against being completely forthright. As was discussed earlier, it is largely for this reason that the prevalence of distraction as a causal factor in crashes is thought to be underestimated. Additionally, crash-based studies are retrospective, so they are limited in their ability to predict trends in or prevent distraction-based crashes that may result from new technologies. In order to predict the potential for driver distraction, experimental studies are frequently conducted.

14.3.1.4 Experimental Studies

Experimental studies can be conducted in the laboratory using driving simulators or in real cars on either closed-course test tracks or the open road. Investigations using each of these forms have been implemented in an effort to study and predict the impact of in-vehicle tasks on driver distraction. Several large-scale efforts including multiple participating agencies include projects entitled HASTE, which stands for Human machine interface And the Safety of Traffic in Europe,[44] and CAMP, which stands for Crash Avoidance Metrics Partnership.[45,46]

These large-scale projects applied different approaches with different primary focuses to arrive at some similar and some disparate conclusions (for a review see Ref. 35). Both HASTE and CAMP projects sought to develop methods and guidelines for assessing the potential distraction of in-vehicle systems. Both projects examined a number of different secondary tasks including both visual-manual and auditory-vocal combinations using laboratory simulations, closed-course test tracks, and open-road driving investigations.

As referred to previously regarding older drivers, HASTE researchers (e.g., Merat et al.[17]) found differences between driving performance in the simulator and on real roads as a function of secondary task distraction.[44] In general, driving performance on real roads was degraded more by secondary tasks than was simulator performance. Carsten and Brookhuis conclude that many of the complexities found in real driving cannot be adequately simulated in the laboratory with current state-of-the-art technologies. Yet, as they maintain, simulator investigations allow systematic control of driving task difficulty, roadway conditions, and other forms of precise control that are simply not possible in real driving situations. Therefore, both forms of experimental investigations have their usefulness.

HASTE investigations also revealed significant differences between visual versus cognitive (i.e., largely auditory) forms of distraction.[44] Visual distractions resulted in degraded steering behavior and reduced lateral position control. Conversely, cognitive distraction resulted in reduced longitudinal control. As they point out, it is important to note that many in-vehicle tasks will potentially result in both visual and cognitive forms of distraction, which may have compounding or interacting effects on driving performance. For example, they discuss the simulator results in some investigations where cognitive distraction resulted in improved steering behavior while degrading longitudinal control, resulting in drivers maintaining reduced levels of temporal headway. They argue that the improved steering performance resulted from drivers adopting a strategy of maintaining lane position by using central versus peripheral vision. This can be thought of as the cognitive load resulting in a form of tunnel vision. Thus, despite their improved lane-keeping performance, they would be unable to detect obstacles and hazards in peripheral vision.

Future research will need to examine the distinct possibility that a given in-vehicle task might result in performance strategy shifts that although improving some aspects of performance, might degrade others. Utilizing multiple measures of performance (i.e., longitudinal and latitudinal driving performance in combination with eye-tracking behavior) and established mental workload techniques (to be discussed in Section 14.3.2.1) are methods of addressing this issue.

CAMP investigations resulted in similar conclusions as HASTE investigations regarding the different patterns of performance consequences of visual-manual versus auditory-vocal distractions.[45] The CAMP project utilized a Multiple Resource Theory (MRT)[47,48] framework to represent the cognitive architecture. MRT is frequently used in human factors investigations to predict time-sharing efficiency or the extent to which two tasks will result in performance decrements if performed together and it is a particularly useful method in driving research.[49-52] MRT is discussed in further detail in Section 14.3.2.2.

CAMP researchers concluded that distraction effects are multidimensional and therefore point out that no single metric will suffice to predict the distraction potential of in-vehicle devices. They recommend a distraction assessment toolkit for use in the laboratory consisting of lateral and longitudinal vehicle control metrics, measures of task completion time, a peripheral detection task, an occlusion task, and a Sternberg memory task. These tasks represented the primary safety critical driver performance aspects identified in the CAMP investigations: eye-glance patterns, lateral and longitudinal lane control, and detection of objects and events.

Unlike the conclusions made by HASTE researchers, CAMP researchers concluded that the cognitive distractions posed by auditory-vocal tasks were of lesser magnitude than those of visual-manual tasks. However, like HASTE researchers, they conclude that monitoring driver eye-glance behavior is critical to making determinations regarding possible distraction.

It is beyond the scope of the current discussion to provide a full account of these large-scale studies. More detailed accounts can be found in the sources referenced here.[44,45,53] For the current purposes, key theoretical concepts essential to understanding and investigating the potential of in-vehicle displays to distract the driver will be discussed.

14.3.2 Key Theoretical Constructs

Understanding the issue of driver distraction and the potential for in-vehicle systems to alter drivers' attention and performance relies on several key theoretical constructs. Two constructs deemed essential to this area have been chosen for discussion here. They are mental workload and multiple resource theory.

14.3.2.1 Mental Workload

Mental workload has been defined in many ways. Essentially, to use the words of W.B. Knowles[54] it asks the question, "How busy is the operator?" Of the many definitions of mental workload that have been presented, they share the assumption that people are limited-capacity processors. In other words, there is an upper limit to the amount of information people can perceive, process, and remember at any given time. If the amount of information necessary to performing a given task exceeds this limited capacity, then performance will suffer. The theoretical relationship between mental workload and driving is illustrated in Figure 14.2. Note that many sources of mental workload (i.e., vehicular control tasks associated with driving as well as the use of in-vehicle displays) place demands on the driver. Performance difficulties occur whenever these demands exceed the driver's mental workload capacity.

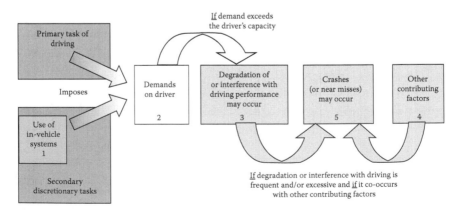

FIGURE 14.2   Driving task and in-vehicle systems workload relationship. (From Angell et al. Driver Workload Metrics Task 2 Final Report, NHTSA 2006, 460.)

Measuring mental workload in different environments has been an essential component of human factors research in the field of transportation for at least fifty years. Dating back to the work of Ivan Brown,[55,56] driver mental workload has been assessed in various driving situations. For example, Brown and Poulton observed that driver mental workload is higher when driving through shopping centers where drivers have to interact with considerably more traffic and pedestrians relative to residential areas where little or no traffic is present.

More recently, researchers such as Recarte and Nunes[34] have examined the effects of various levels of driver mental workload on visual search and discrimination tasks, as well as driver decision making. Mental workload has also been assessed while drivers engage in various extraneous in-vehicle tasks such as conversing on mobile phones[57] using navigational interfaces in both civilian[22] and military settings[58] and other in-vehicle systems.[59,60] Assessment of mental workload is a critical component of ensuring that in-vehicle interfaces and systems do not exceed the information-processing capabilities of the driver.

To the naïve practitioner it may seem at first that measuring various aspects of driver behavior, such as lane-keeping ability, speed maintenance, and steering variability will be sufficient to assess the driver's spare mental capacity. However, further inspection demonstrates that these "primary task" measures of driving performance are insufficient for assessing on-going driver mental workload. As has been documented in numerous investigations,[61–63] primary task measures (such as driving performance) fail to provide an index that is sensitive to the extra effort a person may be putting in as the task gets harder. That is, mental workload may be increasing, but the driver may be able to simply expend greater effort until the point that the driver's capacity is reached. At this point, performance will sharply deteriorate, and in the case of driving excessive risk of crash is present. Rather than accept this rather gross method of classifying mental workload as an either–or, acceptable or unacceptable, a more sensitive index is preferred. Three techniques for assessing mental workload have been developed and well supported in the literature; these include subjective indices, physiological measures, and secondary tasks procedures.[61,64] Each of these techniques has their advantages and disadvantages and may be sensitive to different aspects of the driving situation, leading many researchers to conclude that a battery of assessment procedures using all three techniques is most desirable.[50]

14.3.2.1.1 Subjective Indices

Subjective assessments of mental workload are generally the easiest to administer and for this reason they are frequently used either alone or in combination with other techniques. They have high face validity because the operator him- or herself—or driver in this case—is asked to rate how mentally demanding various aspects of the task are. One might naively assume that the driver is the best person to indicate how difficult a given task is. However, research does not support this assumption.

Several subjective assessments have been developed for the purpose of measuring mental workload in complex environments. One in particular, the NASA Task Load Index, or NASA TLX,[65] is frequently used and has been demonstrated to be moderately effective in assessing some aspects of driver mental workload. However, as mentioned previously, drivers are frequently unaware of how much their performance is affected by the use of in-vehicle displays and therefore subjective indices of workload must be used with caution.

14.3.2.1.2 Physiological Measures

A number of investigations have utilized physiological measures of the driver to examine workload. Physiological measures can often provide quite sensitive indices of workload and may be sensitive to aspects of the driving task that other measures fail to identify. For example, Baldwin and Coyne[49] found that only a physiological measure—specifically event-related potential measures of brain electrical activity—were sensitive to increased driver workload stemming from the presence of heavy fog in a driving simulator scenario. Other physiological measures that have been used successfully as indices of workload in a driving context include measures of heart function[66] and measures of eye movement and function.[33,45]

Physiological indices have several characteristics that may make researchers reluctant to use them. First, the equipment required for data collection is generally more expensive than that required for other forms of workload assessment. It generally requires more training on the part of the experimenter to use the equipment and it may be somewhat obtrusive to the driver. However, tremendous advances have been made in recent years in the area of reducing the obtrusiveness of many physiological recording devices. Dashboard-mounted eye-tracking systems now allow monitoring of eye movements, glance duration, pupil dilation, and percent of time a driver's eyes are closed without any equipment ever needing to come in contact with the driver.

At present, due to the limitations of subjective indices and some of the challenges of physiological techniques for assessing driver mental workload, secondary task measures have been demonstrated to be the preferred and most effective assessment methodology for many investigations.

14.3.2.1.3 Secondary Task Techniques

Secondary task techniques have been demonstrated to provide sensitive indices of mental workload in a wide variety of operational settings.[17,50,67] The technique involves asking a person to perform a secondary, less important task in addition to the primary task of interest. People are instructed to maintain performance on the primary task (i.e., driving) at all times and to devote attention to the secondary task only to the extent that it does not disrupt their performance of the primary task. The reasoning behind this technique is that as the primary task increases in difficulty, performance on the secondary task should deteriorate, even though primary task performance is maintained. That is, rather than the sharp performance drop observed with primary task measures alone, performance on the secondary task should deteriorate gradually as the primary task becomes more difficult.

Verwey and Veltman,[68] for example, found a secondary visual detection task to be sensitive to short periods (10 sec) of elevated workload. Primary task measures such as steering reversals were not sensitive to these short duration peaks in mental workload. Verwey and Veltman's task involved verbally responding to the presence of punctuation marks presented on a display mounted on the center of the dashboard just to the right of the steering wheel. They reasoned that because driving is a heavily visually demanding task, a secondary visual detection task would be most sensitive to short periods of elevated workload. As expected, the number of targets detected dropped significantly as the demands of the driving task increased.

Several other research teams have successfully used different forms of visual detection secondary tasks as indices of driver mental workload and driver distraction. Harms and Patten[2] used a peripheral detection task to investigate the impact of various navigation interfaces on driving performance. They reasoned that the ability to detect important visual information is essential to safe driving. They compared an interface that provided navigation information visually, verbally, or both visually and verbally. Using professional drivers they found that driving performance was not affected by any of the three types of navigation interfaces. Critically though, as discussed in the previous section, this did not mean that using the navigation system did not pose a cognitive challenge to drivers. Both navigation systems that presented information to the driver visually reduced the number of peripheral targets the driver could detect. The interface providing both visual and verbal navigation information also resulted in drivers taking significantly more time to detect peripheral targets relative to the other two interfaces. They note that detection performance while using the verbal interface did not differ significantly from performance when driving through a familiar route, thus not using any guidance system.

Harms and Patton[2] note that the visual peripheral detection task may be somewhat biased or more sensitive to distraction from another visual source, relative to an auditory source. However, they argue that this is justifiable within the context of driving as visual detection is critical to safe driving. Discussion of MRT provides a theoretical justification for this position.

14.3.2.2 Multiple Resource Theory (MRT)

Wickens'[47] MRT was designed to predict time-sharing efficiency in situations where people are expected to perform two tasks at the same time. Wickens and colleagues[47,69] noted that time-sharing efficiency for two tasks performed at the same time (or how well a person could perform the two tasks together) differs by more than simply the difficulty of each task. Difficulty for individual tasks can be assessed by a participant's ability to perform the task by itself, under single task conditions. Wickens observed that two tasks requiring similar processing resources could be time shared less proficiently than a different set of tasks, which although more difficult, relied on different processing resources. To account for the differential time-sharing efficiency of a variety of different tasks, Wickens[47] proposed the MRT model of information processing. MRT has since received considerable empirical support in human factors and other applied literature domains (for reviews see Refs. 48 and 70). According to MRT, time-sharing efficiency, as observed through the performance of each task, is moderated by the information-processing channel (auditory or visual), the type of code (spatial or verbal), and processing stages (encoding, comprehension, response selection) involved in each task. For example, an auditory task will tend to be time-shared more efficiently with a visual task than another auditory task. Likewise, if one task primarily relies on verbal processing (i.e., processing navigational instructions regardless of whether they are presented in text or speech formats) it will tend to be time-shared more efficiently with a task requiring visual-spatial skills (i.e., monitoring a geographic display) rather than another task requiring a verbal processing code (i.e., reading a menu list of nearby restaurant options).

MRT suggests that there are separate pools of resources and the degree to which two tasks interfere with each other (e.g., driving and using a visual map display) will depend on the extent to which the two tasks draw on resources from the same pools. Tasks that draw upon separate resources (e.g., a visual-spatial task like driving combined with an auditory-verbal task like listening to the radio) will interfere less than those that draw upon resources from overlapping pools (i.e., driving while looking at a map). Wickens and colleagues[1,51] have successfully applied MRT to predict patterns of distraction and workload for in-vehicle displays used while driving. Utilizing principles of MRT suggests an important avenue for mitigating the distraction potential of in-vehicle displays.

14.4 Display Design to Mitigate Distraction

The degree of distraction caused by an in-vehicle display is a dynamic interaction of the mental workload and information-processing demands of the system (both in terms of quantity and type required), the driver's workload capacity, and the complexity of the driving task and environment. There is little if anything designers can do to alter the driver's workload capacity. Roadway design, although an important aspect of transportation safety, is beyond the scope of the current review. Attention is focused here on strategies for in-vehicle system design that can mitigate driver distraction. One critical aspect of design is display modality.

14.4.1 Display Modality

Emerging technologies have resulted in the ability to provide drivers with complex visual, auditory, and even haptic or tactile informational displays. Driving is a visually demanding task, particularly when drivers encounter complex traffic and roadway patterns or adverse weather.[41,59] Novice and older drivers have particular difficulty coping with the visual demands of the driving task,[18,19] often even in normal driving conditions. Novice drivers tend to be preoccupied with the vehicle controls and have not yet learned where to effectively direct their visual attention.[13,71] Conversely, older drivers, although they know where to expect potential hazards, take longer to extract and process visual information, longer to respond, and longer to switch their visual attention between two competing sources of visual information.[72–74] Novice drivers tend to need to rely more on foveal vision for lane keeping.[75] Further, with both increasing mental demand (i.e., heavy traffic situations) and age the functional or useful field of view (UFOV) decreases. Driving places heavier mental demands on both novices and older drivers, therefore leaving both high-crash-risk groups more susceptible to visual attention failures. These visual-attentional issues play a significant role in the crash risk of novice and older drivers.[9,76,77] Auditory rather than, or in addition to, visual in-vehicle displays may substantially mitigate driver distraction and appear to be particularly well suited for both novice and older drivers.

14.4.1.1 Auditory Display

In-vehicle auditory displays are increasingly common. They currently range from simple seatbelt reminders and frequency-intensity mapped time-to-contact backup alerts to user-customizable voice navigational guidance systems in a variety of languages and choice of voice styles. In addition to leaving the driver's eyes free to watch the road and roadway hazards, auditory interfaces have a number of other important advantages over visual displays. Auditory displays have excellent attention-getting ability. This characteristic makes them particularly well suited for use in vehicle collision avoidance systems. The driver does not have to be looking at the display in order to be notified by the alert. In fact, the driver can process the alert while keeping his or her eyes on the road or while visually scanning for the hazard. Auditory displays (i.e., voice guidance) have also been found to have beneficial applications in navigational systems, particularly for older and novice drivers.[19,41]

14.4.1.1.1 Auditory Navigation

Commercially available navigation systems such as TomTom, Garmin, and Magellen brands come equipped with voice guidance features in addition to the traditional maps. Most of these systems boast user-customizable voices in several different languages and in the voices of celebrities and other comic features. Some of the voice options are free and others are available for download and purchase on the system's website. So, you can drive around France getting your navigational instructions in English from the voice of Mr. T.

Auditory navigation systems are frequently reported to require less mental workload than visual map-based systems.[78] Additionally, although some drivers have difficulty using visual map displays, auditory guidance instructions are generally found to be accessible to all classes of drivers.

14.4.1.2 Haptic Display

Visual alerts may go unnoticed if drivers' visual attention is not directed at them. Auditory alerts may be missed due to poor signal-to-noise ratio or masking if radios or other infotainment devices are too loud or excessive external noise is present, such as may be found in some urban areas and construction zones. For these and other reasons, haptic alerts are gaining increasing attention as a potential modality to either replace or supplement other alert modalities. In one recent investigation in simulated conditions where participants were asked to simply locate the direction of the alert (it was not tied to an actual threat), providing a haptic alert significantly reduced the time it took to localize the alert (on average by over 250 msec) and dramatically improved localization accuracy relative to auditory alerts alone.[79]

14.4.1.3 Voice Recognition Interfaces

Voice recognition can dramatically decrease distraction relative to using a keyboard. Locating the correct keys, typing or touching in the sequence, and visually inspecting the sequence on a screen for correctness places heavy visual demands on the driver. For this reason considerable effort has been made to develop effective voice recognition technologies. This effort has to be arduous. Humans' ability to comprehend speech is unparalleled. It relies on recognition and interpretation of extremely small temporal and spectral patterns across a wide range and variety of speakers and listening conditions. Nevertheless, tremendous strides have been accomplished and numerous voice recognition systems are currently in use in automated phone networks and most recently in in-vehicle interfaces. Voice recognition software coupled with Bluetooth technology allows drivers to select menu items—changing radio stations, making phone calls and asking for directions simply by speaking to the system. Sound too good to be true? Well, to a large extent it is. User evaluations of the effectiveness of these systems tend to be less than stellar. But progress is still being made.

In terms of distraction, however, voice recognition interfaces do not reduce the distraction potential anywhere near zero. Even voice-based interactions have considerable potential for driver distraction and drivers have a tendency to significantly underestimate the deleterious effects of these distractions on their performance.[30]

14.4.2 Minimizing Mental Workload

The alternative display modalities and response formats just discussed have significant potential to mitigate distraction. Another key determinant to the mental workload and distraction associated with an in-vehicle system and its display will be its level of complexity. More complex messages and menu structures impose greater mental workload and have greater potential for distraction.

14.4.2.1 Complexity

The complexity level of the display, regardless of what modality it is presented in, has a tremendous impact on both mental workload and potential distraction. Federal Highway Administration (FHWA) guidelines based largely on empirical research have been established to aid designers in determining the appropriate complexity level of in-vehicle displays.[80] According to these guidelines, complexity can be thought of as falling along a continuum from high to low. For auditory displays, high-complexity displays contain over nine information units and take over 5 sec to process. Conversely, low-complexity displays contain three to five information units and require less than 5 sec to process.

In in-vehicle routing and navigational systems, message complexity is a critical factor in ensuring that driver's mental workload capabilities are not exceeded.[81] If the visual map is too complex, it may divert the drivers' attention away from the road for an excessive amount of time. In order to make visual information accessible in a single glance, it should contain no more than 20 to 30 character bits.[82] Head-up displays have been found to result in decreased visual hazard scanning performance relative to head-down displays.[83] Further head-down map displays result in better driving performance and hazard scanning behavior relative to text-based turn-by-turn maps even though they are more visually complex.[83]

If the auditory message is too complex, it will likely exceed the working memory capacity limitations of the driver and not be retained (for a review see Ref. 22). Several investigations indicate that auditory route guidance messages should contain no more than three to four informational units or propositional phrases and should take no longer than 5 to 7 sec to present.[81,84–86] When procedural or navigational commands contain more than three propositional phrases, execution errors increase sharply,[87] drivers make more navigational errors,[86] and driving performance becomes less stable.

Srinivasan and Jovanis[88] have suggested that in order to prevent excessive load on a driver's informational processing resources, auditory directional information should be provided in the form of terse commands such as, "Turn left in two blocks onto Park Avenue." However, Reagan and Baldwin[22] have provided evidence that a driver's navigational performance and memory for routes benefits from the addition of a salient landmark in combination with the recommended terse commands without adverse effects on either driving performance or subjective indices of mental workload.

14.4.3 Driver Assistance Systems

Providing the driver with information regarding his or her performance can improve driving and reduce driver distraction.[89] As providing the feedback to the driver may be distracting in and of itself, Donmez and colleagues examined the impact of providing feedback to drivers after a simulated driving session was completed. This subsequent feedback improved drivers' responses to lead vehicle braking events. In other words, when drivers received feedback that they had not exhibited safe driving behaviors in a previous drive and/or that they had a medium to high distraction index they responded more quickly to critical incidents (a car in front of them suddenly braking) in subsequent drives. Participants receiving feedback regarding their performance either subsequent to the drive or both concurrently and subsequently also maintained longer headways between their vehicle and the vehicle in front of them during subsequent drives. Maintaining adequate headways or following distance is critical to safe driving.

Current U.S. law states that drivers must maintain a temporal headway of at least 2 sec between a vehicle and the one in front. However as previous empirical observation indicates,[90,91] many drivers overestimate their temporal headway. Providing these drivers with a temporal headway display has been shown to increase safe following behavior both immediately during field tests and in subsequent drivers up to 6 months later.[90] Results of the Donmez et al.[89] simulator study and Shinar and Schechtman's field study[90] provide support for the position that in-vehicle driver assistance systems can actually improve safe driving behavior.

In sum, emerging in-vehicle systems have the potential to enhance the safety and efficiency of our vehicular transportation network. At the same time, if not designed in accordance with the drivers' mental workload capabilities and sound information-processing principles, they have the potential to dramatically increase driver distraction and lead to decreased safety. Novice and senior drivers are at particular risk of distraction and therefore should be carefully considered in all aspects of the design process. Multiple resource theory offers a sound theoretical basis for developing interfaces that utilize cognitive resources that are not already overburdened in the task of driving. Auditory displays coupled with voice recognition are two formats that may potentially reduce mental workload and mitigate driver distraction. However, as several large-scale investigations have demonstrated, using these alternative formats is not without risk. Care must be taken to ensure that the mental workload required to use emerging in-vehicle systems does not impair the driver's ability to maintain vehicular control, detect hazards, and navigate. With these aspects in mind, the road ahead for in-vehicle systems offers many exciting challenges and ample opportunity for further research.

References

1. Horrey, W.J. and Wickens, C.D., Driving and side task performance: The effects of display clutter, separation, and modality, in Technical Report No. AHFD-02-13/GM-02-2, Savoy, Illinois: University of Illinois at Urbana-Champaign, Aviation Human Factors, Division Institute of Aviation, 2002.
2. Harms, L. and Patten, C., Peripheral detection as a measure of driver distraction. A study of memory-based versus system-based navigation in a built-up area, *Transportation Research Part F: Traffic Psychology and Behaviour*, 6(1), 2003, 23–36.
3. Stutts, J.C., Reinfurt, D.W., Staplin, L., and Rodgman, E.A., The Role of Driver Distraction in Traffic Crashes, Technical Report, AAA Foundation, Washington, DC, 2001.
4. Dingus, T.A. et al., The 100-Car Naturalistic Driving Study, Phase II—Results of the 100-Car Field Experiment, No. DOT HS 810 593, National Highway Traffic Safety Administration, Washington, DC, 2006.
5. Stutts, J. et al., Driver's exposure to distractions in their natural driving environment, *Accident Analysis and Prevention*, 37(6), 2005, 1093–1101.
6. Stutts, J. et al., Distractions in Everyday Driving, AAA Foundation for Traffic Safety, Washington, DC, 2003.
7. Young, M.S. and Stanton, N.A., Miles away: Determining the extent of secondary task interference on simulated driving, *Theoretical Issues in Ergonomics Science*, 8(3), 2007, 233–253.
8. Parasuraman, R. and Riley, V., Humans and automation: Use, misuse, disuse, abuse, *Human Factors*, 39(2), 1997, 230–253.
9. Williams, A.F. and Shabanova, V.I., Responsibility of drivers, by age and gender, for motor-vehicle crash deaths, *Journal of Safety Research*, 34(5), 2003, 527–531.
10. Waller, P.F., Elliott, M.R., Shope, J.T., Raghunathan, T.E., and Little, R.J.A., Changes in young adult offence and crash patterns over time, *Accident Analysis and Prevention*, 33(1), 2001, 117–128.
11. Williams, A.F., Teenage drivers: Patterns of risk, *Journal of Safety Research*, 34(1), 2003, 5–15.
12. Charness, N. and Bosman, E.A., Human factors and age, in *Handbook of Aging and Cognition* (eds Craik, F.I.M. and Salthouse, T.A.), Lawrence Erlbaum, Hillsdale, NJ, 1992.
13. Fisher, D.L. et al., Use of a fixed-base driving simulator to evaluate the effects of experience and PC-based risk awareness training on drivers' decisions, *Human Factors*, 44(2), 2002, 287–302.
14. Burns, P.C., Navigation and the mobility of older drivers, *Journals of Gerontology Series B: Psychological Sciences and Social Sciences*, 54(1), 1999, 49–55.
15. Kline, D.W. et al., Vision, aging, and driving: The problems of older drivers, *Journal of Gerontology*, 47(1), 1992, 27.

16. Owsley, C. et al., Visual/cognitive correlates of vehicle accidents in older drivers, *Psychology and Aging*, 6(3), 1991, 403–415.
17. Merat, N., Anttila, V., and Luoma, J., Comparing the driving performance of average and older drivers: The effect of surrogate in-vehicle information systems, *Transportation Research Part F: Traffic Psychology and Behaviour*, 8(2), 2005, 147–166.
18. Baldwin, C.L., Designing in-vehicle technologies for older drivers: Application of sensory-cognitive interaction theory, *Theoretical Issues in Ergonomics Science*, 3(4), 2002, 307–329.
19. Dingus, T.A., Hulse, M.C., Mollenhauer, M.A., and Fleischman, R.N., Effects of age, system experience, and navigation technique on driving with an Advanced Traveler Information System, *Human Factors*, 39(2), 1997, 177–199.
20. Baldwin, C.L., May, J.F., and Reagan, I., Auditory In-vehicle Messages and Older Drivers, paper presented at the Human Factors and Ergonomics Society, San Francisco, CA, 2006, October 2007.
21. Ferris, R.D., Baldwin, C.L., and Freund, B.M., Age Differences in Appropriate Navigational Command Distance: A Driving Simulation Study, paper presented at the Virginia Academy of Sciences, 80th Annual Meeting, May 2002.
22. Reagan, I. and Baldwin, C.L., Facilitating route memory with auditory route guidance systems, *Journal of Environmental Psychology*, 26(2), 2006, 146–155.
23. Baldwin, C.L. and Schieber, F., Age Differences in Mental Workload with Implications for Driving, paper presented at the 39th Annual Conference of the Human Factors and Ergonomics Society, San Diego, CA, 1995.
24. Consiglio, W., Driscoll, P., Witte, M., and Berg, W.P., Effect of cellular telephone conversations and other potential interference on reaction time in a braking response, *Accident Analysis and Prevention*, 35(4), 2003, 494–500.
25. McCarley, J.S. et al., Conversation disrupts change detection in complex traffic scenes, *Human Factors*, 46(3), 2004, 424–436.
26. Redelmeier, D.A. and Tibshirani, R.J., Association between cellular-telephone calls and motor vehicle collisions, *New England Journal of Medicine*, 336(7), 1997, 453–458.
27. Strayer, D.L., Drews, F.A., and Johnston, W.A., Cell phone-induced failures of visual attention during simulated driving, *Journal of Experimental Psychology: Applied*, 9(1), 2003, 23–32.
28. Strayer, D.L. and Johnston, W.A., Driven to distraction: Dual-task studies of simulated driving and conversing on a cellular telephone, *Psychological Science*, 12(6), 2001, 462–466.
29. Horrey, W.J., Lesch, M.F., and Garabet, A., Assessing the awareness of performance decrements in distracted drivers, *Accident Analysis and Prevention*, 40, 2008, 675–682.
30. Lee, J.D., Caven, B., Haake, S., and Brown, T.L., Speech-based interaction with in-vehicle computers: The effect of speech-based e-mail on drivers' attention to the roadway, *Human Factors*, 43(4), 2001, 631–640.
31. Lesch, M.F. and Hancock, P.A., Driving performance during concurrent cell-phone use: Are drivers aware of their performance decrements? *Accident Analysis and Prevention*, 36(3), 2004, 471–480.

32. Posner, M.I., Orienting of attention, *The Quarterly Journal of Experimental Psychology*, 32(1), 1980, 3–25.
33. Recarte, M.A., and Nunes, L.M., Effects of verbal and spatial-imagery tasks on eye fixations while driving, *Journal of Experimental Psychology: Applied*, 6(1), 2000, 31–43.
34. Recarte, M.A. and Nunes, L.M., Mental workload while driving: Effects on visual search, discrimination, and decision making, *Journal of Experimental Psychology: Applied*, 9(2), 2003, 119–137.
35. NHTSA, *Driver distraction: A review of the current state-of-knowledge*, No. DOT HS 810 787, National Highway Traffic Safety Administration, Department of Transportation, 2008.
36. Strayer, D.L. and Drews, F.A., Cell-phone-induced driver distraction, *Current Directions in Psychological Science*, 16(3), 2007, 128–131.
37. Parker, D., McDonald, L., Rabbitt, P., and Sutcliffe, P., Elderly drivers and their accidents: The aging driver questionnaire, *Accident Analysis and Prevention*, 32(6), 2000, 751–759.
38. Reason, J.T., Manstead, A., Stradling, S., and Baxter, J.S., Errors and violations on the roads: A real distinction? *Ergonomics*, 33(10), 1990, 1315–1332.
39. Streeter, L.A., Vitello, D., and Wonsiewicz, S.A., How to tell people where to go: Comparing navigational aids, *International Journal of Man Machine Studies*, 22(5), 1985, 549–562.
40. Belz, S.M., Robinson, G.S., and Casali, J.G., A new class of auditory warning signals for complex systems: Auditory icons, *Human Factors*, 41(4), 1999, 608–618.
41. Dingus, T.A., Hulse, M.C., and Barfield, W., Human-system interface issues in the design and use of advanced traveler information systems, in Human Factors in Intelligent Transportation Systems (eds Barfield, W. and Dingus, T.A.), 1998, 359–395.
42. Sodnik, J., Dicke, C., Tomazic, S.O., and Billinghurst, M. A user study of auditory versus visual interfaces for use while driving, *International Journal of Human-Computer Studies*, 66(5), 2008, 318–332.
43. Baldwin, C.L., Struckman-Johnson, C., and Struckman-Johnson, D., 2005 South Dakota Statewide Seatbelt Survey, Technical Report, South Dakota Office of Highway Safety, Pierre, SD, 2005.
44. Carsten, O., and Brookhuis, K., Issues arising from the HASTE experiments, *Transportation Research Part F: Traffic Psychology and Behaviour*, 8(2), 2005, 191–196.
45. Angell, L. et al., Driver Workload Metrics Task 2 Final Report, National Highway Traffic Safety Administration, 460, Washington, DC, 2006.
46. NHTSA, Annual Report of the Crash Avoidance Metrics Partnership, April 2001–March 2002, No. DOT HS 809 531, National Highway Safety Administration, Washington, DC, 2002.
47. Wickens, C.D., Processing resources in attention, in Varieties of Attention (eds Parasuraman, R. and Davies, R.), 63–101, Academic Press, Orlando, FL, 1984.
48. Wickens, C.D., Multiple resources and performance prediction, *Theoretical Issues in Ergonomics Science*, 3(2), 2002, 159–177.

49. Baldwin, C.L. and Coyne, J.T., Mental workload as a function of traffic density in an urban environment: Convergence of physiological, behavioral, and subjective indices, a paper presented at the Human Factors in Driving Assessment 2003 Symposium, Park City, UT, July 2003.

50. Baldwin, C.L. and Coyne, J.T., Dissociable aspects of mental workload: Examinations of the P300 ERP component and performance assessments. *Psychologia*, 48, 2005, 102–119.

51. Horrey, W.J. and Wickens, C.D., Multiple resources modeling of task interference in vehicle control, hazard awareness, and in-vehicle task performance a paper presented at the International Driving Symposium on Human Factors in Driver Assessment, Training, and Vehicle Design, Park City, UT, July 2003.

52. Horrey, W.J. and Wickens, C.D., Driving and side task performance: The effects of display clutter, separation, and modality, *Human Factors*, 46(4), 2004, 611–624.

53. Carsten, O. and Brookhuis, K., Editorial, The relationship between distraction and driving performance: Towards a test regime for in-vehicle information systems, *Transportation Research Part F: Traffic Psychology and Behaviour*, 8(2), 2005, 75–77.

54. Knowles, W.B., Operator loading tasks, *Human Factors*, 9(5), 1963, 155–161.

55. Brown, I.D., A comparison of two subsidiary tasks used to measure fatigue in car drivers, *Ergonomics*, 8(4), 1965, 467–473.

56. Brown, I.D. and Poulton, E.C., Measuring the spare "mental capacity" of car drivers by a subsidiary task, *Ergonomics*, 4, 1961, 35–40.

57. Nunes, L. and Recarte, M.A., Cognitive demands of hands-free-phone conversation while driving, *Transportation Research Part F: Traffic Psychology and Behaviour*, 5(2), 2002, 133–144.

58. Leggatt, A.P. and Noyes, J.M., Navigation aids: Effects of crew workload and performance, *Military Psychology*, 12(2), 2000, 89–104.

59. Lansdown, T.C., Fowkes, M., Wierwille, W.W., and Tijerina, L., Workload demands of in-vehicle displays (ed Gale, A.G.), 373, Brown, 1998.

60. Lansdown, T.C., Brook-Carter, N., and Kersloot, T., Distraction from multiple in-vehicle secondary tasks: Vehicle performance and mental workload implications, *Ergonomics*, 47(1), 2004, 91–104.

61. Eggemeier, F., Properties of workload assessment techniques, in Human Mental Workload Advances in Psychology (eds Hancock, P.A. and Meshkati, N.), 52, 41–62, North-Holland, Oxford, England, 1988.

62. Eggemeier, F.T., Wilson, G.F., Kramer, A.F., and Damos, D.L., Workload assessment in multi-task environments, in Multiple-Task Performance (ed Damos, D.L.), 207–216, Taylor & Francis, London, 1991.

63. Gopher, D. and Donchin, E., Workload—An examination of the concept, in Handbook of Perception and Human Performance (eds Boff, K.R., Kaufman, L., and Thomas, J.P.), Vol. II, Cognitive Processes and Performance, 41-41–41-49, John Wiley & Sons, New York, 1986.

64. O'Donnell, R.D. and Eggemeier, F.T., Workload assessment methodology, in Handbook of Perception and Human Performance (eds Boff, K.R., Kaufman, L., and Thomas, J.P.), Vol. II, Cognitive Processes and Performance, 42-41–42-49, John Wiley & Sons, New York, 1986.

65. Hart, S.G. and Staveland, L.E., Development of NASA–TLX (Task Load Index): Results of empirical and theoretical research, in Human Mental Workload (eds Hancock, P.A. and Meshkati, N.), 239–250, North Holland Press, Amsterdam, 1988.

66. Backs, R.W., An autonomic space approach to the psychophysiological assessment of mental workload, in Stress, Workload, and Fatigue (eds Hancock, P.A. and Desmond, P.A.), 279–289, Lawrence Erlbaum, Mahwah, NJ, 2001.

67. Ogden, G.D., Levine, J.M., and Eisner, E.J., Measurement of workload by secondary tasks, *Human Factors*, 21(5), 1979, 529–548.

68. Verwey, W.B. and Veltman, H.A., Detecting short periods of elevated workload: A comparison of nine workload assessment techniques, *Journal of Experimental Psychology: Applied*, 2(3), 1996, 270–285.

69. Wickens, C., Kramer, A., Vanasse, L., and Donchin, E., Performance of concurrent tasks: A psychophysiological analysis of the reciprocity of information-processing resources, *Science*, 221(4615), 1983, 1080–1082.

70. Wickens, C.D., Processing resources and attention, in *Multiple-Task Performance* (ed. Damos, D.L.), 3–34, Taylor & Francis, London, 1991.

71. Fisher, D.L., Narayanaan, V., Pradhan, A., and Pollatsek, A., Using eye movements in driving simulators to evaluate effects of pc-based risk awareness training, a paper presented at the Proceedings of the Human Factors and Ergonomics Society 48th Annual Meeting, 2004.

72. Cerella, J., Information processing rates in the elderly, *Psychological Bulletin*, 98(1), 1985, 67–83.

73. Maquestiaux, F., Hartley, A.A., and Bertsch, J., Can practice overcome age-related differences in the psychological refractory period effect? *Psychology and Aging*, 19(4), 2004, 649–667.

74. Verhaeghen, P. et al., Cognitive efficiency modes in old age: Performance on sequential and coordinative verbal and visuospatial tasks, *Psychology and Aging*, 17(4), 2002, 558–570.

75. Summala, H., Nieminen, T., and Punto, M., Maintaining lane position with peripheral vision during in-vehicle tasks, *Human Factors*, 38(3), 1996, 442–451.

76. Ball, K. and Rebok, G.W., Evaluating the driving ability of older adults, *Journal of Applied Gerontology*, 13(1), 1994, 20–38.

77. Owsley, C. et al., Visual processing impairment and risk of motor vehicle crash among older adults, *JAMA: Journal of the American Medical Association*, 279(14), 1998, 1083–1088.

78. Furukawa, H., Baldwin, C.L., and Carpenter, E.M., Supporting drivers' area-learning task with visual geo-centered and auditory ego-centered guidance: Interference or improved performance? in Human Performance, Situation Awareness, and Automation: Current Research and Trends, HPSAA II (eds Vincenzi, D.A., Mouloua, M., and Hancock, P.A.), 124–129, Daytona Beach, FL, 2004.

79. Fitch, G.M., Kiefer, R.J., Hankey, J.M., and Kleiner, B.M., Toward developing an approach for alerting drivers to the direction of a crash threat, *Human Factors*, 49(4), 2007, 710–720.

80. Campbell, J.L., Richman, J.B., Carney, C., and Lee, J.D., In-vehicle display icons and other information elements volume I: Guidelines, No. FHWA-RD-03-065, Federal Highway Administration, Office of Safety Research and Development, McClean, VA, 2004.

81. Kimura, K., Marunaka, K., and Sugiura, S., Human factors considerations for automotive navigation systems—Legibility, comprehension, and voice guidance, in Ergonomics and Safety of Intelligent Driver Interfaces (ed Noy, Y.I.), 153–167, Lawrence Erlbaum, Mahwah, NJ, 1997.

82. Kimura, K., Osumi, Y., and Nagai, Y., CRT display visibility in automobiles, *Ergonomics*, 33(6), 1990, 707–718.

83. Srinivasan, R. and Jovanis, P.P., Effect of selected in-vehicle route guidance systems on driver reaction times, *Human Factors*, 39(2), 1997, 200–215.

84. Barshi, I., Effects of linguistic properties and message length on misunderstandings in aviation communication, unpublished doctoral dissertation, University of Colorado, Boulder, 1997.

85. Green, P.A. American human factors research on in-vehicle navigation systems, Technical Report UMTRI-92-47, Transportation Research Institute, University of Michigan, Ann Arbor, MI, 1992.

86. Walker, J., Alicandri, E., Sedney, C., and Roberts, K., In-vehicle navigation devices: Effects on the safety of driver performance, Technical Report, No. FHWA-RD-90-053, Federal Highway Administration, Washington, DC, 1990.

87. Scerbo, M.W., Risser, M.R., Baldwin, C.L., and McNamara, D.S., The effects of task interference and message length on implementing speech and simulated data link commands, a paper presented at the Proceedings of the Human Factors and Ergonomics Society 47th Annual Meeting, 2003.

88. Srinivasan, R. and Jovanis, P.P., Effect of in-vehicle route guidance systems on driver workload and choice of vehicle speed: Findings from a driving simulator experiment, in Ergonomics and Safety of Intelligent Driver Interfaces (ed Noy, Y.I.), 97–114, Lawrence Erlbaum, Mahwah, NJ, 1997.

89. Donmez, B., Boyle, L.N., and Lee, J.D., Mitigating driver distraction with retrospective and concurrent feedback, *Accident Analysis and Prevention*, 40(2), 2008, 776–786.

90. Shinar, D. and Schechtman, E., Headway feedback improves intervehicular distance: A field study, *Human Factors*, 44(3), 2002, 474–481.

91. Taieb-Maimon, M. and Shinar, D., Minimum and comfortable driving headways: Reality versus perception, *Human Factors*, 43(1), 2001, 159–172.

92. Cerrelli, E., Crash data and rates for age–sex groups of drivers, 1996, Research Note, National Highway Traffic Safety Administration, Washington, DC, 1998.

Index

H

I

Printed and bound by CPI Group (UK) Ltd, Croydon, CR0 4YY

21/10/2024

01777083-0016